Thomas Jeffrey Parker

Lessons in elementary biology

Thomas Jeffrey Parker

Lessons in elementary biology

ISBN/EAN: 9783337215064

Printed in Europe, USA, Canada, Australia, Japan

Cover: Foto ©berggeist007 / pixelio.de

More available books at **www.hansebooks.com**

LESSONS
IN
ELEMENTARY BIOLOGY

LESSONS IN ELEMENTARY BIOLOGY

BY

T. JEFFERY PARKER, D.Sc., F.R.S.

PROFESSOR OF BIOLOGY IN THE UNIVERSITY OF OTAGO, DUNEDIN, NEW ZEALAND

WITH EIGHTY-EIGHT ILLUSTRATIONS

London
MACMILLAN AND CO.
AND NEW YORK
1893

RICHARD CLAY AND SONS, LIMITED.
LONDON AND BUNGAY.

First Edition, 1891.
Second Edition Revised, 1893.

PREFACE TO THE FIRST EDITION

IN his preface to the new edition of the well-known *Practical Biology*, Professor Huxley gives his reasons for beginning the study of organized nature with the higher forms of animal life, to the abandonment of his earlier method of working from the simpler to the more complex organisms. He says in effect that experience has taught him the unwisdom of taking the beginner at once into the new and strange region of microscopic life, and the advantage of making him commence his studies with a subject of which he is bound to know something—the elementary anatomy and physiology of a vertebrate animal.

Most teachers will probably agree with the general truth of this opinion. The first few weeks of the beginner in natural science are so fully occupied in mastering an unfamiliar and difficult terminology and in acquiring the art of using his eyes and fingers, that he is simply incapable for a time of grasping any of the principles of the science; and, this being the case, the more completely his new work can be connected with any knowledge of the subject, however vague, he may already possess, the better for his progress.

On the other hand, the advantage to logical treatment of proceeding from the simple to the complex—of working upwards from protists to the higher plants and animals—is so immense that it is not to be abandoned without very good and sufficient reasons.

In my own experience I have found that the difficulty may be largely met by a compromise, namely, by beginning the work of the class by a comparative study of one of the higher plants (flowering plant or fern) and of one of the higher animals (rabbit, frog, or crayfish). If there were no limitations as to time, and if it were possible to avoid altogether the valley of the shadow of the coming examination, this preliminary work might be extended with advantage, and made to include a fairly complete although elementary study of animal physiology, with a minimum of anatomical detail, and a somewhat extensive study of flowering plants with special reference to their physiology and to their relations to the rest of nature.

In any case by the time this introductory work is over, the student of average intelligence has overcome preliminary difficulties, and is ready to profit by the second and more systematic part of the course in which organisms are studied in the order of increasing complexity.

It is such a course of general elementary biology which I have attempted to give in the following Lessons, my aim having been to provide a book which may supply in the study the place occupied in the laboratory by "Huxley and Martin," by giving the connected narrative which would be out of place in a practical handbook. I also venture to hope that the work may be of some use to students who have studied zoology and botany as separate subjects, as well as to that large class of workers whose services to English science often receive but scant recognition—I mean amateur microscopists.

As to the general treatment of the subject I have been guided by three principles. Firstly, that the main object of teaching biology as part of a liberal education is to familiarize the student not so much with the facts as with the ideas of science. Secondly, that such ideas are best understood, at least by beginners, when studied in connection with concrete types of animals and plants. And, thirdly, that the types chosen should illustrate without unnecessary complication the particular grade of organization they are intended to typify, and that exceptional cases are out of place in an elementary course.

The types have therefore been selected with a view of illustrating all the more important modifications of structure and the chief physiological processes in plants and animals; and, by the occasional introduction of special lessons on such subjects as biogenesis, evolution, &c., the entire work is so arranged as to give a fairly connected account of the general principles of biology. It is in obedience to the last of the principles just enunciated that I have described so many of the Protozoa, omitted all but a brief reference to the development of Hydra and to the so-called sexual process in Penicillium, and described Nitella instead of Chara, and Polygordius instead of the earthworm. The last-named substitution is of course only made possible by the book being intended for the study and not for the laboratory, but I feel convinced that the student who masters the structure of Polygordius, even from figures and descriptions alone, will be in a far better position to profit by a practical study of one of the higher worms.

Lessons XXVII. and XXX. are mere summaries, and can only be read profitably by those who have studied the organisms described, or allied forms, in some detail. Such abstracts were however necessary to the plan of the book, in order to show how all the higher animals and plants may be described, so to speak, in terms of Polygordius and of the fern.

For many years I have been convinced of the urgent need for a simplification of nomenclature in biology, and have now attempted to carry out a consistent scheme, as will be seen by referring to the definitions in the glossary. Many of Mr. Harvey Gibson's suggestions are adopted and three new words are introduced—phyllula, gamobium, and agamobium. I expect and perhaps deserve to be criticised, or, what is worse, let alone, for the somewhat extreme step of using the word *ovary* in its zoological sense throughout the vegetable kingdom; and for describing as the *venter* of the pistil the so-called ovary of Angiosperms. I would only beg my critics before finally pronouncing judgment to try and look at the book, from the point of view of the beginner, as a graduated course of instruction, and to consider the effect upon the entire scheme of using a term of fundamental importance in two utterly different senses.

A large proportion of the figures are copied either from original sources or from my own drawings—the latter when no authority is mentioned. The majority, even of those which have previously appeared in text-books, have been specially engraved for the work, the draughtsman being my brother, Mr. M. P. Parker. In order to facilitate reference the illustrations referring to each subject have, as far as possible, been grouped together, so that the actual is considerably larger than the nominal number of figures. Full descriptions are given instead of mere lists of reference-letters: these will, I hope, be found useful as abstracts of the subjects illustrated.

I have to thank my friends Mr. A. Dillon Bell and Professor J. H. Scott, M.D., for constant and valuable help in criticising the manuscript. To Dr. Paul Meyer, of the Zoological Station, Naples, I am indebted for specimens of Polygordius; and to Professer Sale, of this University, Professor Haswell, of Sydney, Professor Thomas, of Auckland, and Professors Howes and D. H. Scott, of South Kensington, for important information and criticism on special points. My brother, Professor W. Newton Parker, has kindly promised to undertake a final revision for the press.

DUNEDIN, N.Z.,
August 1890.

PREFACE TO THE SECOND EDITION

IN addition to a thorough revision, Lessons VI. and XXIV. have been largely re-written. Figs. 9, 10, 52, 60, 64, and 66 are new, and Figs. 9, 10, 11, 64, 66, and 67 of the first edition have been withdrawn.

I have received valuable help from Professors W. N. Parker and G. B. Howes, Miss M. Greenwood, and Mr. J. E. S. Moore. Much of the proof-correcting has, as before, fallen upon my brother.

March 1893.

TABLE OF CONTENTS

	PAGE
PREFACE TO THE FIRST EDITION	v
PREFACE TO THE SECOND EDITION	xi
LIST OF ILLUSTRATIONS	xix

LESSON I.
AMŒBA 1

LESSON II.
HÆMATOCOCCUS 23

LESSON III.
HETEROMITA 36

LESSON IV.
EUGLENA 44

LESSON V.
PROTOMYXA 49
THE MYCETOZOA 52

LESSON VI.

A COMPARISON OF THE FOREGOING ORGANISMS WITH CERTAIN CONSTITUENT PARTS OF THE HIGHER ANIMALS AND PLANTS 56

    ANIMAL AND PLANT CELLS 56

    MINUTE STRUCTURE AND DIVISION OF CELLS AND NUCLEI 62

    OVA OF ANIMALS AND PLANTS 68

LESSON VII.

SACCHAROMYCES 71

LESSON VIII.

BACTERIA 82

LESSON IX.

BIOGENESIS AND ABIOGENESIS 95
HOMOGENESIS AND HETEROGENESIS 102

LESSON X.

PARAMŒCIUM 106
STYLONYCHIA 116
OXYTRICHA 120

LESSON XI.

OPALINA 121

LESSON XII.

VORTICELLA 126
ZOOTHAMNIUM 135

LESSON XIII.

SPECIES AND THEIR ORIGIN: THE PRINCIPLES OF CLASSIFICATION 137

LESSON XIV.

THE FORAMINIFERA 148
THE RADIOLARIA 152
THE DIATOMACEÆ 155

LESSON XV.

MUCOR 158

LESSON XVI.

VAUCHERIA 169
CAULERPA 175

LESSON XVII.

THE DISTINCTIVE CHARACTERS OF ANIMALS AND PLANTS . . . 176

LESSON XVIII.

PENICILLIUM 184
AGARICUS 191

LESSON XIX.

SPIROGYRA 194

LESSON XX.

MONOSTROMA 201
ULVA 203
LAMINARIA, &c. 203

LESSON XXI.

	PAGE
NITELLA	206

LESSON XXII.

HYDRA	221

LESSON XXIII.

HYDROID POLYPES	237
BOUGAINVILLEA, &C.	237
DIPHYES	250
PORPITA	253

LESSON XXIV.

SPERMATOGENESIS AND OOGENESIS	255
THE MATURATION AND IMPREGNATION OF THE OVUM	259
THE CONNECTION BETWEEN UNICELLULAR AND DIPLOBLASTIC ANIMALS	264

LESSON XXV.

POLYGORDIUS	271

LESSON XXVI.

POLYGORDIUS (*continued*)	293

LESSON XXVII.

THE GENERAL CHARACTERS OF THE HIGHER ANIMALS	307
THE STARFISH	309
THE CRAYFISH	314
THE FRESH-WATER MUSSEL	320
THE DOGFISH	324

TABLE OF CONTENTS   xvii

LESSON XXVIII.
                                              PAGE
MOSSES   332

LESSON XXIX.
FERNS   344

LESSON XXX.
THE GENERAL CHARACTERS OF THE HIGHER PLANTS 363
    EQUISETUM 366
    SALVINIA   368
    SELAGINELLA 371
    GYMNOSPERMS 373
    ANGIOSPERMS 378

SYNOPSIS 385

INDEX AND GLOSSARY 395

LIST OF ILLUSTRATIONS

FIG.		PAGE
1.	*Amœba*, various species	2
2.	*Protamœba primitiva*	9
3.	*Hæmatococcus pluvialis* and *H. lacustris*	24
4.	*Heteromita rostrata*	38
5.	*Euglena viridis*	45
6.	*Protomyxa aurantiaca*	50
7.	*Badhamia* and *Chondrioderma*	53
8.	Typical animal and vegetable cells	57
9.	Animal and plant cells, detailed structure	62
10.	Stages in the binary fission of a cell	64
11.	Ova of *Carmarina* and *Gymnadenia*	69
12.	*Saccharomyces cerevisiæ*	72
13.	*Bacterium termo*	83
14.	*Bacterium termo*, showing flagella	84
15.	*Micrococcus*	86
16.	*Bacillus subtilis*	87
17.	*Vibrio serpens*, *Spirillum tenue*, and *S. volutans*	88
18.	*Bacillus anthracis*	90
19.	Beaker with culture-tubes	100

FIG.		PAGE
20.	*Paramœcium aurelia*	108
21.	*Paramœcium aurelia*, conjugation	115
22.	*Stylonychia mytilus*	117
23.	*Oxytricha flava*	120
24.	*Opalina ranarum*	122
25.	*Vorticella*	127
26.	*Zoothamnium arbuscula*	134
27.	*Zoothamnium*, various species	138
28.	Diagram illustrating the Origin of the Species of *Zoothamnium* by Creation	142
29.	Diagram illustrating the Origin of the Species of *Zoothamnium* by Evolution	144
30.	*Rotalia*	149
31.	Diagrams of *Foraminifera*	150
32.	*Alveolina quoii*	151
33.	*Lithocircus annularis*	152
34.	*Actinomma asteracanthion*	153
35.	Diagrams of a Diatom and shells of *Navicula* and *Aulacodiscus*	156
36.	*Mucor mucedo* and *M. stolonifer*	159
37.	Moist Chamber	163
38.	*Vaucheria*	170
39.	*Caulerpa scalpelliformis*	174
40.	*Penicillinm glaucum*	186
41.	*Agaricus campestris*	192
42.	*Spirogyra*	195
43.	*Monostroma bullosum* and *M. laceratum*	202
44.	*Laminaria claustoni* and *Lessonia fuscescens*	204
45.	*Nitella*, general structure	207
46.	*Nitella*, terminal bud	212
47.	*Nitella*, spermary	215
48.	*Nitella*, ovary	217
49.	*Chara*, pro-embryo	219
50.	*Hydra viridis* and *H. fusca*, external form	222
51.	*Hydra*, minute structure	226

LIST OF ILLUSTRATIONS

FIG.		PAGE
52.	*Hydra*, nematocyst and nerve-cell	228
53.	*Hydra viridis*, ovum	235
54.	*Bougainvillea ramosa*	238
55.	Diagrams illustrating derivation of *Medusa* from *Hydranth*	242
56.	*Eucopella campanularia*, muscle fibres and nerve-cells	245
57.	*Laomedea flexuosa* and *Eudendrium ramosum*, development	249
58.	*Diphyes campanulata*	252
59.	*Porpita pacifica* and *P. mediterranea*	253
60.	Spermatogenesis in the Mole-Cricket	256
61.	Ovum of *Toxopneustes lividus*	259
62.	Maturation and impregnation of the animal ovum	260
63.	The gastrula	265
64.	*Pandorina morum*	266
65.	*Volvox globator*	268
66.	*Volvox globator*	269
67.	*Polygordius neapolitanus*, external form	272
68.	*Polygordius neapolitanus*, anatomy	274
69.	*Polygordius neapolitanus*, nephridium	285
70.	*Polygordius*, diagram illustrating the relations of the nervous-system	287
71.	*Polygordius neapolitanus*, reproductive organs	294
72.	*Polygordius neapolitanus*, larva in the trochosphere stage	296
73.	Diagram illustrating the origin of the trochosphere from the gastrula	298
74.	*Polygordius neapolitanus*, advanced trochosphere	300
75.	*Polygordius neapolitanus*, larva in a stage intermediate between the trochosphere and the adult	303
76.	Starfish, diagrammatic sections	310
77.	Crayfish, diagrammatic sections	316
78.	Mussel, diagrammatic sections	321
79.	Dogfish, diagrammatic sections	326
80.	Mosses, various genera, anatomy and histology	333
81.	*Funaria*, reproduction and development	338
82.	*Pteris* and *Aspidium*, anatomy and histology	346
83.	Ferns, various genera, reproduction and development	356

FIG.		PAGE
84.	*Equisetum*, reproduction and development	367
85.	*Salvinia*, reproduction and development	369
86.	*Selaginella*, reproduction and development	372
87.	*Gymnosperms*, reproduction and development	374
88.	*Angiosperms*, reproduction and development	379

LESSONS
IN
ELEMENTARY BIOLOGY

LESSONS IN ELEMENTARY BIOLOGY

LESSON I

AMŒBA

It is hardly possible to make a better beginning of the systematic study of Biology than by a detailed examination of a microscopic animalcule often found adhering to weeds and other submerged objects in stagnant water, and known to naturalists as *Amœba*.

Amœbæ are mostly invisible to the naked eye, rarely exceeding one-fourth of a millimetre ($\frac{1}{100}$ inch) in diameter, so that it is necessary to examine them entirely by the aid of the microscope. They can be seen and recognized under the low power of an ordinary student's microscope which magnifies from twenty-five to fifty diameters; but for accurate examination it is necessary to employ a far higher power, one in fact which magnifies about 300 diameters.

Seen under this power, an Amœba appears like a little

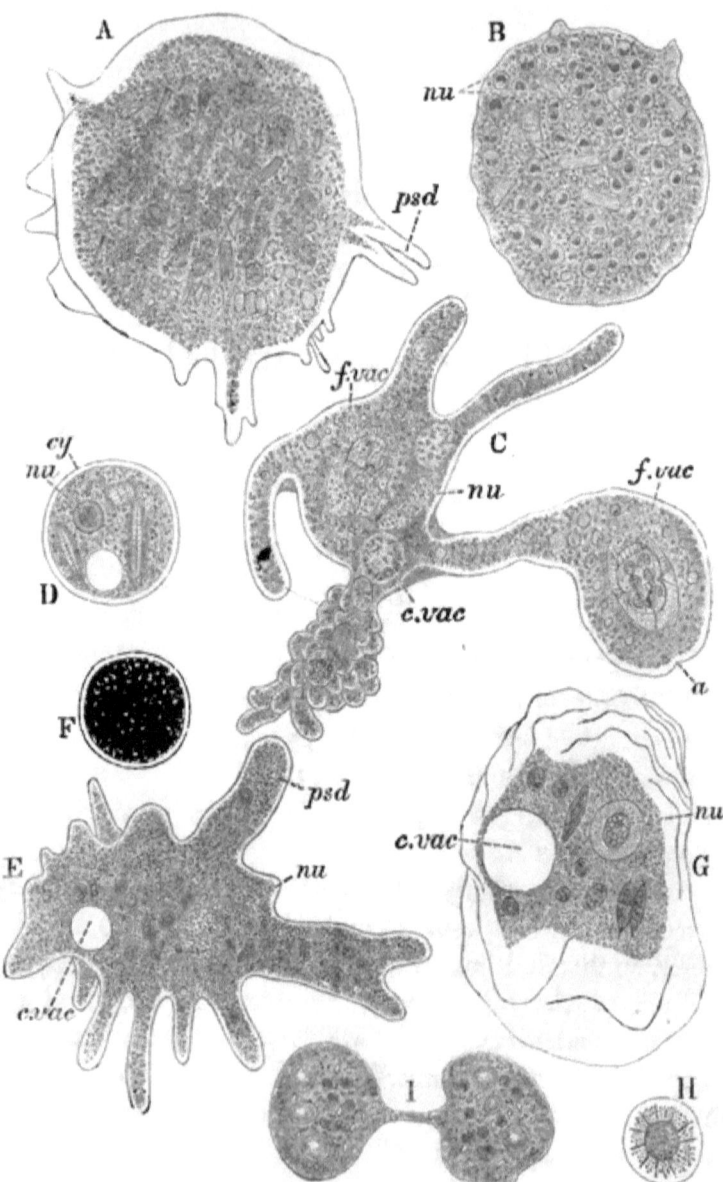

Fig 1.—A. *Amœba quarta*, a living specimen, showing granular endosarc surrounded by clear ectosarc, and several pseudopods (*psd*),

some formed of ectosarc only, others containing a core of endosarc. The larger bodies in the endosarc are mostly food-particles (× 300).[1]

B. The same species, killed and stained with carmine to show the numerous nuclei (*nu*) (× 300).

C. *Amœba proteus*, a living specimen, showing large irregular pseudopods, nucleus (*nu*), contractile vacuole (*c.vac*), and two food vacuoles (*f.vac*), each containing a small infusor (see Lesson X.) which has been ingested as food. The letter *a* to the right of the figure indicates the place where two pseudopods have united to inclose the food vacuole. The contractile vacuole in this figure is supposed to be seen through a layer of granular protoplasm, whereas in the succeeding figures (D, E, and G) it is seen in optical section, and therefore appears clear.

D. An encysted Amœba, showing cell-wall or cyst (*cy*), nucleus (*nu*), clear contractile vacuole (*c.vac*), and three diatoms (see Lesson XIV.) ingested as food.

E. *Amœba proteus*, a living specimen, showing several large pseudopods (*psd*), single nucleus (*nu*), and contractile vacuole (*c.vac*), and numerous food-particles embedded in the granular endosarc (× 330).

F. Nucleus of the same after staining, showing a ground substance or achromatin, containing deeply-stained granules of chromatin, and surrounded by a distinct membrane (× 1010).

G. *Amœba verrucosa*, living specimen, showing wrinkled surface, nucleus (*nu*), large contractile vacuole (*c.vac*) and several ingested organisms (× 330).

H. Nucleus of the same, stained, showing the chromatin aggregated in the centre to form a nucleolus (× 1010).

I. *Amœba proteus*, in the act of multiplying by binary fission (× 500).

(A, B, E, F, G, and H after Gruber; C and I after Leidy; D after Howes.)

shapeless blob of jelly, nearly or quite colourless. The central part of it (Fig. 1, A, C, and E) is granular and semitransparent—something like ground glass—while surrounding this inner mass is a border of perfectly transparent and colourless substance. So clear, indeed, is this outer layer that it is easily overlooked by the beginner, who is apt to take the granular internal substance for the whole Amœba. If in any way the creature can be made to turn over, or if a number of specimens are examined in various positions, these two constitutents will always be found to have the

---

[1] A number preceded by the sign of multiplication indicates the number of diameters to which the object is magnified.

same relations, whence we conclude that an Amœba consists of a granular substance the *endosarc*, completely surrounded by a clear transparent layer or *ectosarc*.

One very noticeable thing about Amœba is that it is never of quite the same shape for long together. Often the changes of form are so slow as to be almost imperceptible, like the movements of the hour-hand of a watch, but by examining it at successive intervals the alteration becomes perfectly obvious, and at the end of half an hour it will probably have altered so much as to be hardly like the same thing.

In an active specimen the way in which the changes of form are brought about is easily seen. At a particular point the ectosarc is pushed out in the form of a small pimple-like elevation (Fig. 1, A, left side) : this increases in size, still consisting of ectosarc only, until at last granules from the endosarc stream into it, and the projection or pseudopod (A, C, E, *psd*) comes to have the same structure as the rest of the Amœba. It must not be forgotten that the animal does not alter perceptibly in volume during the process, every pseudopod thus protruded from one part of the body necessitating the withdrawal of an equal volume from some other part.

This peculiar mode of movement may be illustrated by taking an irregular lump of clay or putty and squeezing it between the fingers. As it is compressed in one direction it will elongate in another, and the squeezing process may be regulated so as to cause the protrusion of comparatively narrow portions from the solid lump, when the resemblance to the movements described in the preceding paragraph will be fairly close. Only it must be borne in mind that in Amœba there is no external compression, the "squeezing" being done by the animalcule itself.

COMPOSITION OF PROTOPLASM

The occurrence of these movements is alone sufficient to show that Amœba is an *organism* or living thing, and no mere mass of dead matter.

The jelly-like substance of which Amœba is composed is called *protoplasm*. It is shown by chemical analysis [1] to consist mainly of certain substances known as *proteids*, bodies of extreme complexity in chemical constitution, the most familiar example of which is white of egg or albumen. They are compounds of carbon, hydrogen, oxygen, nitrogen, and sulphur, the five elements being combined in the following proportions :—

Carbon . .	from	51·5	to	54·5	*per cent.*
Hydrogen .	,,	6·9	,,	7·3	,, ,,
Oxygen .	,,	20·9	,,	23·5	,, ,,
Nitrogen .	,,	15·2	,,	17·0	,, ,,
Sulphur .	,,	0·3	,,	2·0	,, ,,

Besides proteids, protoplasm contains small proportions of mineral matters, especially phosphates and sulphates of potassium, calcium, and magnesium. It also contains a considerable quantity of water which, being as essential a constituent of it as the proteids and the mineral salts, is called *water of organization*.

Protoplasm is dissolved by prolonged treatment with weak acids or alkalies. Strong alcohol coagulates it, *i.e.*, causes it to shrink by withdrawal of water and become comparatively hard and opaque. Coagulation is also produced by raising the temperature to about 40° C.; the reader will remember how the familiar proteid white of egg is coagulated and rendered hard and opaque by heat.

[1] Accurate analyses of the protoplasm of Amœba have not been made, but the various micro-chemical tests which can be applied to it leave no doubt that it agrees in all essential respects with the protoplasm of other organisms, the composition of which is known (see p. 7).

There is another important property of proteids which is tested by the instrument called a dialyser. This consists essentially of a shallow vessel, the bottom of which is made of bladder, or vegetable parchment, or some other organic (animal or vegetable) membrane. If a solution of sugar or of salt is placed in a dialyser and the instrument floated in a larger vessel of distilled water, it will be found after a time that some of the sugar or salt has passed from the dialyser into the outer vessel through the membrane. On the other hand, if a solution of white of egg is placed in the dialyser no such transference to the outer vessel will take place.

The dialyser thus allows us to divide substances into two classes : *crystalloids*—so called because most of them, like salt and sugar, are capable of existing in the form of crystals—which, in the state of solution, will diffuse through an organic membrane ; and *colloids* or glue-like substances which will not diffuse. Protoplasm, like the proteids of which it is largely composed, is a colloid, that is, is non-diffusible.

Another character of proteids is their *instability*. A lump of salt or of sugar, a piece of wood or of chalk, may be preserved unaltered for any length of time, but a proteid if left to itself very soon begins to *decompose;* it acquires an offensive odour, and breaks up into simpler and simpler compounds, the most important of which are water (H_2O), carbon dioxide or carbonic acid (CO_2), ammonia (NH_3), and sulphuretted hydrogen (H_2S)[1]. In this character of instability or readiness to decompose protoplasm notoriously agrees with its constituent proteids ; any dead organism will,

[1] For a more detailed account of the phenomena of putrefaction see Lesson VIII., in which it will be seen that the above statement as to the instability of (dead) proteids requires qualification ; as a matter of fact they only decompose in the presence of living Bacteria.

unless special means are taken to preserve it, undergo more or less speedy decomposition.

Many of these properties of protoplasm can hardly be verified in the case of Amœba, owing to its minute size and the difficulty of isolating it from other organisms (waterweeds, &c.) with which it is always associated; but there are some tests which can be readily applied to it while under observation beneath the microscope.

One of the most striking of these micro-chemical tests depends upon the avidity with which protoplasm takes up certain colouring matters. If a drop of a neutral or slightly alkaline solution of carmine or logwood, or of some aniline dye, or a weak solution of iodine, is added to the water containing Amœba, the animalcule is killed, and at the same time becomes more or less deeply stained. The theory is that protoplasm has a slightly acid reaction, and thus produces precipitation of the colouring matter from the neutral or alkaline solution.

The staining is, however, not uniform. The endosarc, owing to the granules it contains, appears darker than the ectosarc, and there is usually to be seen, in the endosarc, a rounded spot more brightly stained than the rest. This structure, which can sometimes be seen in the living Amœba (Fig. 1, C, E, and G, *nu*), while frequently its presence is revealed only by staining (comp. A and B), is called the *nucleus*.

But when viewed under a sufficiently high power, the nucleus itself is seen to be unequally stained. It has lately been shown, in many Amœbæ, to be a globular body, enclosed in a very delicate membrane, and made up of two constituents, one of which is deeply stained by colouring matters, and is hence called *chromatin*, while the other, the *nuclear matrix* or *achromatin*, takes a lighter tint (Fig. 1, F). The relative arrangement of chromatin and matrix varies in different Amœbæ: sometimes there are granules of chromatin in an achromatic ground substance (F); sometimes the chromatin is collected towards the surface or periphery of the nucleus; sometimes, again, it becomes aggregated in the centre (G, H). In the latter case the nucleus is seen to have a deeply-stained central portion, which is then distinguished as the *nucleolus*.

When it is said that Amœbæ sometimes have one kind of nucleus and sometimes another, it must not be inferred that the same animalcule varies in this respect. What is meant is that there are found in stagnant water many kinds or *species* of Amœba which are distinguished from one another, amongst other things, by the character of their nuclei, just as the various species of *Felis*—the cat, lion, tiger, lynx, &c.—are distinguished from one another, amongst other things, by the colour and markings of their fur. According to the method of *binomial nomenclature* introduced into biology by Linnæus, the same *generic name* is applied to all such closely allied species, while each is specially distinguished by a second or *specific name* of its own. Thus under the *genus* Amœba are included *Amœba proteus* (Fig. 1, C, E, and F), with long lobed pseudopods and a nucleus containing evenly-disposed granules of chromatin; *A. quarta* (A and B), with short pseudopods and numerous nuclei; *A. verrucosa* (G and H) with crumpled or folded surface, no well-marked pseudopods, and a nucleus with a central aggregation of chromatin, or nucleolus; and many others.

Besides the nucleus, there is another structure frequently visible in the living Amœba. This is a clear, rounded space in the ectosarc (C, E, and G, *c. vac*), which periodically disappears with a sudden contraction and then slowly re-appears, its movements reminding one of the beating of a minute colourless heart. It is called the *contractile vacuole*, and consists of a cavity in the ectosarc containing a watery fluid.

Occasionally Amœbæ—or more strictly Amœba-like organisms—are met with which have neither nucleus[1] nor contractile vacuole, and are therefore placed in the separate genus *Protamœba* (Fig. 2). They may be looked upon as the simplest of living things.

The preceding paragraphs may be summed up by saying that Amœba is a mass of protoplasm produced into temporary processes or pseudopods, divisible into ectosarc and

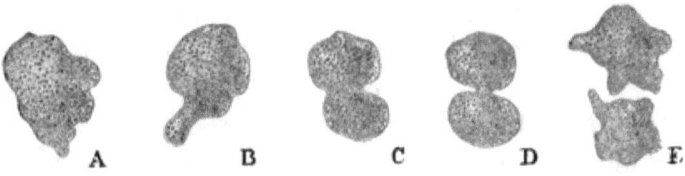

FIG. 2—*Protamœba primitiva*; A, B, the same specimen drawn at short intervals of time, showing changes of form.
C—E. Three stages in the process of binary fission. (After Haeckel.)

endosarc, and containing a nucleus and a contractile vacuole : that the nucleus consists of two substances, chromatin and achromatin, enclosed in a distinct membrane : and that the contractile vacuole is a mere cavity in the protoplasm containing fluid. All these facts come under the head of *Morphology*, the division of biology which treats of form and structure : we must now study the *Physiology* of our animalcule—that is, consider the actions or *functions* it is capable of performing.

[1] Judging from the analogy of the Infusoria it seems very probable that such apparently non-nucleate forms as Protamœba contain chromatin diffused in the form of minute granules throughout their substance (see end of Lesson X., p. 118), or that they are forms which have lost their nuclei.

First of all, as we have already seen, it moves, the movement consisting in the slow protrusion and withdrawal of pseudopods. This may be expressed generally by saying that Amœba is *contractile*, or that it exhibits *contractility*. But here it must be borne in mind that contraction does not mean the same thing in biology as in physics. When it is said that a red-hot bar of iron contracts on cooling, what is meant is that there is an actual reduction in volume, the bar becoming smaller in all dimensions. But when it is said that an Amœba contracts, what is meant is that it diminishes in one dimension while increasing in another, no perceptible alteration in volume taking place: each time a pseudopod is protruded an equivalent volume of protoplasm is withdrawn from some other part of the body.

We may say then that contractility is a function of the protoplasm of Amœba—that is, that it is one of the actions which the protoplasm is capable of performing.

A contraction may arise in one or other of two ways. In most cases the movements of an Amœba take place without any obvious external cause; they are what would be called in the higher animals voluntary movements—movements dictated by the will and not necessarily in response to any external stimulus. Such movements are called *automatic*. On the other hand, movements may be induced in Amœba by external stimuli, by a sudden shock, or by coming into contact with an object suitable for food: such movements are the result of *irritability* of the protoplasm, which is thus both automatic and irritable—that is, its contractility may be set in action either by internal or by external stimuli.

Under certain circumstances an Amœba temporarily loses its power of movement, draws in its pseudopods, and becomes a globular mass around which is formed a thick, shell-like coat, called the *cyst* or *cell-wall* (Fig. 1, D, *cy*). The composition of this is not known; it is certainly not protoplasmic, and very probably consists of some nitrogenous substance allied in composition to horn and to the chitin which forms the external shell of crustacea, insects, &c. After remaining in this *encysted* condition for a time, the Amœba escapes by the rupture of its cell-wall, and resumes its active life.

Very often an Amœba in the course of its wanderings comes in contact with a still smaller organism, such as a diatom (see Lesson XIV., Fig. 35) or a small infusor (see Lessons X.—XII.). When this happens the Amœba may be seen to send out pseudopods which gradually creep round the prey, and finally unite on the far side of it, as in Fig. 1, C, *a*. The diatom or other organism becomes in this way completely enclosed in a cavity or *food-vacuole* (*f. vac*), which also contains a small quantity of water necessarily included with the prey. The latter is taken in by the Amœba as food: so that another function performed by the animalcule is the *reception* of food, the first step in the process of *nutrition*. It is to be noted that the reception of food takes place in a particular way, viz. by *ingestion*— *i.e.* it is enclosed raw and entire in the living protoplasm. It has been noticed that Amœba usually ingests at its hinder end—that is, the end directed backwards in progression.

Having thus ingested its prey, the Amœba continues its course, when, if carefully watched, the swallowed organism will be seen to undergo certain changes. Its protoplasm is slowly dissolved; if it contains chlorophyll—the green colouring matter of plants—this is gradually turned to brown; and finally nothing is left but the case or cell-wall in which many minute organisms, such as diatoms, are enclosed.

Finally, the Amœba as it creeps slowly on leaves this empty cell-wall behind, and thus gets rid of what it has no further use for. It is thus able to *ingest* living organisms as food; to dissolve or *digest* their protoplasm; and to *egest* or get rid of any insoluble materials they may contain. Note that all this is done without either ingestive aperture (mouth), digestive cavity (stomach), or egestive aperture (anus); the food is simply taken in by the flowing round it of pseudopods, digested as it lies enclosed in the protoplasm, and got rid of by the Amœba flowing away from it.

It has just been said that the protoplasm of the prey is dissolved or digested: we must now consider more particularly what this means.

The stomachs of the higher animals—ourselves, for instance—produce in their interior a fluid called *gastric juice*. When this fluid is brought into contact with albumen or any other proteid a remarkable change takes place. The proteid is dissolved and at the same time rendered diffusible, so as to be capable, like a solution of salt or sugar, of passing through an organic membrane (see p. 6). The diffusible proteids thus formed by the action of gastric juice upon ordinary proteids are called *peptones*: the transformation is effected through the agency of a constituent of the gastric juice called *pepsin*.

There can be little doubt that the protoplasm of Amœba is able to convert that of its prey into a soluble and diffusible form, possibly by the agency of some substance analogous to pepsin, and that the dissolved matters diffuse through the body of the Amœba until the latter is, as it were, soaked through and through with them. Under these circumstances the Amœba may be compared to a sponge which is allowed to absorb water, the sponge itself representing the living protoplasm, the water the solution of proteids which permeates it. It has been proved by experiment that proteids are the only class of food which Amœba can make use of: it is unable to digest either starch or fat—two very important constituents of the food of the higher animals. Mineral matters must, however, be taken with the food in the form of a weak watery solution, since the water in which the animalcule lives is never absolutely pure.

The Amœba being thus permeated, as it were, with a nutrient solution, a very important process takes place. The elements of the solution, hitherto arranged in the form of peptones, mineral salts, and water, become re-arranged in such a way as to form new particles of living protoplasm, which are deposited among the pre-existing particles. In a word, the food is *assimilated* or converted into the actual living substance of the Amœba.

One effect of this formation of new protoplasm is obvious: if nothing happens to counteract it, the Amœba must *grow*, the increase in size being brought about in much the same way as that of a heap of stones would be by continually thrusting new pebbles into the interior of the heap. This mode of growth—by the interposition of new particles among old ones—is called growth by *intussusception*, and is very characteristic of the growth of protoplasm. It is necessary to distinguish it, because there is another mode of growth which is characteristic of minerals and occurs also in some organized structures. A crystal of alum, for instance, suspended in a strong solution of the same substance grows, but the increase is due to the deposition of successive layers on the surface of the original crystal, in much the same way as a candle might be made to grow by repeatedly dipping it into melted grease. This can be proved by colouring the crystal with logwood or some other dye before suspending it, when a gradually-increasing colourless layer will be deposited round the coloured crystal: if growth took place by intussusception we should have a gradual weakening of the tint as the crystal increased in size. This mode of growth—by the deposition of successive layers—is called growth by *accretion*.

It is probable that the cyst of Amœba referred to above (p. 11) grows by accretion. Judging from the analogy of other organisms it would seem that, after rounding itself off, the surface of the sphere of protoplasm undergoes a chemical change resulting in the formation of a thin superficial layer of non-protoplasmic substance. The process is repeated, new layers being continually deposited within the old ones until the cell-wall attains its full thickness. The cyst is therefore a substance separated or *secreted* from the protoplasm; it is the first instance we have met with of a *product of secretion*.

From the fact that Amœba rarely attains a greater diameter than $\frac{1}{4}$ mm., it follows that something must happen to counteract the constant tendency to grow, which is one of the results of assimilation. We all know what happens in our own case: if we take a certain amount of exercise—walk ten miles or lift a series of heavy weights—we undergo a loss of substance manifested by a diminution in weight and by the sensation of hunger. Our bodies have done a certain amount of work, and have undergone a proportional amount of waste, just as a fire every time it blazes up consumes a certain weight of coal.

Precisely the same thing happens on a small scale with Amœba. Every time it thrusts out or withdraws a pseudopod, every time it contracts its vacuole, it does a certain amount of work—moves a definite weight of protoplasm through a given space. And every movement, however slight, is accompanied by a proportional waste of substance, a certain fraction of the protoplasm becoming oxidized, or in other words undergoing a process of low temperature combustion.

When we say that any combustible body is burnt what we usually mean is that it has combined with oxygen, forming certain products of combustion due to the chemical union of the oxygen with the substance burnt. For instance, when carbon is burnt the product of combustion is carbon dioxide or carbonic acid ($C + O_2 = CO_2$) : when hydrogen is burnt, water ($H_2 + O = H_2O$). The products of the slow combustion which our own bodies are constantly undergoing are these same two bodies—carbon dioxide given off mainly in the air breathed out, and water given off mainly in the form of perspiration and urine—together with two compounds containing nitrogen, urea (CH_4N_2O) and uric acid $C_5H_4N_4O_3$), both occurring mainly in the urine. In some animals urea and uric acid are replaced by other compounds such as guanin ($C_5H_5N_5O$), but it may be taken as proved that in all living things the product of combustion are carbon dioxide, water, and some nitrogenous substance of simpler constitution than proteids, and allied to the three just mentioned.

With this breaking down of proteids the vital activity of all organisms are invariably connected. Just as useful mechanical work may be done by the fall of a weight from a given height to the level of the ground, so the work done by the organism is a result of its complex proteids falling, so to speak, to the level of simpler substances. In both instances potential energy or energy of position is converted into kinetic or actual energy.

In the particular case under consideration we have to rely upon analogy and not upon direct experiment. We may, however, be quite sure that the products of combustion or waste matters of Amœba include carbon dioxide, water and some comparatively simple (as compared with proteids) compound of nitrogen.

These waste matters or *excretory products* are given off partly from the general surface of the body, but partly, it would seem, through the agency of the contractile vacuole. It appears that the water taken in with the food, together in all probability with some of that formed by oxidation of the protoplasm, makes its way to the vacuole, and is expelled by its contraction. We have here another function performed by Amœba, that of *excretion*, or the getting rid of waste matters.

In this connection the reader must be warned against a possible misunderstanding arising from the fact that the word excretion is often used in two senses. We often hear, for instance, of solid and liquid "excreta." In Amœba the solid excreta, or more correctly *fæces*, consist of such things as the indigestible cell-walls, starch-grains, &c., of the organisms upon which it feeds ; but the rejection of these is no more a process of excretion than the spitting out of a cherry-stone, since they are simply parts of the food which have never been assimilated—never formed part and parcel of the organism. True excreta, on the other hand, are invariably products of the waste or decomposition of protoplasm.

The statement just made that the protoplasm of Amœba constantly undergoes oxidation presupposes a constant supply of oxygen. The water in which the animalcule lives invariably contains that gas in solution : on the other hand, as we have seen, the protoplasm is continually forming carbon dioxide. Now when two gases are separated from one another by a porous partition, an interchange takes place between them, each diffusing into the space occupied by the other. The same process of gaseous diffusion is continually going on between the carbon dioxide in the interior of Amœba and the oxygen in the surrounding water, the protoplasm acting as the porous partition. In this way the carbon dioxide is got rid of, and at the same time a supply of oxygen is obtained for further combustion.

The taking in of oxygen might be looked upon as a kind of feeding process, the food being gaseous instead of solid or liquid, just as we might speak of "feeding" a fire both with coals and with air. Moreover, as we have seen, the giving out of carbon dioxide is a process of excretion. It is, however, usual and convenient to speak of this process of exchange of gases as *respiration* or breathing, which is therefore another function performed by the protoplasm of Amœba.

The oxidation of protoplasm in the body of an organism, like the combustion of wood or coal in a fire, is accompanied by an *evolution of heat*. That this occurs in Amœba cannot be doubted, although it has never been proved. The heat thus generated is, however, constantly being lost to the surrounding water, so that the temperature of Amœba, if we could but measure it, would probably be found, like that of a frog or a fish, to be very little if at all above that of the medium in which it lives.

We thus see that a very elaborate series of chemical processes is constantly going on in the interior of Amœba. These processes are divisible into two sets: those which begin with the digestion of food and end with the manufacture of living protoplasm, and those which have to do with the destruction of protoplasm and end with excretion.

The whole series of processes are spoken of collectively as *metabolism*. We have, first of all, digested food diffused through the protoplasm and finally converted into fresh

C

living protoplasm : these are processes of *constructive metabolism* or *anabolism*. Next we have the protoplasm gradually breaking down and undergoing conversion into excretory products : this is the process of *destructive metabolism* or *katabolism*. There can be little doubt that both are processes of extreme complexity : it seems probable that after the food is once dissolved there ensues the successive formation of numerous bodies of gradually increasing complexity (*anabolic mesostates* or *anastates*), culminating in protoplasm ; and that the protoplasm, when once formed, is decomposed into a series of substances of gradually diminishing complexity (*katabolic mesostates* or *katastates*), the end of the series being formed by the comparatively simple products of excretion. The granules in the endosarc are probably to be looked upon as various mesostates imbedded in the protoplasm proper.

Living protoplasm is thus the most unstable of substances ; it is never precisely the same thing for two consecutive seconds : it "decomposes but to recompose," and recomposes but to decompose ; its existence, like that of a waterfall or a fountain, depends upon the constant flow of matter into it and away from it.

It follows from what has been said that if the income of an Amœba, *i.e.*, the total weight of substances taken in (food *plus* oxygen *plus* water) is greater than its expenditure or the total weight of substances given out (fæces *plus* excreta proper *plus* carbon dioxide) the animalcule will grow : if less it will dwindle away : if the two are equal it will remain of the same weight or in a state of physiological equilibrium.

We see then that the fundamental condition of existence of the individual Amœba is that it should be able to form new protoplasm out of the food supplied to it. But something more than this is necessary. Amœbæ are subject to all sorts of casualties; they may be eaten by other organisms or the pool in which they live may be dried up; in one way or another they are constantly coming to an end. From which it follows that if the race of Amœbæ is to be preserved there must be some provision by which the individuals composing it are enabled to produce new individuals. In other words Amœba must, in addition to its other functions, perform that of *reproduction*.

An Amœba reproduces itself in a very simple way. The nucleus first divides into two: then the whole organism elongates, the two nuclei at the same time travelling away from one another: next a furrow appears across the middle of the drawn-out body between the nuclei (Fig. 1, I; fig. 2, C, D): the furrow deepens until finally the animalcule separates into two separate Amœbæ (Fig. 2, E), which henceforward lead an independent existence.

This, the simplest method of reproduction known, is called *simple* or *binary fission*. Notice how strikingly different it is from the mode of multiplication with which we are familiar in the higher animals. A fowl, for instance, multiplies by laying eggs at certain intervals, in each of which, under favourable circumstances, and after a definite lapse of time, a chick is developed: moreover, the parent bird, after continuing to produce eggs for a longer or shorter time, dies. An Amœba, on the other hand, simply divides into two Amœbæ, each exactly like itself, and in doing so ceases to exist as a distinct individual. Instead of the successive production of offspring from an ultimately dying parent, we have the simultaneous production of offspring by the division of the parent, which does not die, but becomes simply merged in its progeny. There can be no better instance of the fact that reproduction is discontinuous growth.

From this it seems that an Amœba, unless suffering a violent death, is practically immortal, since it divides into two completely organized individuals, each of which begins life with half of the entire body of its parent, there being therefore nothing left of the latter to die. It would appear, however, judging from the analogy of the Infusoria (see Lesson X.) that such organisms as Amœba cannot go on multiplying indefinitely by simple fission, and that occasionally two individuals come into contact and undergo complete fusion. A *conjugation* of this kind has been observed in Amœba, but has been more thoroughly studied in other forms (see Lessons III. and X.). Whether it is a necessary condition of continued existence in our animalcule or not, it appears certain that "death has no place as a natural recurrent phenomenon" in that organism.

If an Amœba does happen to be killed and to escape being eaten it will undergo gradual decomposition, becoming converted into various simple substances of which carbon dioxide, water, and ammonia are the chief. (See p. 90.)

In conclusion, a few facts may be mentioned as to the conditions of life of Amœba—the circumstances under which it will live or die, flourish or otherwise.

In the first place, it will live only within certain limits of temperature. In moderately warm weather the temperature to which it is exposed may be taken as about 15° C. If gradually warmed beyond this point the movements at first show an increased activity, then become more and more sluggish, and at about 30°—35° C. cease altogether, recommencing, however, when the temperature is lowered. If the heating is continued up to about 40° C. the animalcule is killed by the coagulation of its protoplasm (see p. 5): it is then said to suffer *heat-rigor* or death-stiffening produced by heat. Similarly when it is cooled below the ordinary temperature the movements become slower and slower, and at the freezing point (0° C.) cease entirely. But freezing, unlike over-heating, does not kill the protoplasm, but only renders it temporarily inert; on thawing, the movements recommence. We may therefore distinguish an *optimum* temperature at which the vital actions are carried on with the greatest activity; *maximum* and *minimum* temperatures above and below which respectively they cease; and an *ultra-maximum* temperature at which death ensues. There is no definite ultra-minimum temperature known in the case of Amœba.

The quantity of water present in the protoplasm—as water of organization (see p. 5)—is another matter of importance. The water in which Amœba lives, although fresh, always contains a certain percentage of salts in solution, and the protoplasm is affected by any alteration in the density of the surrounding medium; for instance, by replacing it by distilled water and so reducing the density, or by adding salt and so increasing it. The addition of common salt, (sodium chloride) to the amount of 2 per cent. causes Amœba to withdraw its pseudopods and undergo a certain amount of shrinkage: it is then said to pass into a condition of *dry-rigor*. Under these circumstances it may be restored to its normal condition by adding a sufficient proportion of water to bring back the fluid to its original density.

In this connection it is interesting to notice that the deleterious effects of an excess of salt are produced only when the salt is added suddenly. By the very gradual addition of sodium chloride Amœbæ have been brought to live in a 4 per cent. solution, *i.e.*, one twice as strong as would, if added suddenly, produce dry-rigor.

From what has been said above on the subject of respiration (p. 17) it follows that free oxygen is necessary for the existence of Amœba. Light, on the other hand, appears to be unnecessary, amœboid movements having been shown to go on actively in darkness.

LESSON II

HÆMATOCOCCUS

THE rain-water which collects in puddles, open gutters, &c., is frequently found to have a green colour. This colour is due to the presence of various organisms—plants or animals—one of the commonest of which is called *Hæmatococcus* (or as it is sometimes called *Protococcus* or *Sphærella*) *pluvialis*.

Like Amœba, Hæmatococcus is so small as to require a high power for its examination. Magnified three or four hundred diameters it has the appearance (Fig. 3, A) of an ovoidal body, somewhat pointed at one end, and of a bright green colour, more or less flecked with equally bright red.

Like Amœba, moreover, it is in constant movement, but the character of the movement is very different in the two cases. An active Hæmatococcus is seen to swim about the field of the microscope in all directions and with considerable apparent rapidity. We say *apparent* rapidity because the rate of progression is magnified to the same extent as the organism itself, and what appears a racing speed under the microscope is actually a very slow crawl when divided by 300. It has been found that such organisms as Hæmatococcus travel at the rate of one foot in from a quarter of an hour to an hour: or, to express the fact in another and fairer way, that they travel a distance equal to 2½ times their own diameter in one second. In swimming the pointed end is always directed forwards and

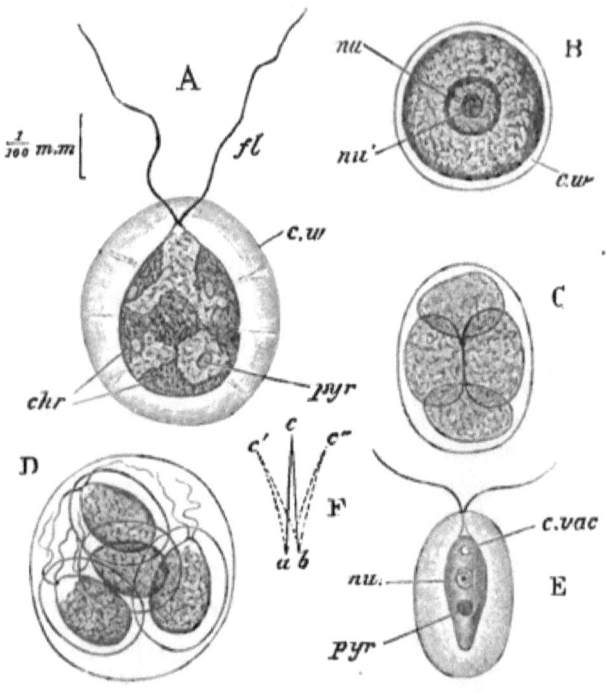

FIG. 3.—A. *Hæmatococcus pluvialis*, motile phase. Living specimen, showing protoplasm with chromatophores (*chr*) and pyrenoids (*pyr*), cell-wall (*c.w*) connected to cell-body by protoplasmic filaments, and flagella *fl*. The scale to the left applies to Figs. A—D.

B. Resting stage of the same, showing nucleus (*nu*) with nucleolus (*nu'*), and thick cell-wall (*c.w*) in contact with protoplasm.

C. The same, showing division of the cell-body in the resting stage into four daughter-cells.

D. The same, showing the development of flagella and detached cell-wall by the daughter-cells before their liberation from the inclosing mother-cell-wall.

E. *Hæmatococcus lacustris*, showing nucleus (*nu*), single large pyrenoid (*pyr*), and contractile vacuole (*c.vac*),

F. Diagram illustrating the movement of a flagellum: *ab*, its base; *c*, *c'*, *c"*, different positions assumed by its apex. (E, after Bütschli.)

the forward movement is accompanied by a rotation of the organism upon its longer axis.

Careful watching shows that the outline of a swimming Hæmatococcus does not change, so that there is evidently no protrusion of pseudopods, and at first the cause of the movement appears rather mysterious. Sooner or later, however, the little creature is sure to come to rest, and there can then be seen projecting from the pointed end two excessively delicate colourless threads (Fig. 3, A, *fl*), each about half as long again as the animalcule itself: these are called *flagella* or sometimes *cilia*.[1] In a Hæmatococcus which has come to rest these can often be seen gently waving from side to side: when this slow movement is exchanged for a rapid one the whole organism is propelled through the water, the flagella acting like a pair of extremely fine and flexible fins or paddles. Thus the movement of Hæmatococcus is not *amœboid*, *i.e.*, produced by the protrusion and withdrawal of pseudopods, but is *ciliary*, *i.e.*, due to the rapid vibration of cilia or flagella.

The flagella are still more clearly seen by adding a drop of iodine solution to the water: this immediately kills and stains the organism, and the flagella are seen to take on a distinct yellow tint. By this and other tests it is shown that Hæmatococcus, like Amœba, consists of protoplasm, and that the flagella are simply filamentous processes of the protoplasm.

It was mentioned above that in swimming the pointed end

[1] The word *cilium* is sometimes used as a general term to include any delicate vibratile process of protoplasm: often, however, it is used in a restricted sense for a rhythmically vibrating thread, of which each cell bears a considerable number (see Fig. 8, E, and Fig. 21); a flagellum is a cilium having a whip-lash-like movement, and each cell bearing only a limited number—one or two, or occasionally as many as four.

with the flagella goes first; this may therefore be distinguished as the anterior extremity, the opposite or blunt end being posterior. So that, as compared with Amœba, Hæmatococcus exhibits a *differentiation of structure :* an anterior and a posterior end can be distinguished, and a part of the protoplasm is differentiated or set apart as flagella.

The green colour of the body is due to the presence of a special pigment called *chlorophyll*, the substance to which the colour of leaves is due. That this is something quite distinct from the protoplasm may be seen by treatment with alcohol, which simply kills and coagulates the protoplasm, but completely dissolves out the chlorophyll, producing a clear green solution. The solution, although green by transmitted light, is red under a strong reflected light, and is hence *fluorescent :* when examined through the spectroscope it has the effect of absorbing the whole of the blue and violet end of the spectrum as well as a part of the red. The red colour which occurs in so many individuals, sometimes almost replacing the green, is due to a colouring matter closely allied in its properties to chlorophyll and called *hæmatochrome*.

At first sight the chlorophyll appears to be evenly distributed over the whole body, but accurate examination under a high power shows it to be lodged in a variable number of irregular structures called *chromatophores* (Fig. 3, A, *chr.*), which together form a layer immediately beneath the surface. Each chromatophore consists of a protoplasmic substance impregnated with chlorophyll.

After solution of the chlorophyll with alcohol a nucleus (B, *nu.*) can be made out; like the nucleus of Amœba it is stained by iodine, magenta, &c. Other bodies which might easily be mistaken for nuclei are also visible in the living organism. These are small ovoidal structures (A, *pyr.*), with clearly defined outlines occurring in varying numbers in the chromatophores. When treated with iodine they assume a deep, apparently black but really dark blue, colour. The assumption of a blue colour with iodine is the characteristic test of the well-known substance *starch*, as can be seen by letting a few drops of a weak solution of iodine fall upon some ordinary washing starch. The bodies in question have been found to consist of a proteid substance covered with a layer of starch, and are called *pyrenoids*. Starch itself is a definite chemical compound belonging to the group of *carbo-hydrates*, *i.e.*, bodies containing the elements carbon, hydrogen, and oxygen: its formula is $C_6 H_{10} O_5$.

In Hæmatococcus pluvialis there is no contractile vacuole, but in another species, *H. lacustris*, this structure is present as a minute space near the anterior or pointed end (Fig. 3, E, *c. vac.*).

There is still another characteristic structure to which no reference has yet been made. This appears at the first view something like a delicate haze around the green body, but by careful focusing is seen to be really an extremely thin globular shell (A, *c.w.*) composed of some colourless transparent material and separated by a space containing water from the body to which it is connected by very delicate radiating strands of protoplasm. It is perforated by two extremely minute apertures for the passage of the flagella. Obviously we may consider this shell as a cyst or cell-wall differing from that of an encysted Amœba (Fig. 1, D) in not being in close contact with the protoplasm.

A more important difference, however, lies in its chemical composition. The cyst or cell-wall of Amœba, as stated in the preceding lesson (p. 11) is very probably nitrogenous:

that of Hæmatococcus, on the other hand, is formed of a carbohydrate called *cellulose*, allied in composition to starch, sugar, and gum, and having the formula $C_6 H_{10} O_5$. Many vegetable substances, such as cotton, consist of cellulose, and wood is a modification of the same compound. Cellulose is stained yellow by iodine, but iodine and sulphuric acid together turn it blue, and a similar colour is produced by a solution of iodine and potassium iodide in zinc chloride known as Schulze's solution. These tests are quite easily applied to Hæmatococcus: the protoplasm stains a deep yellowish-brown, around which is seen a sort of blue cloud due to the stained and partly-dissolved cell-wall.

It has been stated that in stagnant water in which it has been cultivated for a length of time Hæmatococcus sometimes assumes an amœboid form. In any case, after leading an active existence for a longer or shorter time it comes to rest, loses its flagella, and throws around itself a thick cell-wall of cellulose (Fig. 3, B), thus becoming encysted. So that, as in Amœba, there is an alternation of an active or motile with a stationary or resting condition.

In the matter of nutrition the differences between Hæmatococcus and Amœba are very marked and indeed fundamental. As we have seen, Hæmatococcus has no pseudopods, and therefore cannot take in solid food after the manner of Amœba: moreover, even in its active condition it is usually surrounded by an imperforate cell-wall, which of course quite precludes the possibility of ingestion. As a matter of observation, also, however long it is watched it is never seen to *feed* in the ordinary sense of the word. Nevertheless it must take in food in some way or other, or the decomposition of its protoplasm would soon bring it to an end.

Hæmatococcus lives in rain-water. This is never pure water, but always contains certain mineral salts in solution, especially nitrates, ammonia salts, and often sodium chloride or common table salt. These salts, being crystalloids, can and do diffuse into the water of organization of the animalcule, so that we may consider its protoplasm to be constantly permeated by a very weak saline solution, the most important elements contained in which are oxygen, hydrogen, nitrogen, potassium, sodium, calcium, sulphur, and phosphorus.

If water containing a large quantity of Hæmatococcus is exposed to sunlight, minute bubbles are found to appear in it, and these bubbles, if collected and properly tested, are found to consist largely of oxygen. Accurate chemical analysis has shown that this oxygen is produced by the decomposition of the carbon dioxide contained in solution in rain-water, and indeed in all water exposed to the air, the gas, which is always present in small quantities in the atmosphere, being very soluble in water.

As the carbon dioxide is decomposed in this way, its oxygen being given off, it is evident that its carbon must be retained. As a matter of fact it is retained by the organism but not in the form of carbon; in all probability a double decomposition takes place between the carbon dioxide absorbed and the water of organization, the result being the liberation of oxygen in the form of gas and the simultaneous production of some extremely simple form of carbohydrate, *i.e.*, some compound of carbon, hydrogen, and oxygen, with a comparatively small number of atoms to the molecule.

The next step seems to be that the carbohydrate thus formed unites with the ammonia salts or the nitrates absorbed from the surrounding water, the result being the formation of some comparatively simple nitrogenous compound, probably belonging to the class of amides, one of the best known of which—asparagin—has the formula $C_4 H_8 N_2 O_3$. Then further combinations take place, substances of greater and greater complexity are produced, sulphur from the absorbed sulphates enters into combination, and proteids are formed. From these, finally, fresh living protoplasm arises.

From the foregoing account, which only aims at giving the very briefest outline of a subject as yet imperfectly understood, it will be seen that, as in Amœba, the final result of the nutritive process is the manufacture of protoplasm, and that this result is attained by the formation of various substances of increasing complexity or anastates (see p. 18). But it must be noted that the steps in this process of constructive metabolism are widely different in the two cases. In Amœba we start with living protoplasm—that of the prey—which is killed and broken up into diffusible proteids, these being afterwards re-combined to form new molecules of the living protoplasm of Amœba. So that the food of Amœba is, to begin with, as complex as itself, and is first broken down by digestion into simpler compounds, these being afterwards re-combined into more complex ones. In Hæmatococcus, on the other hand, we start with extremely simple compounds, such as carbon dioxide, water, nitrates, sulphates, &c. Nothing which can be properly called digestion, *i.e.*, a breaking up and dissolving of the food, takes place, but its various constituents are combined into substances of gradually increasing complexity, protoplasm, as before, being the final result.

To express the matter in another way: Amœba can only make protoplasm out of proteids already formed by some other organism: Hæmatococcus can form it out of simple liquid and gaseous inorganic materials.

Speaking generally, it may be said that these two methods

of nutrition are respectively characteristic of the two great groups of living things. Animals require solid food containing ready-made proteids, and cannot build up their protoplasm out of simpler compounds. Green plants, *i.e.*, all the ordinary trees, shrubs, weeds, &c., take only liquid and gaseous food, and built up their protoplasm out of carbon dioxide, water, and mineral salts. The first of these methods of nutrition is conveniently distinguished as *holozoic*, or wholly-animal, the second as *holophytic*, or wholly-vegetal.

It is important to note that only those plants or parts of plants in which chlorophyll is present are capable of holophytic nutrition. Whatever may be the precise way in which the process is effected, it is certain that the decomposition of carbon dioxide which characterizes this form of nutrition is a function of chlorophyll, or to speak more accurately, of chromatophores, since there is reason for thinking that it is the protoplasm of these and not the actual green pigment which is the active agent in the process.

Moreover, it must not be forgotten that the decomposition of carbon dioxide is carried on only during daylight, so that organisms in which holophytic nutrition obtains are dependent upon the sun for their very existence. While Amœba derives its energy from the breaking down of the proteids in its food (see p. 12), the food of Hæmatococcus is too simple to serve as a source of energy, and it is only by the help of sunlight that the work of constructive metabolism can be carried on. This may be expressed by saying that Hæmatococcus, in common with other organisms, containing chlorophyll, is supplied with kinetic energy (in the form of light or radiant energy) directly by the sun.

As in Amœba, destructive metabolism is constantly going on side by side with constructive. The protoplasm becomes oxidized, water, carbon dioxide, and nitrogenous waste matters being formed and finally got rid of. Obviously, then, absorption of oxygen must take place, or in other words, respiration must be one of the functions of the protoplasm of Hæmatococcus as of that of Amœba. In many green, *i.e.*, chlorophyll-containing, plants, this has been proved to be the case; respiration, *i.e.*, the taking in of oxygen and giving out of carbon dioxide, is constantly going on, but during daylight is obscured by the converse process—the taking in of carbon dioxide for nutritive purposes and the giving out of the oxygen liberated by its decomposition. In darkness, when this latter process is in abeyance, the occurrence of respiration is more readily ascertained.

Owing to the constant decomposition, during sunlight, of carbon dioxide, a larger volume of oxygen than of carbon dioxide is evolved; and if an analysis were made of all the ingesta of the organism (carbon dioxide *plus* mineral salts *plus* respiratory oxygen) they would be found to contain less oxygen than the egesta (oxygen from decomposition of carbon dioxide *plus* water, excreted carbon dioxide and nitrogenous waste); so that the nutritive process in Hæmatococcus is, as a whole, a process of deoxidation. In Amœba, on the other hand, the ingesta (food *plus* respiratory oxygen) contain more oxygen than the egesta (fæces *plus* carbon dioxide, water, and nitrogenous excreta), the nutritive process being therefore on the whole one of oxidation. This difference is, speaking broadly, characteristic of plants and animals generally; animals, as a rule, take in more free oxygen than they give out, while green plants always give out more than they take in.

But destructive metabolism is manifested not only in the formation of waste products, but in that of substances simpler than protoplasm which remain an integral part of the organism, viz., cellulose and starch. The cell-wall is probably formed by the conversion of a thin superficial layer of protoplasm into cellulose, the cyst attaining its final thickness by frequent repetition of the process (see p. 14). The starch of the pyrenoids is apparently formed by a similar process of decomposition or destructive metabolism of protoplasm, growth taking place in both instances by accretion and not by intussusception.

We see then that destructive metabolism may result in the formation of (*a*) *waste products* and (*b*) *plastic products*, the former being got rid of as of no further use, while the latter remain an integral part of the organism.

Let us now turn once more to the movements of Hæmatococcus, and consider in some detail the manner of their performance.

Each flagellum (Fig. 3, A, *fl*) is a thread of protoplasm of uniform diameter except at its distal or free end where it tapers to a point. The lashing movements are brought about by the flagellum bending successively in different directions; for instance, if in Fig. 3 F, *abc* represents it in the position of rest, *abc'* will show the form assumed when it is deflected to the left, and *abc"* when the bending is towards the right. In the position *abc* the two sides *ab*, *ac* are obviously equal to one another, but in the flexed positions it is equally obvious that the concave sides *ac'*, *bc"* are shorter than the convex sides *bc'*, *ac"*; in other words, as the flagellum bends to the left side *ac* becomes shortened, as it bends to the right the side *bc*.

This may be otherwise expressed by saying that, in bending to the left the side *ac* contracts (see p. 10), in bending to the right the side *bc*, or that the movement is performed by the alternate contraction of opposite sides of the flagellum.

Thus the ciliary movement of Hæmatococcus, like the amœboid movement of Amœba, is a phenomenon of *contractility*. Imagine an Amœba to draw in all its pseudopods but two, and to protrude these two until they became mere threads; imagine further these threads to contract regularly and rapidly instead of irregularly and slowly; the result would be the substitution of pseudopods by flagella, *i.e.*, of temporary slow-moving processes of protoplasm by permanent rapidly-moving ones.

To put the matter in another way: in Amœba the function of contractility is performed by the whole organism; in Hæmatococcus it is discharged by a small part only, viz., the flagella, the rest of the protoplasm being incapable of movement. We have therefore in Hæmatococcus a *differentiation of structure* accompanied by a *differentiation of function* or *division of physiological labour*.

The expression "division of physiological labour" was invented by the great French physiologist, Henri Milne-Edwards, to express the fact that a sort of rough correspondence exists between lowly and highly organized animals and plants on the one hand, and lowly and highly organized human societies on the other. In primitive communities there is little or no division of labour: every man is his own butcher, baker, soldier, doctor, &c., there is no distinction between "classes" and "masses," and each individual is to a great extent independent of all the rest. Whereas in complex civilized communities society is differentiated into politicians, soldiers, professional men, mechanics, labourers, and so on, each class being to a great extent dependent on every other. This comparison of an advanced society with a high organism is at least as old as Æsop, who gives expression to it in the well-known fable of "the Belly and Members."

We see the very first step towards a division of labour in the minute organism now under consideration. If we could cut off a pseudopod of Amœba the creature would be little or none the worse, since every part would be capable of sending off similar processes, and so movement would be in no way hindered. But if we could amputate the flagella of Hæmatococcus its movements would be absolutely stopped.

Hæmatococcus multiplies only in the resting condition (p. 28, and Fig. 3, B); as in Amœba its protoplasm undergoes simple or binary fission, but with the peculiarity that the process is immediately repeated, so that four *daughter-cells* are produced within the single mother-cell-wall (Fig. 3 C). By the rupture of the latter the daughter-cells are set free as the ordinary motile form; sometimes they acquire their flagella and detached cell-wall before making their escape (D).

Under certain circumstances the resting form divides into eight instead of four daughter-cells, and these when liberated are found to be smaller than the ordinary motile form, and to have no cell-wall. Hæmatococcus is therefore *dimorphic*, *i.e.*, occurs, in the motile condition, under two distinct forms: the larger or ordinary form with detached cell-wall is called a *megazooid*, the smaller form without a cell-wall a *microzooid*.

LESSON III

HETEROMITA

WHEN animal or vegetable matter is placed in water and allowed to stand at the ordinary temperature, the well-known process called decomposition sooner or later sets in, the water becoming turbid and acquiring a bad smell. A drop of it examined under the microscope is then found to teem with minute organisms. To one of these, called "the Springing Monad," or in the language of zoology, *Heteromita rostrata*, we must now direct our attention; it is found in infusion of cod's head which has been allowed to stand for two or three months.

Heteromita (Fig. 4, A) is considerably smaller than either Amœba or Hæmatococcus, being only $\frac{1}{120}$ mm. ($\frac{1}{3000}$ inch) in average length. It has a certain resemblance in general form to Hæmatococcus, being somewhat ovoidal and pointed at one end. Like Hæmatococcus also it has two flagella, but only one of these (*fl.* 1) proceeds from its beak-like anterior end and is directed forwards as the creature swims; the other (*fl.* 2) springs a short distance from the beak, and in the ordinary swimming position is trailed after the organism as in A^2 and F^4. Thus in Heteromita, besides an anterior and a posterior end, we may distinguish a *ventral* surface which is directed downwards in the ordinary position, and bears the second or trailing flagellum, and an opposite or *dorsal* surface directed upwards.

Often instead of swimming freely in the fluid a Heteromita is found anchored as it were to a bit of the decomposing substance by its ventral flagellum as in A^1. Under these circumstances it is in constant movement, springing backwards and forwards by alternately coiling and uncoiling the attached ventral flagellum. The general character of the movement will be readily understood from the figure, in which A^1 shows the monad with coiled flagellum, A^2 after it has sprung forward to the full extent of the flagellum. It is from this curious habit that the name "springing monad" is derived.

Towards the posterior end of the body is a nucleus (*nu*), and at the anterior end a contractile vacuole (*c. vac*). There is no trace of an investing membrane or cell-wall, and the protoplasm is colourless. Also, as is invariably the case with organisms devoid of chlorophyll, there is no starch.

In considering the nutrition of Heteromita it is necessary, first of all, to take into consideration the precise nature of its surroundings. It lives, as already stated, in decomposing infusions of animal matter. Such infusions contain proteids in solution, in part split up by the process of decomposition into simpler compounds some of which are diffusible; this process is due, as we shall see hereafter (Lesson VIII.), to the action of the minute organisms known as Bacteria, which are always present in vast numbers in putrescent substances.

As Heteromita contains no chlorophyll its nutrition is obviously not holophytic. Observation seems to show pretty conclusively that it is not holozoic; apart from the

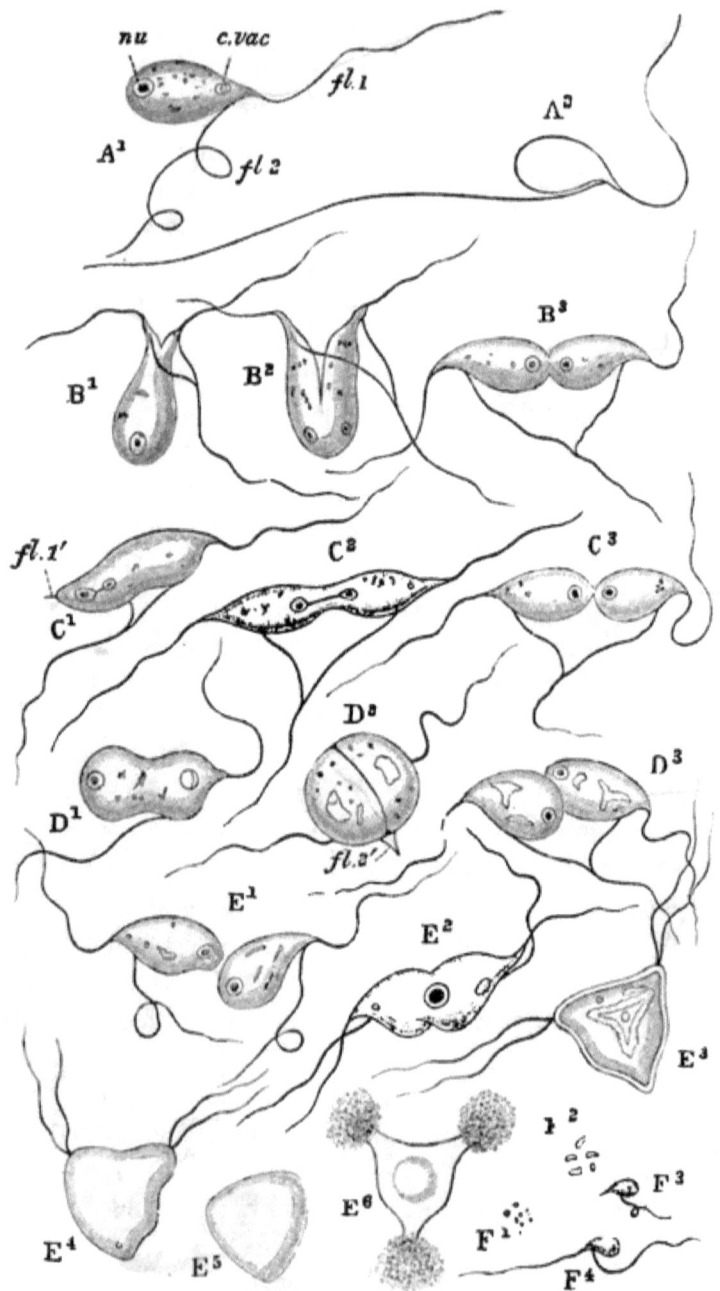

FIG. 4.—*Heteromita rostrata.*
A[1], the living organism, showing nucleus (*nu*), contractile vacuole

(*c. vac*), anterior flagellum (*fl.* 1), and coiled ventral flagellum (*fl.* 2) by which the organism is anchored; A^2 shows the position at the forward limit of the spring, the ventral flagellum being fully extended.

B^1—B^3, three stages in the longitudinal fission of the anchored form.

C^1—C^3. Three stages in the transverse fission of the same : *fl.* 1^1, rudiment of newly formed anterior flagellum.

D^1—D^3, three stages in the fission of the free-swimming form : *fl.* 2^1, rudiment of the newly-formed ventral flagella.

E^1, free-swimming and anchored forms about to conjugate : E^2, commencement of conjugation : E^3, E^4, two stages in the development of the zygote : E^5, the fully formed zygote : E^6, dehiscence of the zygote and emission of spores.

F^1—F^4, four stages in the development of the spores.
(After Dallinger.)

fact that it possesses neither mouth nor pseudopods, examples have been kept under observation for hours together by trained microscopists, and have never been observed to ingest the bacteria or other particles, dead or alive, contained in the fluid. There remains only one way in which nutrition can take place, namely, by absorption of the proteids and other nutrient substances in the solution, *i.e.*, by these substances diffusing into the water of organization of the monad. Whether the proteids are rendered diffusible by the process of decomposition alone, *i.e.*, by the action of bacteria (see p. 91), or whether a kind of surface digestion takes place, the protoplasm of Heteromita converting the proteids in immediate contact with it into peptones or allied compounds, is not certain.

Thus Heteromita feeds neither by taking solid proteinaceous food into its interior (holozoic nutrition) nor by decomposing carbon dioxide and combining the carbon with water and mineral salts (holophytic nutrition), but by absorbing decomposing proteids and other nutrient substances in the liquid form ; this is the *saprophytic* mode of nutrition. It will be seen that the main difference between saprophytic and holozoic nutrition is that in the former digestion, *i.e.*, the process of rendering food-stuffs soluble and diffusible, takes place outside the body so that constructive metabolism can begin at once.

It is worthy of notice that while the process of feeding is strictly intermittent in Amœba, which only takes in food at intervals, and largely intermittent in Hæmatococcus, in which the decomposition of carbon dioxide takes place only during daylight, in Heteromita it is continuous, the organism living in a solution of putrefying proteids which it is constantly absorbing. It may be said to live immersed in an immense cauldron of broth which it is for ever imbibing, not by its mouth, for it has none, but by the whole surface of its body.

Respiration and excretion probably take place in the same manner as in Amœba. It has been shown that the optimum temperature for saprophytic monads is about 18° C., the ultra-maximum or thermal death-point about 60° C. But it is an interesting fact that by very slowly increasing the temperature, Dr. Dallinger was able in the course of several months to accustom some of these forms—not Heteromita itself but closely allied genera—to live at a temperature exceeding 68° C.

The ordinary method of reproduction is by simple fission, the process affecting not only the body but the flagella as well. In Fig 4, B^1, the commencement of fission is shown; the anterior flagellum has undergone complete longitudinal division, while the split has only extended about a third of the length of the body and ventral flagellum. In B^2 the process has gone further, and in B^3 the products of division are on the point of separating.

More frequently, however, fission instead of being longitudinal, *i.e.*, in the direction of the long axis of the monad, is transverse, *i.e.*, at right angles to the long axis. This process is shown in c^1—c^3, and is seen to differ from that described in the preceding paragraph in the circumstance that the anterior flagellum of the parent form is unaffected, and becomes without alteration the anterior flagellum of one of the daughter-forms—that to the right in the figures. The anterior flagellum of the other product of division—that to the left—is a new structure formed as an outgrowth from the body: its commencement is shown in c^1, *fl.* 1'.

These two modes of fission—longitudinal and transverse—both occur in the anchored form of Heteromita, *i.e.*, in individuals attached by the ventral flagellum. The free-swimming form presents a third variety of the process. It comes to rest, loses its regular outline (D^1), becoming almost amœboid in form and finally (D^2) globular. Division then takes place: the flagella of the parent become each the anterior flagellum of one of the daughter-cells (compare D^1, D^2, and D^3), while their ventral flagella are formed by the splitting of a little outgrowth of the dividing body (D^2, *fl.* 2').

As in Amœba fission is invariably preceded by division of the nucleus.

But in Heteromita fission is not the only mode of reproduction. Under certain circumstances a free-swimming form approaches an anchored form, and applies itself to it in such a way that the posterior ends of the two are in contact (E^1). The two individuals then fuse with one another as completely as two drops of gum on a plate unite when brought into contact. Fusion of the nuclei also takes place, and there is formed an irregular body (E^2) with a single nucleus and with two flagella at each end. This swims about freely, and as it does so the last trace of distinction between the two monads of which it is formed is lost, and a triangular form is assumed (E^3), the two pairs of cilia being situated at two of the angles. Still later the protoplasm of this triangular body loses all trace of nucleus, granules, &c., and becomes perfectly clear (E^4): then it comes to rest and loses its flagella, appearing as a clear, homogeneous, three-cornered sac with slightly convex sides (E^5). This body, formed by the *conjugation* of the two monads, is called a *zygote*, the two conjugating individuals being distinguished as *gametes*.

The zygote remains quiescent for some time, and then, after undergoing wave-like movements of its surface, bursts at its three angles (E^6), its contents escaping in the form of granules called *spores*, so minute as to be barely visible even under the highest powers of the best modern microscopes. They are formed by the protoplasm of the zygote dividing into an immense number of separate masses, a process known as *multiple fission*.

Carefully watched, these almost ultra-microscopic particles (F^1) are found to grow into clear visibility and to take on a distinctly oval shape (F^2). Still increasing in size they develop a ventral flagellum (F^3) which is at first quite quiescent: finally, the pointed end sends out a process which becomes an anterior flagellum (F^4). The spore has now become a Heteromita resembling the parent form in all but size.

It will be seen that this remarkable mode of multiplication by conjugation differs from multiplication by fission in the fact that it requires the co-operation of two individuals which undergo complete fusion. As we shall see more plainly later on (Lessons XV. and XVI.) conjugation is the simplest case of *sexual reproduction*, differing from the sexual reproduction of the higher organisms in that the two conjugating bodies or gametes are each an entire individual, and in the further circumstance that the gametes resemble one another in form and size, so that there is no distinction of sex,[1] but each takes an equal and similar share in the production of the zygote. Binary fission, on the other hand, is an example of *asexual* reproduction.

[1] It might perhaps be allowable to consider the active, free-swimming monad which seeks and attaches itself to the anchored form as a male, and the passive anchored form as a female gamete (see Lesson XII.).

Notice also another important fact. The spores when first emitted from the ruptured zygote are mere granules of protoplasm, approaching as nearly as anything in nature to the mathematical definition of a point, "without parts and without magnitude." And during its growth a spore increases not only in size but also in complexity, in other words undergoes a progressive differentiation or *development*. This is an instance of the principle known as Von Baer's law, according to which "development is a progress from the simple to the complex, from the general to the special, from the homogeneous to the heterogeneous." In Heteromita, then, we have our first instance of development, since in simple fission there is no development, each product of division being from the first similar to the parent in all but size.

Lastly, Heteromita is the first instance we have had of an organism with a definite *life-history*. It multiplies asexually by simple fission producing free-swimming and anchored forms: these conjugate in pairs forming a zygote, in which, by multiple fission, numerous spores are formed: the spores develop into the adult form, asexual multiplication begins once more, and so the cycle of existence is completed.

It must be borne in mind that further researches may reveal the occurrence of a true sexual process in Amœba and Hæmatococcus.

LESSON IV

EUGLENA

THE rain-water collected in puddles by the road-side, on roofs, &c., is often found to have a bright green colour: this is sometimes due to the presence of delicate water weeds visible to the naked eye (Lesson XVI.), but frequently the water when held up to the light in a glass vessel appears uniformly green, no suspended matter being visible to the unaided sight. Under these circumstances the green colour is frequently due to the presence of vast numbers of an organism known as *Euglena viridis*.

Although microscopic, Euglena is considerably larger than either Hæmatococcus or Heteromita, its length varying from $\frac{1}{24}$ mm. to $\frac{1}{6}$ mm. The body is spindle-shaped, wide in the middle and narrow at both ends (Fig. 5, A—E): one extremity is blunter than the other, and from it proceeds a single long flagellum (*fl*) by the action of which the organism swims with great rapidity, the flagellum being, as in Hæmatococcus, directed forwards. Besides its rapid swimming movements Euglena frequently performs slow movements of contraction and expansion, something like those of a short worm, the body becoming broadened out first at the anterior end, then in the middle, then at the posterior end, twisting to the right and left, and so on (Fig. 5, A—D). These movements are so characteristic of the genus that the name *euglenoid* is applied to them.

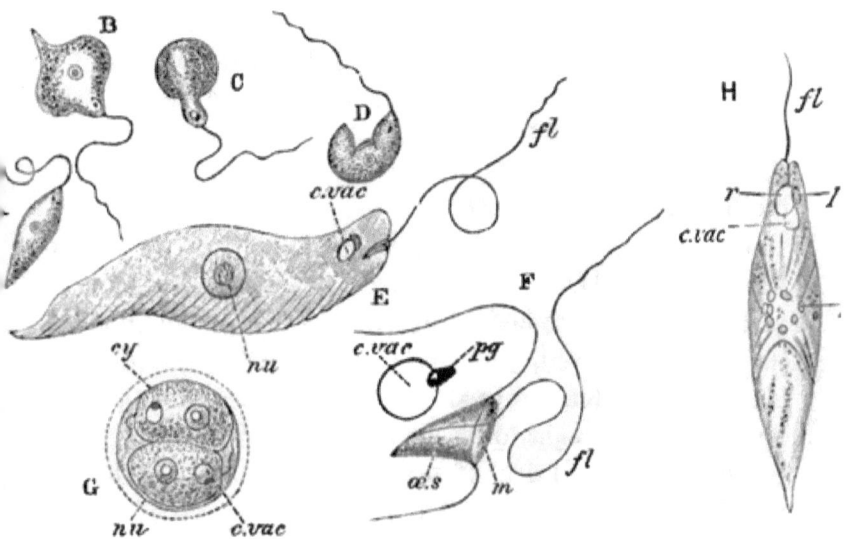

FIG. 5.—*Euglena viridis*.

A—D, four views of the living organism, showing the changes of form produced by the characteristic euglenoid movements.

E, enlarged view, showing the nucleus (*nu*), reservoir of the contractile vacuole (*c.vac*), with adjacent pigment spot, and gullet with a single flagellum springing from it.

F, enlarged view of the anterior end of E, showing pigment-spot (*pg*) and reservoir (*c. vac*), mouth (*m*), gullet (*œ. s*), and origin of flagellum (*fl*).

G, resting form after binary fission, showing cyst or cell-wall (*cy*), and the nuclei (*nu*) and reservoirs (*c. vac*) of the daughter-cells.

H, active form showing contractile vacuole (*c. vac*), reservoir (*r*), and paramylum-bodies (*p*).

(A—G, after Saville Kent : H, from Bütschli after Klebs.)

The body consists of protoplasm covered with a very delicate skin or *cuticle* which is often finely striated, and is to be looked upon as a superficial hardening of the protoplasm. The green colour is due to the presence of chlorophyll, which tinges all the central part of the body, the two ends being colourless. It is difficult to make out whether the chlorophyll is lodged in one chromatophore or in several.

In Hæmatococcus we saw that chlorophyll was associated with starch (p. 27). In Euglena there are, near the middle of the body, a number of grains of *paramylum* (H, *p*), a carbohydrate of the same composition as starch ($C_6H_{10}O_5$), but differing from it in remaining uncoloured by iodine.

Water containing Euglena gives off bubbles of oxygen in sunlight: as in Hæmatococcus the carbon dioxide in solution in the water is decomposed in the presence of chlorophyll, its oxygen evolved, and its carbon combined with the elements of water and used in nutrition. For a long time Euglena was thought to be nourished entirely in this way, but there is a good deal of reason for thinking that this is not the case.

When the anterior end of a Euglena is very highly magnified it is found to have the form shown in Fig. 5, F. It is produced into a blunt snout-like extremity at the base of which is a conical depression (*œ. s*) leading into the soft internal protoplasm:—just the sort of depression one could make in a clay model of Euglena by thrusting one's finger or the end of a pencil into the clay. From the bottom of this tube the flagellum arises, and by its continual movement gives rise to a sort of whirlpool in the neighbourhood. By the current thus produced minute solid food-particles are swept down the tube and forced into the soft internal protoplasm, where they doubtless become digested in the same way as the substances ingested by an Amœba. That solid particles are so ingested by Euglena has been proved by diffusing finely powdered carmine in the water, when the coloured particles were seen to be swallowed in the way described.

The depression in question is therefore a *gullet*, and its external aperture or margin (*m*) is a *mouth*. Euglena, like Amœba, takes in solid food, but instead of ingesting it at almost any part of the body, it can do so only at one particular point where there is a special ingestive aperture or mouth. This is clearly a case of specialization or differentiation of structure : in virtue of the possession of a mouth and gullet Euglena is more highly organized than Amœba.

It thus appears that in Euglena nutrition is both holozoic and holophytic : very probably it is mainly holophytic during daylight and holozoic in darkness.

Near the centre of the body or somewhat towards the posterior end is a nucleus (E, *nu*) with a well-marked nucleolus, and at the anterior end is a clear space (*c. vac*) looking very like a contractile vacuole. It has been shown, however, that this space is in reality a non-contractile cavity or reservoir (H, *r*) into which the true contractile vacuole (*c. vac*) opens, and which itself discharges into the gullet.

In close relation with the reservoir is found a little bright red speck (*pg*) called the *pigment spot* or *stigma*. It consists of hæmatochrome (see p. 26) and is curiously like an eye in appearance, so much so that it is sometimes known as the eye-spot. There seems, however, to be no reason for assigning a visual function to it : indeed it has been shown that the greatest sensitiveness to light is manifested by the colourless anterior end of the body.

As in Hæmatococcus a resting condition alternates with the motile phase : the organism loses its flagellum and surrounds itself with a cyst of cellulose (Fig. 5, G, *cy*), from which, after a period of rest, it emerges to resume active life.

Reproduction takes place by simple fission of the resting form, the plane of division being always longitudinal (G). Sometimes each product of division or daughter-cell divides again: finally the two, or four, or sometimes even eight daughter-cells emerge from the cyst as active Euglenæ. A process of multiple fission (p. 42) has also been described, numerous minute active spores being produced which gradually assume the ordinary form and size.

LESSON V

PROTOMYXA AND THE MYCETOZOA

WHEN Professor Haeckel was investigating the zoology of the Canary Islands more than twenty years ago he discovered a very remarkable organism which he named *Protomyxa aurantiaca*. It was found in sea-water attached to a shell called *Spirula*, and was at once noticeable from the bright orange colour which suggested its specific name. Apparently no one has since been fortunate enough to find it.

In its fully developed stage Protomyxa is the largest of all the organisms we have yet studied, being fully 1mm. ($\frac{1}{25}$ inch) in diameter, and therefore visible to the naked eye as a small orange speck. In general appearance (Fig. 6, A) it is not unlike an immense Amoeba, the chief difference lying in the fact that the pseudopods (*psd*) instead of being short, blunt processes, few in number (comp. Fig. 1, p. 2) are very numerous, slender, branching threads which often unite with one another so as to form networks. No nucleus was observed [1] and no contractile vacuole, but it is quite possible that a renewed examination might prove the presence of one or both of these structures.

The figure (A) is enough to show that nutrition is holozoic,

[1] See p. 9, note.

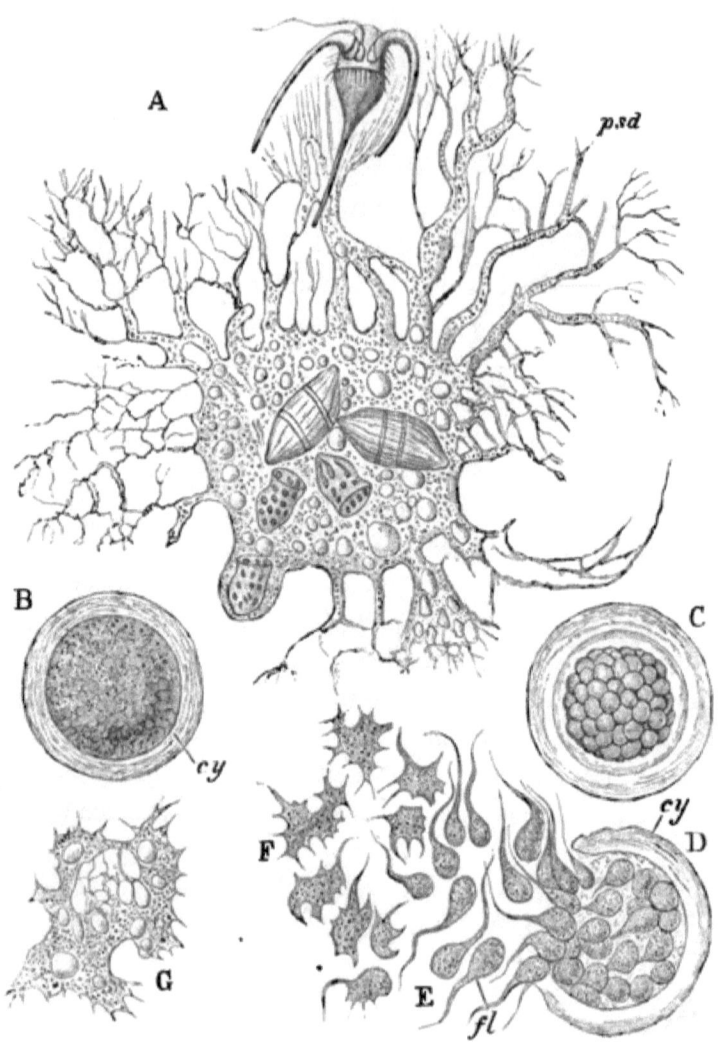

FIG. 6.—*Protomyxa aurantiaca*.

A, the living organism (plasmodium), showing fine branched pseudopods (*psd*) and several ingested organisms.
B, the same, encysted: *cy* the cell-wall.
C, the protoplasm of the encysted form breaking up into spores.
D, dehiscence of the cyst and emergence of
E, flagellulæ which afterwards become converted into
F, amœbulæ.
G, amœbulæ uniting to form a plasmodium. (After Haeckel.)

the specimen has ingested several minute organisms and is in the act of capturing another.

But the main interest of Protomyxa lies in its very curious and complicated life-history. After crawling over the Spirula shell for a longer or shorter time it draws in its pseudopods, comes to rest, and surrounds itself with a cyst (B, *cy*). The composition of the cyst is not known, but it is apparently not cellulose, since it is not coloured by iodine and sulphuric acid (p. 28).

Next, the encysted protoplasm undergoes multiple fission, dividing into a number of spores (C): soon the cyst bursts and its contents emerge (D) as bodies which differ utterly in appearance from the amœboid form from which we started. Each spore has in fact become a little ovoid body of an orange colour, provided with a single flagellum (E, *fl*) by the lashing of which it swims through the water after the manner of a monad.

It is convenient to have a name by which to distinguish these flagellate bodies, just as we have special names for the young of the higher animals, such as tadpoles or kittens. From the fact of their distinguishing character being the possession of a flagellum they are called *flagellulæ*; the same name will be applied to the flagellate young of various other organisms which we shall study hereafter.

After swimming about actively for a time each flagellula settles down on some convenient substratum and undergoes a remarkable change: its movements become sluggish, its outline irregular, and its flagellum short and thick, until it finally takes on the form of a little Amœba (F). For this stage also a name is required: it is not an Amœba but an amœboid phase in the life-history of a totally different organism: it is called an *amœbula*.

The process just described may be taken as a practical

proof of the statement made in a previous Lesson (p. 34) that a flagellum is nothing more than a delicate and relatively permanent pseudopod. In Protomyxa we have a flagellula directly converted into an amœbula, the flagellum of the former becoming one of the pseudopods of the latter.

The amœbulæ thus formed may simply increase in size and send out numerous delicate pseudopods, thus becoming converted into the ordinary Protomyxa-form. Frequently, however, they attain this form by a very curious process: they come together in twos and threes until they are in actual contact with one another, when they undergo complete and permanent fusion (G). In this case the Protomyxa-form is produced not by the development of a single amœbula but by the conjugation or fusion of a variable number of amœbulæ. A body formed in this way by the fusion of amœbulæ is called a *plasmodium*, so that in the life-history of Protomyxa we can distinguish an encysted, a ciliated or flagellate, an amœboid, and a plasmodial phase.

The nature of a plasmodium will be made clearer by a short consideration of the strange group of organisms known as *Mycetozoa* or sometimes "slime-fungi." They occur as gelatinous masses on the bark of trees, on the surface of tan-pits, and sometimes in water. It must be remembered that *Mycetozoa* is the name not of a genus but of a *class* in which are included several genera, such as *Badhamia*, *Chondrioderma*, &c. (see Fig. 7): a general account of the class is all that is necessary for our present purpose.

The Mycetozoa consists of sheets or networks of protoplasm which may be as much as 30 cm. (1ft.) in diameter, and throughout the substance of which are found numerous nuclei. In this condition they creep about over bark or some

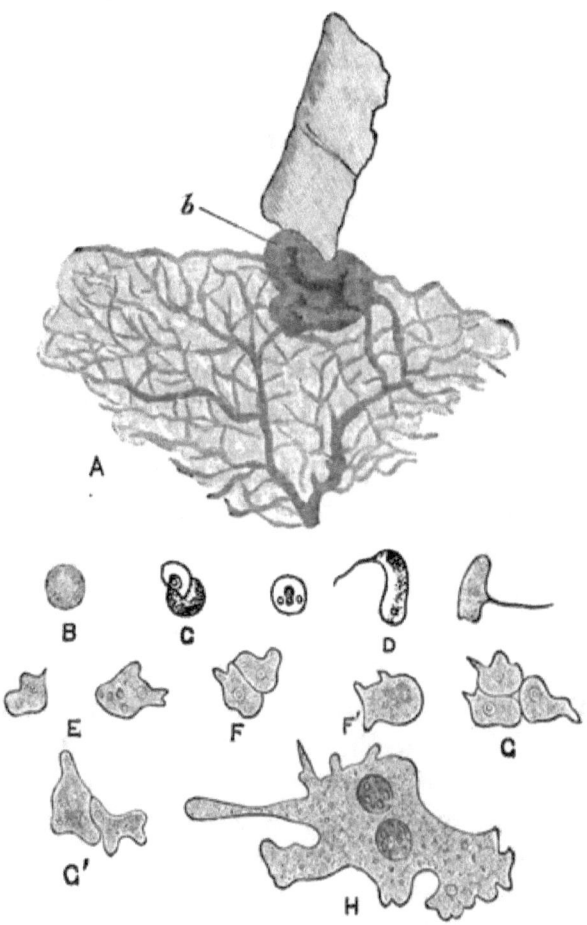

Fig. 7.—A, part of the plasmodium of *Badhamia* ($\times 3\frac{1}{2}$); *b*, a short pseudopod enclosing a bit of mushroom stem.

B, spore of *Chondrioderma*.
C, the same, undergoing dehiscence.
D, flagellulæ liberated from spores of the same.
E, amœbulæ formed by metamorphosis of flagellulæ.
F, two amœbulæ about to fuse : F', the same after complete union.
G, G', two stages in the formation of a three-celled plasmodium.
H, a small plasmodium.
(A, after Lister : B—H, from Sachs after Cienkowski.)

other substance: and in doing so ingest solid food (Fig. 7, A). It has been proved that they digest protoplasm: and in one genus pepsin—the constituent of our own gastric juice by which the digestion of proteids is effected (see p. 12)—has been found. They can also digest starch which has been swollen by a moderate heat—as in our own bread and rice-puddings—but are unable to make use of raw starch.

After living in this free condition, like a gigantic terrestrial Amœba, for a longer or shorter time, either a part or the whole of the protoplasm becomes encysted[1] and breaks up into spores. These (B) consist of a globular mass of protoplasm covered with a wall of cellulose: the cysts are also formed of cellulose.

By the rupture of the cell-wall of the spore (C) the protoplasm is liberated as a flagellula (D) provided with a nucleus and a contractile vacuole, and frequently exhibiting amœboid as well as ciliary movements. After a time the flagellulæ lose their cilia and pass into the condition of amœbulæ (E), which finally fuse to form the plasmodium with which we started (F—H). In the young plasmodia (G^1) the nuclei of the constituent amœbulæ are clearly visible, and from them the nuclei of the fully developed plasmodia are probably derived. It would seem, therefore, that in the fusion of amœbulæ to form the plasmodium of Mycetozoa the cell-bodies (protoplasm) alone coalesce, not the nuclei.

There is a suggestive analogy between this process of plasmodium-formation and that of conjugation as seen in Heteromita. Two Heteromitæ fuse and form a zygote the

[1] The process of formation of the cyst or sporangium is a complicated one, and will not be described here. See De Bary, *Fungi, Mycetozoa, and Bacteria* (Oxford, 1887).

protoplasm of which divides into spores. In Protomyxa and the Mycetozoa not two but several amœbulæ unite to form a plasmodium which after a time becomes encysted and breaks up into spores. So that we might look upon the conjugation of Heteromita as an extremely simple plasmodial phase in its life-history, or upon the formation of a plasmodium by Protomyxa and the Mycetozoa as a process of multiple conjugation.

There is, however, an important difference between the two cases by reason of which the analogy is far from complete. In Heteromita the nuclei of the two gametes are no longer visible (p. 41): they coalesce during conjugation, and the product of their union subsequently, in all probability, breaks up to form the nuclei of the spores. In the Mycetozoa neither fusion nor apparent disappearance of the nuclei of the amœbulæ has been observed.

LESSON VI

A COMPARISON OF THE FOREGOING ORGANISMS WITH CERTAIN CONSTITUENT PARTS OF THE HIGHER ANIMALS AND PLANTS

WHEN a drop of the blood of a crayfish, lobster, or crab is examined under a high power, it is found to consist of a nearly colourless fluid, the *plasma*, in which float a number of minute solid bodies, the *blood-corpuscles* or *leucocytes*. Each of these (Fig. 8, A) is a colourless mass of protoplasm, reminding one at once of an Amœba, and on careful watching the resemblance becomes closer still, for the corpuscle is seen to put out and withdraw pseudopods (A^1—A^4) and so gradually to alter its form completely. Moreover the addition of iodine, logwood, or any other suitable colouring matter reveals the presence of a large nucleus (A^5, A^6, *nu*) : so that, save for the absence of a contractile vacuole in the leucocyte, the description of Amœba in Lesson I. would apply almost equally well to it.

The blood of a fish, a frog (B^1), a reptile, or a bird contains quite similar leucocytes, but in addition there are found in the blood of these red-blooded animal bodies called *red corpuscles*. They are flat oval discs of protoplasm (B^5, B^6)

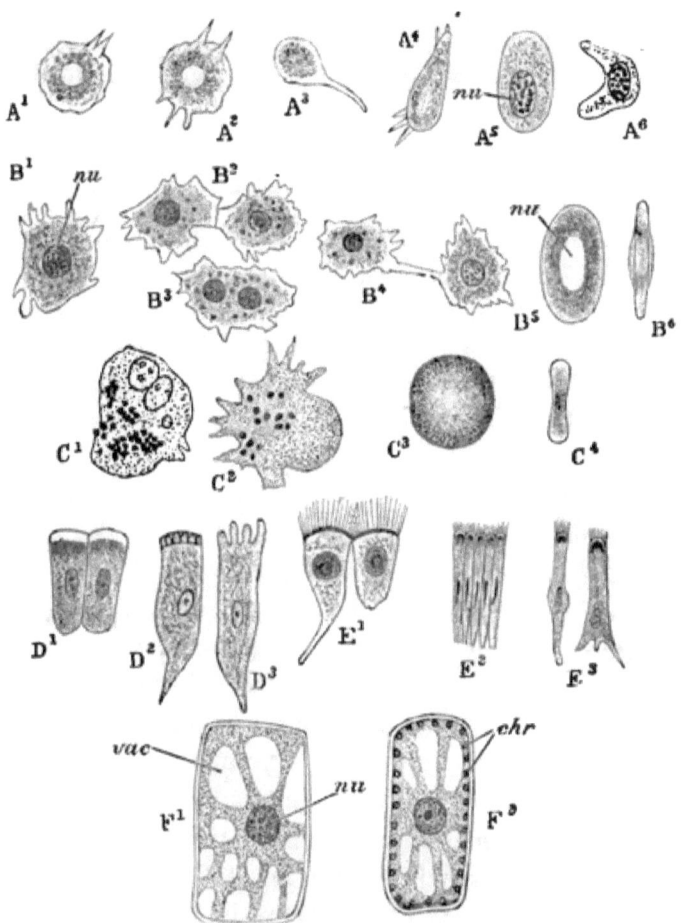

FIG. 8.—Typical Animal and Vegetable Cells.

A^1—A^4, living leucocyte (blood corpuscle) of a crayfish showing amœboid movements: A^5, A^6, the same, killed and stained, showing the nucleus (nu).

B^1, leucocyte of the frog, nu the nucleus; B^2, two leucocytes beginning to conjugate: B^3, the same after conjugation, a binucleate plasmodium being formed: B^4, a leucocyte undergoing binary fission: B^5, surface view and B^6, edge view of a red corpuscle of the same, nu, the nucleus.

C^1, C^2, leucocytes of the newt; in C^1 particles of vermilion, represented by black dots, have been ingested.

C^3, surface view and C^4, edge view of a red corpuscle of man.

D^1, columnar epithelial cells from intestine of frog: D^2, a similar cell showing striated distal border from which in D^3 pseudopods are protruded.

E^1, ciliated epithelial cell from mouth of frog: E^2, E^3, similar cells from windpipe of dog.

F^1, parenchyma cell from root of lily, showing nucleus (*nu*), vacuoles (*vac*), and cell-wall: F^2, a similar cell from leaf of bean, showing nucleus, vacuoles, cell-wall and chromatophores (*chr*).

(B, D^1, and E^1, after Howes: C, E^2, and E^3, after Klein and Noble Smith: D^2, D^3, after Wiedersheim: F^1, after Sachs: F^2, after Behrens.)

coloured by a pigment called *hæmoglobin*, and provided each with a large nucleus (*nu*) which, when the corpuscle is seen from the edge, produces a bulging of its central part. These bodies may be compared to Amœbæ which have drawn in their pseudopods, assumed a flattened form, and become coloured with hæmoglobin.

In the blood of mammals, such as the rabbit, dog, or man, similar leucocytes occur, but their red blood corpuscles (c^3, c^4) have the form of biconcave discs, and are devoid of nuclei.

In many animals the leucocytes have been observed to ingest solid particles (c^1), to multiply by simple fission (B^4) and to coalesce with one another forming plasmodia (B^2) (p. 52).

The stomach and intestines of animals are lined with a sort of soft slimy skin called *mucous membrane*. If a bit of the surface of this membrane—in a frog or rabbit for instance—is snipped off and "teased out," *i.e.*, torn apart with needles, it is found when examined under a high power to be made up of an immense number of microscopic bodies called *epithelial cells*, which in the living animal, lie close to one another in the inner layer of mucous membrane in something the same way as the blocks of a wood pavement lie on the surface of a road. An epithelial cell (D^1, D^2) consists of a rod-like mass of protoplasm, containing a large nucleus, and is therefore comparable to an elongated Amœba without pseudopods. In some animals the resemblance is still closer : the epithelial cells have been observed to throw out pseudopods from their free surfaces (D^3), that is, from the only part where any such movement is possible, since they are elsewhere in close contact with their fellow cells.

The mouth of the frog and the trachea or windpipe of air-breathing vertebrates such as reptiles, birds, and mammals, are also lined with mucous membrane, but the epithelial cells which constitute its inner layer differ in one important respect from those of the stomach and intestine. If examined quite fresh each is found to bear on its free surface, *i.e.*, the surface which bounds the cavity of the mouth or windpipe, a number of delicate protoplasmic threads or cilia ($E^1—E^3$) which are in constant vibratory movement. In the process of teasing out the mucous membrane some of the cells are pretty sure to become detached, and are then seen to swim about in the containing fluid by the action of their cilia. These *ciliated epithelial cells* remind one strongly of Heteromita, except for the fact that they bear numerous cilia in constant rhythmical movement instead of two only—in this case distinguished as flagella—presenting an irregular lashing movement.

Similar ciliated epithelial cells are found on the gills of oysters, mussels, &c., and in many other situations.

The stem or root of an ordinary herbaceous plant, such as a geranium or sweet-pea, is found when cut across to consist of a central mass of pith, around which is a circle of woody substance, and around this again a soft greenish material called the *cortex*. A thin section shows the latter to be made up of innumerable polyhedral bodies called *parenchyma cells*, fitting closely to one another like the bricks in a wall.

A parenchyma cell examined in detail (F^1) is seen to consist of protoplasm hollowed out internally into one or more cavities or vacuoles (*vac*) containing a clear fluid. These vacuoles differ from those of Amœba, Heteromita, or Euglena in being non-contractile; they are in fact mere cavities in the protoplasm containing a watery fluid: the layer of protoplasm immediately surrounding them is denser than the rest. Sometimes there is only one such space occupying the whole interior of the cell, sometimes, as in the example figured, there are several, separated from one another by delicate bands or sheets of protoplasm. The cell contains a large nucleus (*nu*) and is completely enclosed in a moderately thick cell-wall composed of cellulose.

The above description applies to the cells composing the deeper layers of the cortex, *i.e.*, those nearest the woody layer: in the more superficial cells, as well as in the internal cells of a leaf, there is something else to notice. Imbedded in the protoplasm, just within the cell wall, are a number of minute ovoid bodies of a bright green colour (F^2, *chr*). These are chromatophores or chlorophyll corpuscles; they consist of protoplasm coloured with chlorophyll which can be proved experimentally to have the same properties as the chlorophyll of Hæmatococcus and Euglena.

Such a green parenchyma cell is clearly comparable with an encysted Hæmatococcus or Euglena, the main differences being that in the plant cell the form is polyhedral owing to the pressure of neighbouring cells and that the chromatophores are relatively small and numerous. Similarly a colourless parenchyma cell resembles an encysted Amœba.

The pith, the epidermis or thin skin which forms the outer surface of herbaceous plants, the greater part of the leaves and other portions of the plant may be shown to consist of an aggregation of cells agreeing in essential respects with the above description.

We come therefore to a very remarkable result. The higher animals and plants are built up—in part at least—of elements which resemble in their essential features the minute and lowly organisms studied in previous lessons. Those elements are called by the general name of *cells*; hence the higher organisms, whether plants or animals, are *multicellular* or are to be considered as *cell-aggregates*, while in the case of such beings as Amœba, Hæmatococcus, Heteromita, or Euglena, the entire organism is a single cell, or is *unicellular*.

Note further that the cells of the higher animals and plants, like entire unicellular organisms, may occur in either the amœboid (Fig. 8, A, B^1 C^1,) the ciliated (E), or the encysted (F) condition, and that a plasmodial phase (B^2) is sometimes produced by the union of two or more amœboid cells.

One of the most characteristic features in the unicellullar organisms described in the preceding lessons is the constancy of the occurrence of binary fission as a mode of multiplication. The analogy between these organisms and the cells of the higher animals and plants becomes still closer when we find that in the latter also simple fission is the normal mode of multiplication, the increase in size of growing parts being brought about by the continual division of their constituent cells.

The process of division in animal and vegetable cells is frequently accompanied by certain very characteristic and complicated changes in the nucleus to which we must now direct our attention. First of all, however, it will be necessary to describe the exact microscopic structure of cells and their nuclei as far as it is known at present.

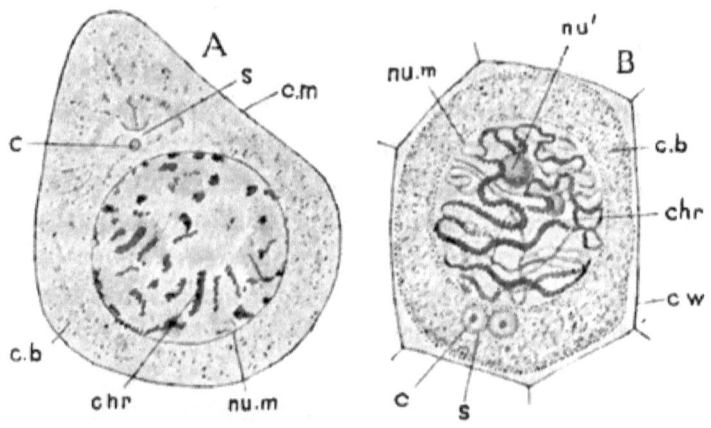

FIG. 9.—A, Cell from the genital ridge of a young salamander, showing cell-membrane (*c. m*), protoplasm or cell-body (*c. b*) with directive sphere (*s*) and central particle (*c*), and nucleus with membrane (*nu. m*) and irregular network of chromatin (*chr*).

B. Cell from the immature stamen of a lily, showing cell-wall (*c. w*), protoplasm with two directive spheres (*s*), and nucleus as in A.

Both figures very highly magnified.

(A, from a drawing by Mr. J. E. S. Moore : B, after Guignard.)

There seems to be a good deal of variation in the precise structure of various animal and plant cells, but the more recent researches show that in the cell-body or protoplasm (Fig. 9, *c. b*) two constituents may be distinguished, a clear semi-fluid substance, traversed by a delicate sponge-work. Now under the microscope the whole cell is not seen at once but only an *optical section* of it, that is all the parts which are in focus at one time : by altering the focus we view the object at successive depths, each view being practically a slice parallel to the lenses of the instrument. This being the case, protoplasm presents the microscopic appearance of a clear or slightly granular matrix traversed by a delicate network. In the epithelial cells of animals the protoplasm is bounded externally by a *cell-membrane* (Fig. 9, A, *c. m*) of extreme tenuity, in plants by a cell-wall (B, *c. w*) of cellulose: in amœboid cells the ectosarc or transparent non-granular portion of the cell consists of clear protoplasm only, the granular endosarc alone possessing the sponge work. In the majority of full-grown plant cells (Fig. 8, F) and in some animal cells the protoplasm is more or less extensively vacuolated, but in the young growing parts as well as in the ordinary cells of animals the foregoing description holds good. It is quite possible that the reticular character of the cell may be merely the optical expression of an extensive but minute vacuolation, or may be due to the presence of innumerable minute granules developed in the protoplasm as products of metabolism.

The nucleus is usually spherical in form: it is enclosed in a delicate *nuclear membrane* (*n.m*) and contains, as in Amœba (p. 7) two constituents, the *nuclear matrix* and the *chromatin* which exhibit far more striking differences than the two constituents of the cell-body. The nuclear matrix is a homogeneous semi-fluid substance which forms the ground-work of the nucleus: it resembles the clear cell-protoplasm in its general characters, amongst other things in being unaffected by dyes. The chromatin (*chr*) takes the form of a network or sponge-work of very variable form, and is distinguished from all other constituents of the cell by its strong affinity for aniline and other dyes. Frequently one or more minute globular structures, the *nucleoli* (B, *nu'*), occur in the nucleus either connected with the network or lying freely in its meshes: they also have a strong affinity for dyes although they often differ considerably from the chromatin in their micro-chemical reactions.

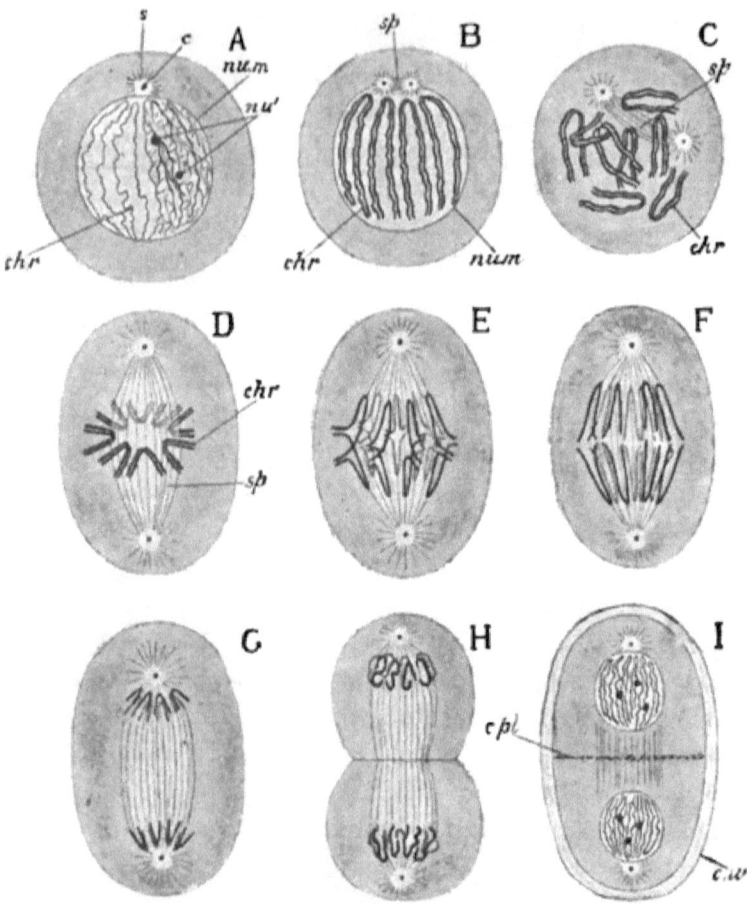

FIG. 10.—Diagrams illustrating the process of indirect cell division or karyokinesis.

A, The resting cell: the nucleus shows a nuclear membrane (*nu.m*), chromatin (*chr*) arranged in loops united into a network (the latter shown on the right side only), and two nucleoli (*nu'*): near the nucleus is a directive sphere (*s*), containing a centrosome (*c*) and surrounded by radiating protoplasmic filaments.

B, The chromatin has resolved itself into distinct loops or chromosomes (*chr*) which have divided longitudinally: the nuclear membrane has begun to disappear: there are two directive spheres and between them is seen the commencement of the nuclear spindle (*sp*).

C, The nuclear membrane has disappeared: the chromosomes are arranged irregularly: the spindle has increased in size and is situated definitely within the nuclear area.

D, The chromosomes are arranged round the equator of the fully formed nuclear spindle.

E, The daughter-loops of the chromosomes are passing in opposite directions towards the poles of the spindle, each having a spindle-fibre attached to it.

F, Later stage of the same process.

G, The chromosomes are now arranged in two distinct groups one at each pole of the spindle.

H, The daughter-cells are partly separated by constriction and the chromosomes of each group are uniting to form the network of the daughter-nucleus.

I, Shows the division of a plant cell by the formation of a cell-plate (*c. pl*): the daughter nuclei are fully formed.

(Altered from Flemming, Rabl, &c.)

In the body of some cells and possibly of all there is found a globular body, surrounded by a radiating arrangement of the protoplasm and called the *directive sphere* (s): it lies close to the nucleus, and contains a minute granule known as the *central particle* or *centrosome* (c). In many plant cells two directive spheres have been found in each cell (B, s).

The precise changes which take place during the fission of a cell are, like the structure of the cell itself, subject to considerable variation. We will consider what may probably be taken as a typical case (Fig. 10).

First of all, the directive sphere divides (B, s) and the products of its division gradually separate from one another (c), ultimately passing to opposite poles of the nucleus (D). At the same time the network of chromatin divides into a number of separate filaments called *chromosomes* (B, *chr*), the number of which appears to be constant in any given species of animal or plant, although it may vary in different species from two to twenty-four. Soon after this the nuclear membrane and the free nucleoli disappear (B, C) and the nucleus is seen to contain a spindle-shaped body (*sp*) formed of excessively delicate fibres which converge at each pole to the corresponding directive sphere. The precise origin of this nuclear spindle is uncertain : it may arise either from the nuclear matrix or, more probably, from the protoplasm of the cell : it is not affected by colouring matters.

At the same time each chromosome splits, sometimes transversely, but usually along its whole length so as to form two parallel rods or loops in close contact with one another (B) : in this way the number of chromosomes is doubled, each one being now represented by a pair.

The divided chromosomes now pass to the equator of the spindle (D) and assume the form either of V- shaped loops, or of short rods, which arrange themselves in a radiating manner so as to present a star-like figure when the cell is viewed in the direction of the long axis of the spindle. Everything is now ready for division to which all the foregoing processes are preparatory.

The two chromosomes of each pair now gradually pass to opposite poles of the spindle (E, F), two distinct groups being thus produced (G) and each chromosome of each group being the twin of one in the other group. Probably the fibres of the spindle are the active agents in this process, the chromosomes being dragged in opposite directions by their contraction.

After reaching the poles of the spindle the chromosomes of each group unite with one another to form a network (H) around which a nuclear membrane finally makes its appearance (I). In this way two nuclei are produced within a single cell, the chromosomes of the *daughter-nuclei*, as well as their attendant directive spheres, being formed by the binary fission of those of the *mother-nucleus*.

But *pari passu* with this process of nuclear division, fission of the cell-body is also going on. This may take place by a simple process of constriction (H)—in much the same way as a lump of clay or dough would divide if a loop of string were tied round its middle and then tightened—or by the formation of what is known as a *cell-plate*. This arises as a row of granules formed from the equatorial part of the nuclear spindle (I): the granules extend until they form a complete equatorial plate dividing the cell-body into two halves: fission then takes place by the cell-plate splitting into two along a plane parallel with its flat surfaces.[1] In plants the cell-plate gives rise to a partition wall of cellulose which divides the two daughter-cells from one another.

In some cases the dividing nucleus instead of going through the complicated processes just described divides by simple constriction. We have therefore to distinguish between *direct* and *indirect nuclear division*. To the latter very elaborate method the name *karyokinesis* is often applied.

In this connection the reader will not fail to note the extreme complexity of structure revealed in cells and their nuclei by the highest powers of the microscope. When the constituent cells of the higher animals and plants were discovered, during the early years of the present century, by Schleiden and Schwann, they were looked upon as the *ultima Thule* of microscopic analysis. Now the demonstration of the cells themselves is an easy matter, the problem is to make out their ultimate constitution. What would be the

---

[1] It must not be forgotten that the cells which are necessarily represented in such diagrams as Fig. 10 as planes are really solid bodies, and that consequently the cell-plate represented in the figures as a line is actually a plane at right angles to the plane of the paper.

result if we could get microscopes as superior to those of to-day as those of to-day are to the primitive instruments of eighty or ninety years ago, it is impossible even to conjecture. But of one thing we may feel confident—of the enormous strides which our knowledge of the constitution of living things is destined to make during the next half century.

The striking general resemblance between the cells of the higher animals and plants and entire unicellular organisms has been commented on as a very remarkable fact: there is another equally significant circumstance to which we must now advert.

All the higher animals begin life as an egg, which is either passed out of the body of the parent as such, as in most fishes, frogs, birds, &c., or undergoes the first stages of its development within the body of the parent, as in sharks, some reptiles, and nearly all mammals.

The structure of the egg is, in essential respects, the same in all animals from the highest to the lowest. In a jelly-fish, for instance, it consists (Fig. 11, A) of a globular mass of protoplasm (ga), in which are deposited granules of a proteinaceous substance known as *yolk-spherules*. Within the protoplasm is a large clear nucleus ($g.v.$), the chromatin of which is aggregated into a central mass or nucleolus ($g.m.$). An investing membrane may or may not be present. In other words the egg is a cell: it is convenient, for reasons which will appear immediately, to speak of it as the *ovum* or *egg-cell*.

The young or immature ova of all animals present this structure, but in many cases certain modifications are undergone before the egg is mature, *i.e.*, capable of development into a new individual. For instance, the protoplasm may throw out pseudopods, the egg becoming amœboid (see

Fig. 53); or the surface of the protoplasm may secrete a thick cell-wall (see Fig. 61). The most extraordinary modification takes place in some Vertebrata, such as birds. In a hen's egg, for instance, the yolk-spherules increase immensely, swelling out the microscopic ovum until it becomes what we know as the "yolk" of the egg: around this layers of albumen or "white" are deposited, and finally the shell membrane and the shell. Hence we have to distinguish carefully in eggs of this character between the entire "egg" in the ordinary acceptation of the term, and the ovum or egg-cell.

But complexities of this sort do not alter the fundamental

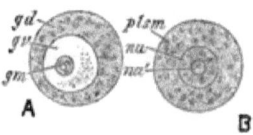

FIG. 11.—A, ovum of an animal (*Carmarina hastata*, one of the jelly fishes), showing protoplasm (*gd*), nucleus (*gv*), and nucleolus (*gm*).
B, ovum of a plant (*Gymnadenia conopsea*, one of the orchids), showing protoplasm (*plsm*), nucleus (*nu*), and nucleolus (*nu'*).
(A, from Balfour after Haeckel: B, after Marshall Ward.)

fact that all the higher animals begin life as a single cell, or in other words that multicellular animals, however large and complex they may be in their adult condition, originate as unicellular bodies of microscopic size.

The same is the case with all the higher plants. The pistil or seed-vessel of an ordinary flower contains one or more little ovoidal bodies, the so-called "ovules" (more accurately megasporangia (see Lesson XXX., and Fig. 89), which, when the flower withers, develop into the seeds. A section of an ovule shows it to contain a large cavity, the embryo-sac or megaspore (see Fig. 89, D), at one end of which is a microscopic cell (*ov*, and Fig. 12 B), consisting as usual of protoplasm (*plsm*), nucleus (*nu*), and nucleolus (*nu'*). This is the ovum or egg-cell of the plant: from it the new plant, which springs from the germinating seed, arises. Thus the higher plants, like the higher animals, are, in their earliest stage of existence, microscopic and unicellular.

LESSON VII

SACCHAROMYCES

EVERY one is familiar with the appearance of the ordinary brewer's yeast—the light-brown, muddy, frothing substance which is formed on the surface of the fermenting vats in breweries and is used in the manufacture of bread to make the dough "rise."

Examined under the microscope yeast is seen to consist of a fluid in which are suspended immense numbers of minute particles, the presence of which produces the muddiness of the yeast. Each of these bodies is a unicellular organism, the *yeast-plant*, or in botanical language *Saccharomyces cerevisiæ*.

Saccharomyces consists of a globular or ellipsoidal mass of protoplasm (Fig. 12), about $\frac{1}{100}$ mm. in diameter, and surrounded with a delicate cell-wall of cellulose (c, *c.w.*). In the protoplasm are one or more non-contractile vacuoles (*vac*)—mere spaces filled with fluid and varying according to the state of nutrition of the cell. Granules also occur in the protoplasm which are products of metabolism, some of them being of a proteid material, others fat globules. Under ordinary circumstances no nucleus is to be seen: but recently, by the employment of a special mode of staining, a small rounded nucleus has been shown to exist near the centre of the cell.

The cell-wall is so thin that it is difficult to be sure of its presence unless very high powers are employed. It can however be easily demonstrated by staining yeast with

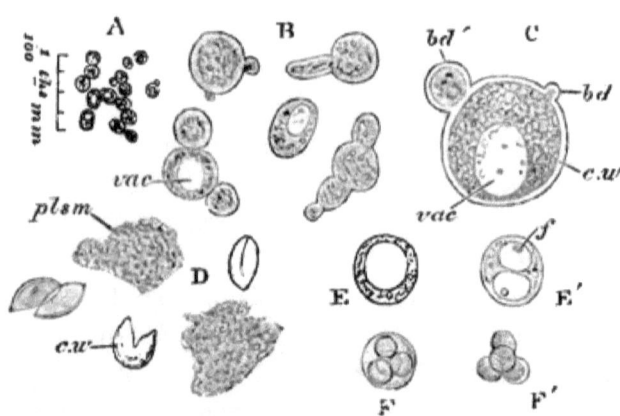

FIG. 12.—*Saccharomyces cerevisiae.*

A, a group of cells under a moderately high power. The scale to the left applies to this figure only.

B, several cells more highly magnified, showing various stages of budding, *vac*, the vacuole.

C, a single cell with two buds (*bd, bd'*) still more highly magnified : *c.w*, cell-wall : *vac*, vacuole.

D, cells, crushed by pressure : *c.w*, the ruptured cell-walls : *plsm*, the squeezed out protoplasm.

E, E', starved cells, showing large vacuoles and fat globules (*f*).

F, F', formation of spores by fission of the protoplasm of a starved cell : in F the spores are still enclosed in the mother-cell-wall, in F' they are free.

magenta, and then applying pressure to the cover-glass so as to crush the cells. Under this treatment the cell-walls are burst and appear as crumpled sacs, split in various ways and unstained by the magenta (D, *c.w*), while the squeezed-out protoplasm is seen in the form of irregular masses (*plsm*) stained pink by the dye.

The mode of multiplication of Saccharomyces is readily made out in actively fermenting yeast, and is seen to differ from anything we have met with hitherto. A small pimple-like elevation (C, *bd*) appears on the surface of a cell and gradually increases in size: examined under a high power this *bud* is found to consist of an offshoot of the protoplasm of the parent cell covered with a very thin layer of cellulose: it is formed by the protoplasm growing out into an offshoot —like a small pseudopod—which pushes the cell-wall before it. The bud increases in size (*bd'*) until it forms a little globular body touching the parent cell at one pole: then a process of fission takes place along the plane of junction, the protoplasm of the bud or daughter-cell becoming separated from that of the mother-cell and a cellulose partition being secreted between the two. Finally the bud becomes completely detached as a separate yeast-cell.

It frequently happens that a Saccharomyces buds in several places and each of its daughter-cells buds again, before detachment of the buds takes place. In this way chains or groups of cells are produced (B), such *cell-colonies* consisting of two or more generations of cells, the central one standing in relation of parent, grandparent, or great-grandparent to the others.

It must be observed that this process of budding or *gemmation* is after all only a modification of simple fission. In the latter the two daughter-cells are of equal size and are both smaller than the parent-cell, while in gemmation one—the mother-cell—is much larger than the other— the daughter-cell or bud—and is of the same size as, indeed is practically identical with, the original dividing-cell. Hence in budding, the parent form does not, as in simple fission, lose its individuality, becoming wholly merged in its twin offspring, but merely undergoes separation of a small portion of its substance in the form of a bud, which by assimilation of nutriment gradually grows to the size of its parent, the latter thus retaining its individuality and continuing to produce fresh buds as long as it lives.

Multiplication by budding goes on only while the Saccharomyces is well supplied with food: if the supply of nutriment fails, a different mode of reproduction obtains. Yeast can be effectually starved by spreading out a thin layer of it on a slab of plaster-of-Paris kept moist under a bell-jar: under these circumstances the yeast is of course supplied with nothing but water.

In a few days the yeast-cells thus circumstanced are found to have altered in appearance: large vacuoles appear in them (Fig. 12, E, E′) and numerous fat-globules (f) are formed. The protoplasm has been undergoing destructive metabolism, and, there being nothing to supply new material, has diminished in quantity, and at the same time been partly converted into fat. Both in plants and in animals it is found that fatty degeneration, or the conversion of protoplasm into fat by destructive metabolism, is a constant phenomenon of starvation.

After a time the protoplasm collects towards the centre of the cell and divides simultaneously into four masses arranged like a pyramid of four billiard balls, three at the base and one above (F). Each of these surrounds itself with a thick cellulose coat and becomes a *spore*, the four spores being sooner or later liberated by the rupture of the mother-cell wall (F′).

The spores being protected by their thick cell-walls are able to withstand starvation and drought for a long time; when placed under favourable circumstances they develop into the ordinary form of Saccharomyces. So that reproduction by multiple fission appears to be, in the yeast-plant, a last effort of the organism to withstand extinction.

The physiology of nutrition of Saccharomyces has been studied with great care by several men of science and notably by Pasteur, and is in consequence better known than that of any other low organism. For this reason it will be advisable to consider it somewhat in detail.

The first process in the manufacture of beer is the preparation of a solution of malt called "sweet-wort." Malt is barley which has been allowed to germinate or sprout, *i.e.*, the young plant is allowed to grow to a certain extent from the seed. During germination the starch which forms so large a portion of the grain of barley is partly converted into sugar: barley also contains soluble proteids and mineral salts, so that when malt is infused in hot water the sweet-wort formed may be looked upon as a solution of sugar, proteid, and salts.

Into this wort a quantity of yeast is placed. Very soon the liquid begins to froth, the quantity of yeast increasing enormously: this means of course that the yeast-cells are budding actively, as can be readily made out by microscopic examination. If while the frothing is going on a lighted candle is lowered into the vat the flame will be immediately extinguished: if an animal were placed in the same position it would be suffocated.

Chemical examination shows that the extinction of the candle's flame or of the animal's life is caused by a rapid evolution of carbon dioxide from the fermenting wort, the frothing being due to the escape of the gas from the liquid.

After a time the evolution of gas ceases, and the liquid is then found to be no longer sweet but to have acquired what we know as an alcoholic or spirituous flavour. Analysis shows that the sugar has nearly or quite disappeared, while a new substance, *alcohol*, has made its appearance. The sweet-wort has, in fact, been converted into beer.

Expressed in the form of a chemical equation what has happened is this :—

$$C_6H_{12}O_6 = 2(C_2H_6O) + 2(CO_2)$$
Grape sugar.   Alcohol.   Carbon dioxide.

One molecule of sugar has, by the action of yeast, been split up into two molecules of alcohol which remain in the fluid, and two of carbon dioxide which are given off as gas. This is the process known as *alcoholic fermentation*.

It has been shown by accurate analysis that only about 95 per cent. of the sugar is thus converted into alcohol and carbon dioxide : 4 per cent. is decomposed, with the formation of glycerine, succinic acid, and carbon dioxide, and 1 per cent. is used as nutriment by the yeast cells.

For the accurate study of fermentation the sweet-wort of the brewer is unsuitable, being a fluid of complex and uncertain composition, and the nature of the process, as well as the part played in it by Saccharomyces, becomes much clearer if we substitute the artificial wort invented by M. Pasteur, and called after him Pasteur's solution. It is made of the following ingredients :—

Water, H_2O	83·76 per cent.
Cane sugar, $C_{12}H_{22}O_{11}$	15·00 ,, ,,
Ammonium tartrate $(NH_4)_2C_4H_4O_6$	1·00 ,, ,,
Potassium phosphate, K_3PO_4.	0·20 ,, ,,
Calcium phosphate, $Ca_3(PO_4)_2$	0·02 ,, ,,
Magnesium sulphate, $MgSO_4$	0·02 ,, ,,
	100·00

The composition of this fluid is not a matter of guess-work, but is the result of careful experiments, and is determined by the following considerations.

It is obvious that if we are to study alcoholic fermentation sugar must be present,[1] since the essence of the process is the formation of alcohol from sugar.

Then nitrogen in some form as well as carbon, oxygen, and hydrogen must be present, since these four elements enter into the composition of protoplasm, and all but the first-named (nitrogen) into that of cellulose, and they are thus required in order that the yeast should live and multiply. The form in which nitrogen can best be assimilated was found out by experiment. We saw that in the manufacture of beer the yeast cells obtain their nitrogen largely in the form of soluble proteids : green plants obtain theirs largely in the simple form of nitrates. It was found that while proteids are, so to say, an unnecessarily complex food for Saccharomyces, nitrates are not complex enough, and an ammonia compound is necessary, ammonium tartrate being the most suitable. Thus while Saccharomyces can build up the molecule of protoplasm from less complex food-stuffs than are required by Amœba, it cannot make use of such comparatively simple compounds as suffice for Hæmatococcus : moreover it appears to be indifferent whether its nitrogen is supplied to it in the form of ammonium tartrate or in the higher form of proteids.

Then as to the remaining ingredients of the fluid—potassium and calcium phosphate and magnesium sulphate. If a quantity of yeast is burnt, precisely the same thing happens as when one of the higher animals or plants is subjected to the same process. It first chars by the libera-

[1] It is a matter of indifference whether cane-sugar or grape-sugar is used.

tion of carbon, then as the heat is continued the carbon is completely consumed, going off by combination with the oxygen of the air in the form of carbon dioxide; at the same time the nitrogen is given off mostly as nitrogen gas, the hydrogen by union with atmospheric oxygen as water-vapour, and the sulphur as sulphurous acid or sulphur dioxide (SO_2). Finally, nothing is left but a small quantity of white ash which is found by analysis to contain phosphoric acid, potash, lime, and magnesia; *i.e.*, precisely the ingredients of the three mineral constituents of Pasteur's solution with the exception of sulphur, which, as already stated, is given off during the process of burning as sulphur dioxide.

Thus the principle of construction of an artificial nutrient solution such as Pasteur's is that it should contain all the elements existing in the organism it is designed to support; or in other words, the substances by the combination of which the waste of the organism due to destructive metabolism may be made good.

That Pasteur's solution exactly fulfils these requirements may be proved by omitting one or other of the constituents from it, and finding out how the omission affects the well-being of Saccharomyces.

If the sugar is left out the yeast-cells grow and multiply, but with great slowness. This shows that sugar is not necessary to the life of the organism, but only to that active condition which accompanies fermentation. A glance at the composition of Pasteur's solution will show that all the necessary elements are supplied without sugar.

Omission of ammonium tartrate is fatal: without it the cells neither grow nor multiply. This, of course, is just what one would expect since, apart from ammonium tartrate, the fluid contains no nitrogen without which the molecules of protoplasm cannot be built up.

It is somewhat curious to find that potassium and calcium phosphates are equally necessary; although occurring in such minute quantities they are absolutely essential to the well-being of the yeast-cells, and without them the organism, although supplied with abundance of sugar and ammonium tartrate, will not live. This may be taken as proving that phosphorus, calcium, and magnesium form an integral part of the protoplasm of Saccharomyces, although existing in almost infinitesimal proportions.

Lastly, magnesium sulphate must not be omitted if the organism is to flourish: unlike the other two mineral constituents it is not absolutely essential to life, but without it the vital processes are sluggish.

Thus by growing yeast in a fluid of known composition it can be ascertained exactly what elements and combinations of elements are necessary to life, what advantageous though not absolutely essential, and what unnecessary.

The precise effect of the growth and multiplication of yeast upon a saccharine fluid, or in other words the nature of alcoholic fermentation, can be readily ascertained by a simple experiment with Pasteur's solution. A quantity of the solution with a little yeast is placed in a flask the neck of which is fitted with a bent tube leading into a vessel of lime-water or solution of calcium oxide. When the usual disengagement of carbon dioxide (see p. 75) takes place the gas passes through the tube into the lime-water and causes an immediate precipitation of calcium carbonate as a white powder which effervesces with acids. This proves the gas evolved during fermentation to be carbon dioxide since no other converts lime into carbonate. When fermentation is complete the presence of alcohol may be proved by distillation: a colourless, mobile, pungent, and inflammable liquid being obtained.

By experimenting with several flasks of this kind it can be proved that fermentation goes on as well in darkness as in light, and that it is quite independent of free oxygen. Indeed the process does not go on if free oxygen—*i.e.*, oxygen in the form of dissolved gas—is present in the fluid ; from which it would seem that Saccharomyces must be able to obtain the oxygen, which like all other organisms it requires for its metabolic processes, from the food supplied to it.

The process of fermentation goes on most actively, between 28° and 34°C : at low temperatures it is comparatively slow, and at 38°C. multiplication ceases.

If a small portion of yeast is boiled so as to kill the cells, and then added to a flask of Pasteur's solution, no fermentation takes place, from which it is proved that the decomposition of sugar is effected by the living yeast-cells only. There seems to be no doubt that the property of exciting alcoholic fermentation is a function of the living protoplasm of Saccharomyces. The yeast-plant is therefore known as an *organized ferment:* when growing in a saccharine solution it not only performs the ordinary metabolic processes necessary for its own existence, but induces decomposition of the sugar present, this decomposition being unaccompanied by any corresponding change in the yeast-plant itself.

It is necessary to mention in this connection that there is an important group of not-living bodies which produce striking chemical changes in various substances without themselves undergoing any change : these are distinguished as *unorganized ferments.* A well-known example is *pepsin*, which is found in the gastric juice of the higher animals, and has the function of converting proteids into peptones (see p. 12) : its presence has been proved in the Mycetozoa (p. 52), and probably it or some similar *peptonizing* or *proteolytic* ferment effects this change in all organisms which have the power of digesting proteids. Another instance is furnished by *diastase*, which effects the conversion of starch into grape sugar: it is present in germinating barley (see p. 73), and an infinitesimal quantity of it can convert immense quantities of starch. The *ptyalin* of our own saliva has a like action, and probably some similar *diastatic* or *amylolytic* ferment is present in the Mycetozoa which, as we saw (p. 52), are able to digest cooked starch.

LESSON VIII

BACTERIA

IT is a matter of common observation that if certain moist organic substances, such as meat, soup, milk, &c., are allowed to stand at a moderate temperature for a few days—more or fewer according as the weather is hot or cold—they "go bad" or putrefy; *i.e.* they acquire an offensive smell, a taste which few are willing to ascertain by direct experiment, and often a greatly altered appearance.

One of the most convenient substances for studying the phenomena of putrefaction is an infusion of hay, made by pouring hot water on a handful of hay and straining the resultant brown fluid through blotting paper. Pasteur's solution may also be used, or mutton-broth well boiled and filtered, or indeed almost any vegetable or animal infusion.

If some such fluid is placed in a glass vessel, covered with a sheet of glass or paper to prevent the access of dust, the naked-eye appearances of putrefaction will be found to manifest themselves with great regularity. The fluid, at first quite clear and limpid, becomes gradually dull and turbid. The opacity increases and a scum forms on the surface: at the same time the odour of putrefaction arises, and especially in the case of animal infusions, quickly becomes very strong and disagreeable.

The scum after attaining a perceptible thickness breaks up and falls to the bottom, and after this the fluid slowly clears again, becoming once more quite transparent and losing its bad smell. If exposed to the light patches of green appear in it sooner or later, due to the presence of microscopic organisms containing chlorophyll. The fluid has acquired, in fact, the characteristics of an ordinary stagnant pond, and is quite incapable of further putrefaction. The whole series of changes may occupy many months.

Microscopic examination shows that the freshly-prepared fluid is free from organisms, and indeed, if properly filtered,

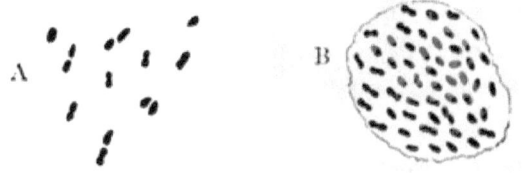

FIG. 13.—*Bacterium termo.* A, motile stage : B, resting stage or zoogloea. (From Klein.)

from particles of any sort. But the case is very different when a drop of infusion in which turbidity has set in is placed under a high power. The fluid is then seen to be crowded with incalculable millions of minute specks, only just visible under a power of 300 or 400 diameters, and all in active movement. These specks are *Bacteria*, or as they are sometimes called, *microbes* or *micro-organisms ;* they belong to the particular genus and species called *Bacterium termo.*

Seen under the high power of an ordinary student's microscope Bacterium termo has the appearance shown in Fig. 13, A : it is like a minute finger-biscuit, *i.e.* has the form of a rod constricted in the middle. But it is only by using the very highest powers of the microscope that its precise form and structure can be satisfactorily made out. It is then seen (Fig. 14) to consist of a little double spindle, showing neither nucleus, vacuole, nor other internal structure. It stains very deeply with aniline dyes, and from this and other circumstances there is reason for thinking that the whole cell consists of chromatin covered with a membrane of extreme tenuity formed of cellulose. It may therefore be considered as a cell consisting of cell-wall and nucleus only, the cell-body being absent. At each end is attached a flagellum about as long as the cell itself.

Bacterium termo is much smaller than any organism we have yet considered, so small in fact that, as it is always

Fig. 14.—*Bacterium termo* (× 4000), showing the terminal flagella. (After Dallinger.)

easier to deal with whole numbers than with fractions, its size is best expressed by taking as a standard the one-thousandth of a millimetre, called a *micromillimetre* and expressed by the symbol μ. The entire length of the organism under consideration is from 1·5 to 2 μ, *i.e.* about the $\frac{1}{500}$ mm. or the $\frac{1}{12500}$ inch. In other words, its entire length is not more than one-fourth the diameter of a yeast-cell or of a human blood-corpuscle. The diameter of the flagellum has been estimated by Dallinger to be about $\frac{1}{8}$ μ or $\frac{1}{204000}$ inch, a smallness of which it is as difficult to form any clear conception as of the distances of the fixed stars.

Some slight notion of these almost infinitely small dimensions may, however, be obtained in the following way. Fig.

14 shows a Bacterium termo magnified 4000 diameters, the scale above the figure representing $\frac{1}{100}$ mm. magnified to the same amount. The height of this book is a little over 18 cm.; this multiplied by 4,000 gives 72,000 cm. = 720 metres = 2362 feet. We therefore get the proportion—*as* 2362 feet, or nearly six times the height of St. Paul's, *is to* the height of the present volume, *so* the length of Fig. 14 *is to* that of Bacterium termo.

It was mentioned above that at a certain stage of putrefaction a scum forms on the surface of the fluid. This film consists of innumerable motionless Bacteria imbedded in a transparent gelatinous substance formed of a proteid material (Fig. 13, B). After continuing in the active condition for a time the Bacteria rise to the surface, lose their flagella, and throw out this gelatinous substance in which they lie imbedded. The bacterial jelly thus formed is called a *zooglœa*. Thus in Bacterium termo, as in so many of the organisms we have studied, there is an alternation of an active with a resting condition.

During the earlier stages of putrefaction Bacterium termo is usually the only organism found in the fluid, but later on other microbes make their appearance. Of these the commonest are distinguished by the generic names *Micrococcus, Bacillus, Vibrio,* and *Spirillum.*

Micrococcus (Fig. 15) is a minute form, the cells of which are about 2μ ($\frac{1}{500}$ mm.) in diameter. It differs from Bacterium in being globular instead of spindle-shaped and in having no motile phase. Like Bacterium it assumes the zooglœa condition (Fig. 15, 4).

Bacillus is commonly found in putrescent infusions in which the process of decay has gone on for some days: as its numbers increase those of Bacterium termo diminish, until Bacillus becomes the dominant form. Its cells (Fig. 16) are rod-shaped and about 6μ ($\frac{1}{170}$ mm.) in length in the commonest species. Both motionless and active forms are found, the latter having a flagellum at each end. The zooglæa condition is often assumed, and the rods are frequently found united end to end so as to form filaments.

Vibrio resembles Bacillus, but the rod-like cells (Fig. 17, A) are wavy instead of straight. They are actively motile and when highly magnified are found to be provided with a

FIG. 15.—*Micrococcus*. 1, single and double (dumb-bell shaped) forms: 2 and 3, chain-forms: 4, a zooglæa.

flagellum at each end. Vibriones vary from 8μ to 25μ in length.

Spirillum is at once distinguished by its spiral form, the cells resembling minute corkscrews (Fig. 17, B & C) and being provided with a flagellum at each end (C). The smaller species, such as S. tenue (B) are from 2 to 5 μ in length, but the larger forms, such as S. volutans (C) attain a length of from 25 to 30μ. In swimming Spirillum appears on a superficial examination to undulate like a worm or a serpent, but this is an optical illusion: the spiral is really a permanent one, but during progression it rotates upon its long axis, like Hæmatococcus (p. 25), and this double movement produces the appearance of undulation.

Most Bacteria are colourless, but three species (*Bacterium viride, B. chlorinum*, and *Bacillus virens*) contain chlorophyll, and several others form pigments of varying tints and often of great intensity. For instance, there are red, yellow, brown, blue, and violet species of Micrococcus which grow

FIG. 16.—*Bacillus subtilis*, showing various stages between single orms and long filaments (*Leptothrix*).

on slices of boiled potato, hard-boiled egg, &c., forming brilliantly coloured patches; and the yellow colour often assumed by milk after it has been allowed to stand for a considerable time is due to the presence of *Bacterium xanthinum*.

All Bacteria multiply by simple transverse fission, the process taking place sometimes during the motile, sometimes during the resting condition. Frequently the daughter-cells do not separate completely from one another but remain loosely attached, forming chains. These are very common in some species of micrococcus (see Fig. 15).

Bacillus when undergoing fission behaves something like Heteromita: the mother-cell divides transversely across the middle, and the two halves gradually wriggle away from one another, but remain connected for a time by a very fine thread

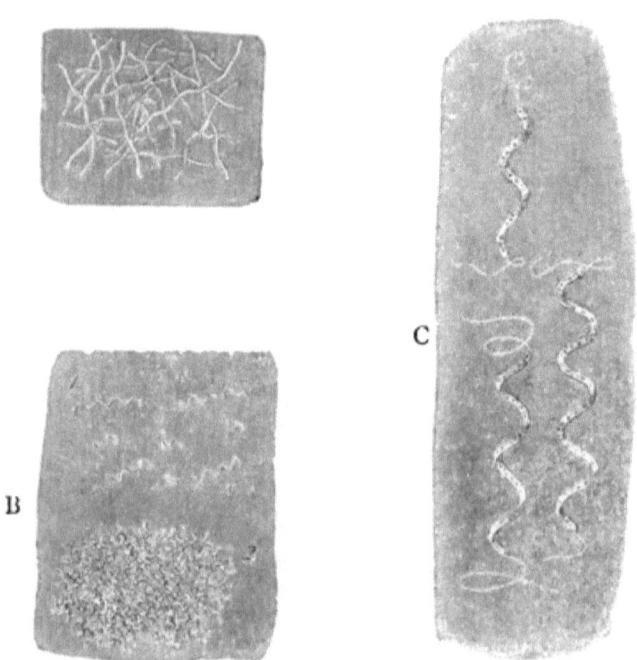

FIG. 17.—A, *Vibrio.* B, *Spirillum tenue.* C, *Spirilum volutans.* (From Klein.)

of protoplasm which extends between their adjacent ends. This is drawn out by the gradual separation of the two cells until it attains twice the length of a flagellum when it snaps in the middle, thus providing each daughter-cell with a new flagellum. Bacillus may, however, divide while in the resting condition and, under certain circumstances, the process is repeated again and again, and the daughter-cells remaining in contact form a long wavy or twisted filament called *Leptothrix* (Fig. 16) the separate elements of which are usually only visible after staining.

Bacillus also multiplies by a peculiar process of spore-formation which may take place either in the ordinary resting form or in a leptothrix filament. A bright dot appears at one place in the protoplasm (Fig. 18): this increases in size, the greater part of the protoplasm being used up in its formation, and finally takes on the form of a clear oval spore which remains for some time enclosed in the cell-wall of the Bacillus, by the rupture of which it is finally liberated. Spores of this kind are termed *endospores*. In other Bacteria spores are formed directly from the ordinary cells, which become thick walled (*arthrospores*). The spores differ from the Bacilli in being unstained by aniline dyes.

After a period of rest the spores, under favourable circumstances, germinate by growing out at one end so as to become rod-like, and thus finally assuming the form of ordinary Bacilli.

There are other genera often included among Bacteria for the description of which the student is referred to the more special treatises.[1] One remark must, however, be made in concluding the present brief account of the morphology of the group. There is a great deal of evidence to show that what have been spoken of as genera (Bacterium, Bacillus, Spirillum, &c.) may merge into one another and are therefore to be looked upon as phases in the life-history of various microbes rather than as true and distinct genera. But this is a point which cannot at present be considered as settled.

The conditions of life of Bacteria are very various. Some live in water, such as that of stagnant ponds, and of these

---

[1] See especially De Bary, *Fungi, Mycetozoa, and Bacteria* (Oxford, 1887), and Klein, *Micro-organisms and Disease* (London, 1886).

three species, as already stated (p. 85), contain chlorophyll. The nutrition of such forms must obviously be holophytic, and in the case of Bacterium chlorinum the giving off of oxygen in sunlight has actually been proved.

But this mode of nutrition is rare among the Bacteria: nearly all of those to which reference has been made are

FIG. 18.—Spore-formation in *Bacillus*. (From Klein.)

saprophytes, that is, live upon decomposing animal and vegetable matters. They are, in fact, nourished in precisely the same way as Heteromita (see p. 37). Many of these forms such as Bacterium termo, and species of Bacillus, Vibrio, &c., will, however, flourish in Pasteur's solution, in which they obtain their nitrogen in the form of ammonium tartrate instead of decomposing proteid. It has also been shown that some Bacteria can go further and make use of nitrates as a source of nitrogen, and of a carbonate or even of carbon dioxide as a source of carbon: in other words, they are able to live upon purely inorganic matter in spite of the fact that they contain no chlorophyll. Some species may even multiply to a considerable extent in distilled water.

But *pari passu* with their ordinary nutritive processes, many Bacteria exert an action on the fluids on which they live comparable to that exerted on a saccharine solution by the yeast-plant. Such microbes are, in fact, organized ferments.

Every one is familiar with the turning sour of milk. This change is due to the conversion of the milk-sugar into lactic acid.

$$C_6H_{12}O_6 = 2(C_3H_6O_3),$$
Sugar. Lactic Acid.

The transformation is brought about by the agency of *Bacterium lactis*, a microbe closely resembling *B. termo*.

Beer and wine are two other fluids which frequently turn sour, there being in this case a conversion of alcohol into acetic acid, represented by the equation—

$$C_2H_6O + O_2 = H_2O + C_2H_4O_2.$$
Alcohol. Oxygen. Water. Acetic Acid.

The ferment in this instance is *Bacterium aceti*, often called *Mycoderma aceti*, or the "vinegar plant." It will be noticed that in this case oxygen enters into the reaction: it is a case of fermentation by oxidation.

Putrefaction itself is another instance of fermentation induced by a microbe. Bacterium termo—the putrefactive ferment—causes the decomposition of proteids into simpler compounds, amongst which are such gases as ammonia

(NH_3), sulphuretted hydrogen (H_2S), and ammonium sulphide ($(NH_4)_2S$), the evolution of which produces the characteristic odour of putrefaction.

The final stage in putrefaction is the formation of nitrates and nitrites. The process is a double one, both stages being due to special forms of Bacteria. In the first place, by the agency of the *nitrous ferment*, ammonia is converted into nitrous acid—

$$NH_3 + 3O = H_2O + HNO_2$$
Ammonia.   Oxygen.   Water.   Nitrous Acid.

The *nitric ferment* then comes into action, converting the nitrous into nitric acid—

$$NHO_2 + O = HNO_3$$
Nitrous Acid.   Oxygen.   Nitric Acid.

This process is one of vast importance, since by its agency the soil is constantly receiving fresh supplies of nitric acid which is one of the most important substances used as food by plants.

Besides holophytes and saprophytes there are included among Bacteria many *parasites*, that is, species which feed not on decomposing but on living organisms. Many of the most deadly infectious diseases, such as tuberculosis, diphtheria, typhoid fever, and cholera, are due to the presence in the tissues or fluids of the body of particular species of microbes, which feed upon the parts affected and give rise to the morbid symptoms characteristic of the disease.

Some Bacteria, like the majority of the organisms previously studied, require free oxygen for their existence, but others, like Saccharomyces during active fermentation (see p. 78), are quite independent of free oxygen and must therefore be able to take the oxygen, without which their metabolic processes could not go on, from some of the compounds contained in the fluid in which they live. Bacteria are for this reason divided into *aërobic* species which require free oxygen, and *anaërobic* species which do not.

As to temperature, common observation tells us that Bacteria flourish only within certain limits. We know for instance that organic substances can be preserved from putrefaction by being kept either at the freezing-point, or at or near the boiling-point. One important branch of modern industry, the trade in frozen meat, depends upon the fact that the putrefactive Bacteria, like other organisms, are rendered inactive by freezing, and every housekeeper knows how easily putrefaction can be staved off by roasting or boiling. Simiarly it is a matter of common observation that a moderately igh temperature is advantageous to these organisms, the heat of summer or of the tropics being notoriously favourable to putrefaction. In the case of Bacterium termo, it has been found that the optimum temperature is from 30° to 35° C., but that the microbe will flourish between 5° and 40° C.

Although fully-formed Bacteria, like other organisms, are usually killed by exposure to heat several degrees below boiling-point, yet the spores of some species will withstand, at any rate for a limited time, a much higher temperature— even one as high as 130° C. On the other hand, putrefactive Bacteria retain their power of development after being exposed to a temperature of −111° C., although during the time of exposure all vital activity is of course suspended.

Bacteria also resemble other organisms in being unable to carry on active life without a due supply of water : no perfectly dry substance ever putrefies. The preservation for ages of the dried bodies of animals in such countries as Egypt and Peru depends at least as much upon the moistureless air as upon the antiseptics used in embalming.

For the most part Bacteria are unaffected by light, since they grow equally well in darkness and in ordinary daylight. Many of them, however, will not bear prolonged exposure to direct sunlight, and it has been found possible to arrest the putrefaction of an organic infusion by *insolation*, or exposure to the direct action of the sun's rays. It has also been proved that it is the light-rays and not the heat-rays which are thus prejudicial to the life of micro-organisms.

LESSON IX

BIOGENESIS AND ABIOGENESIS: HOMOGENESIS AND HETEROGENESIS

THE study of the foregoing living things and especially of Bacteria, the smallest and probably the simplest of all known organisms, naturally leads us to the consideration of one of the most important problems of biology—the problem of the origin of life.

In all the higher organisms we know that each individual arises in some way or other from a pre-existing individual: no one doubts that every bird now living arose by a process of development from an egg formed in the body of a parent bird, and that every tree now growing took its origin either from a seed or from a bud produced by a parent plant. But there have always—until quite recently, at any rate—been upholders of the view that the lower forms of life, bacteria, monads, and the like, may under certain circumstances originate independently of pre-existing organisms: that, for instance, in a flask of hay-infusion or mutton-broth, boiled so as to kill any living things present in it, fresh forms of life may arise *de novo*, may in fact be created then and there.

We have therefore two theories of the lower organisms,

the theory of *Biogenesis*, according to which each living thing, however simple, arises by a natural process of budding, fission, spore-formation, or what not, from a parent organism : and the theory of *Abiogenesis*, or as it is sometimes called *Spontaneous* or *Equivocal Generation*, according to which fully formed living organisms sometimes arise from not-living matter.

In former times the occurrence of abiogenesis was universally believed in. The expression that a piece of meat has " bred maggots " ; the opinion that parasites such as the gall-insects of plants or the tape-worms in the intestines of animals originate where they are found ; the belief still held in some rural districts in the occurrence of showers of frogs, or in the transformation of horse-hairs kept in water into eels ; all indicate a survival of this belief.

Aristotle, one of the greatest men of science of antiquity, explicitly teaches abiogenesis. He states that some animals " spring from putrid matter," that certain insects " spring from the dew which falls upon plants," that thread-worms " originate in the mud of wells and running waters," that fleas " originate in very small portions of corrupted matter," and that " bugs proceed from the moisture which collects on the bodies of animals, lice from the flesh of other creatures."

Little more than 200 years ago one Alexander Ross, commenting on Sir Thomas Browne's doubt as to " whether mice may be bred by putrefaction," says, "so may he doubt whether in cheese and timber worms are generated ; or if beetles and wasps in cow's dung ; or if butterflies, locusts, grasshoppers, shell-fish, snails, eels, and such like, be procreated of putrefied matter, which is apt to receive the form of that creature to which it is by formative power disposed. To question this is to question reason, sense, and experience.

If he doubts of this let him go to Egypt, and there he will find the fields swarming with mice, begot of the mud of Nylus, to the great calamity of the inhabitants."

As accurate inquiries into these matters were made, the number of cases in which equivocal generation was supposed to occur was rapidly diminished. It was a simple matter—when once thought of—to prove, as Redi did in 1638, that no maggots were ever "bred" in meat on which flies were prevented by wire screens from laying their eggs. Far more difficult was the task, also begun in the seventeenth century, of proving that parasites, such as tape-worms, arise from eggs taken in with the food; but gradually this proposition was firmly established, so that no one of any scientific culture continued to believe in the abiogenetic origin of the more highly organized animals any more than in showers of frogs, or in the origin of geese from barnacles.

But a new phase of the question was opened with the invention of the microscope. In 1683, Anthony van Leeuwenhoek discovered Bacteria, and it was soon found that however carefully meat might be protected by screens, or infusions by being placed in well-corked or stoppered bottles, putrefaction always set in sooner or later, and was invariably accompanied by the development of myriads of bacteria, monads, and other low organisms. It was not surprising, considering the rapidity with which these were found to make their appearance, that many men of science imagined them to be produced abiogenetically.

Let us consider exactly what this implies. Suppose we have a vessel of hay-infusion, and in it a single Bacterium. The microbe will absorb the nutrient fluid and convert it into fresh protoplasm: it will divide repeatedly, and, its progeny repeating the process, the vessel will soon contain millions of Bacteria instead of one. This means, of course, that a certain amount of fresh living protoplasm has been formed out of the constituents of the hay-infusion, through the agency in the first instance of a single living Bacterium. The question naturally arises—Why may not the formation of protoplasm take place independently of this insignificant speck of living matter?

It must not be thought that this question is in any way a vain or absurd one. That living protoplasm has at some period of the world's history originated from not-living matter seems a necessary corollary of the doctrine of evolution, and is obviously the very essence of the doctrine of special creation; and there is no *à priori* reason why it should be impossible to imitate the unknown conditions under which this took place. At present, however, we have absolutely no data towards the solution of this fundamental problem.

But however insoluble may be the question as to how life first dawned upon our planet, the origin of living things at the present day is capable of investigation in the ordinary way of observation and experiment. The problem may be stated as follows:—any putrescible infusion,—*i.e.* any fluid capable of putrefaction—will be found after a longer or shorter exposure to swarm with bacteria and monads: do these organisms or the spores from which they first arise reach the infusion from without, or are they generated within it? And the general lines upon which an investigation into the problem must be conducted are simple: given a vessel of any putrescible infusion; let this be subjected to some process which, without rendering it incapable of supporting life, shall kill any living things contained in it; let it then be placed under such circumstances that no living particles, however small, can reach it from without. If, after these two conditions have been rigorously complied with, living organisms appear in the fluid, such organisms must have originated abiogenetically.

To kill any microbes contained in the fluid it is usually quite sufficient to boil it thoroughly. As we have seen, protoplasm enters into heat-rigor at a temperature considerably below the boiling-point of water, so that, with an exception which will be referred to presently, a few minutes' boiling suffices to *sterilize* all ordinary infusions, *i.e.*, to kill any organisms they may contain.

Then as to preventing the entrance of organisms or their spores from without. This may be done in various ways. One way is to take a flask with the neck drawn out into a very slender tube, to boil the fluid in it for a sufficient time, and then, while ebullition is going on, to close the end of the tube by melting the glass in the flame of a Bunsen-burner or spirit-lamp, thus hermetically sealing the flask.

By this method not only organisms and their spores are excluded from the flask but also air. But this is obviously unnecessary: it is evident that air may be admitted to the fluid with perfect impunity if only it can be filtered, that is, passed through some substance which shall retain all solid particles however small, and therefore of course bacteria, monads, and their spores.

A perfectly efficient filter for this purpose is furnished by cotton-wool. A flask or test-tube is partly filled with the infusion: the latter is boiled, and during ebullition cotton-wool is pushed into the mouth of the vessel until a long and firm plug is formed (Fig 19). When the source of heat is removed, and, by the cooling of the fluid, the steam which filled the upper part of the tube condenses, air passes in to supply its place, but as it does so it is filtered of even the smallest solid particles by having to pass through the close meshes of the cotton-wool.

Experiments of this sort conducted with proper care have been known for many years to give negative results in the great majority of cases: the fluids remain perfectly sterile for any length of time. But in certain instances, in spite of the most careful precautions, bacteria were found to appear

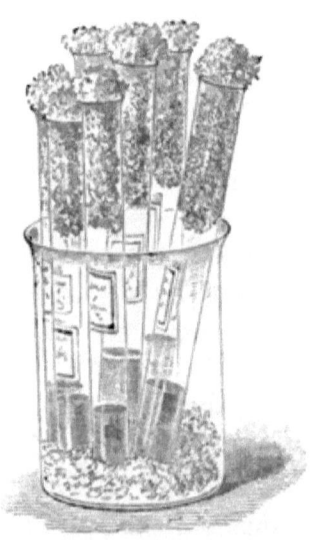

FIG. 19.—A Beaker with a number of test-tubes containing putrescible infusions and plugged with cotton-wool. (From Klein.)

in such fluids, and for years a fierce controversy raged between the biogenists and the abiogenists, the latter insisting that the experiments in question proved the occurrence of spontaneous generation, while the biogenists considered that all such cases were due to defective methods—either to imperfect sterilization of the fluid or to imperfect exclusion of germ-containing atmospheric dust.

The matter was finally set at rest, and the biogenists

proved to be in the right, by the important discovery that the spores of bacteria and monads are not killed by a temperature many degrees higher than is sufficient to destroy the adult forms : that in fact while the fully developed organisms are killed by a few minutes' exposure to a temperature of 70° C. the spores are frequently able to survive several hours' boiling, and must be heated to 130°—150° C. in order that their destruction may be assured. It was also shown that the more thoroughly the spores are dried the more difficult they are to kill, just as well-dried peas are hardly affected by an amount of boiling sufficient to reduce fresh ones to a pulp.

This discovery of the high thermal death-point or ultra-maximum temperature of the spores of these organisms has necessitated certain additional precautions in experiments with putrescible infusions. In the first place the flask and the cotton-wool should both be heated in an oven to a temperature of 150° C., and thus effectually sterilized. The flask being filled and plugged with cotton-wool is well boiled and then kept for some hours at a temperature of 32°—38°C., the optimum temperature for bacteria. The object of this is to allow any spores which have not been killed by boiling to germinate, in other words to pass into the adult condition in which the temperature of boiling water is fatal. The infusion is then boiled again, so as to destroy any such freshly germinated forms it may contain. The same process is repeated once or twice, the final result being that the very driest and most indurated spores are induced to germinate, and are thereupon slain. It must not be forgotten that repeated boiling does not render the fluid incapable of supporting life, as may be seen by removing the cotton-wool plug, when it will in a short time swarm with microbes.

Experiments conducted with these precautions all tell the

same tale : they prove conclusively that in properly sterilized putrescible infusions, adequately protected from the entrance of atmospheric germs, no micro-organisms ever make their appearance. So that the last argument for abiogenesis has been proved to be fallacious, and the doctrine of biogenesis shown, as conclusively as observation and experiment can show it, to be of universal application as far as existing conditions known to us are concerned.

It is also necessary to add that the presence of microbes in considerable quantities in our atmosphere has been proved experimentally. By drawing air through tubes lined with a solid nutrient material Prof. Percy Frankland showed that the air of South Kensington contained about thirty-five micro-organisms in every ten litres, and by exposing circular discs coated with the same substance he was further able to prove that in the same locality 279 micro-organisms fall upon one square foot of surface in one minute.

There is another question intimately connected with that of Biogenesis, although strictly speaking quite independent of it. It is a matter of common observation that, in both animals and plants, like produces like : that a cutting from a willow will never give rise to an oak, nor a snake emerge from a hen's egg. In other words, ordinary observation teaches the general truth of the doctrine of *Homogenesis*.

But there has always been a residuum of belief in the opposite doctrine of *Heterogenesis*, according to which the offspring of a given animal or plant may be something utterly different from itself, a plant giving rise to an animal or *vice versâ*, a lowly to a highly organized plant or animal and so on. Perhaps the most extreme case in which heterogenesis was once seriously believed to occur is that of the "barnacle-geese." Buds of a particular tree growing near the sea were said to produce barnacles, and these falling into the water to develop into geese. This sounds absurd enough, but within the last twenty years two or three men of science have described, as the result of repeated observations, the occurrence of quite similar cases among microscopic organisms. For instance, the blood-corpuscles of the silkworm have been said to give rise to fungi, the protoplasm of the green weed Nitella (see Fig. 45) to Amœba and Infusoria (see p. 107), Euglenæ to thread-worms, and so on.

It is proverbially difficult to prove a negative, and it might not be easy to demonstrate, what all competent naturalists must be firmly convinced of, that every one of these supposed cases of heterogenesis is founded either upon errors of observation or upon faulty inductions from correct observations.

Let us take a particular case by way of example. Many years ago Dr. Dallinger observed among a number of Vorticellæ or bell-animalcules (Fig. 26) one which appeared to have become encysted upon its stalk. After watching it for some time, there was seen to emerge from the cyst a free-swimming ciliated Infusor called *Amphileptus*, not unlike a long-necked Paramœcium (Fig. 20, p. 108). Many observers would have put this down as a clear case of heterogenesis: Dallinger simply recorded the observation and waited. Two years later the occurrence was explained: he found the same two species in a pond, and watched an Amphileptus seize and devour a Vorticella, and, after finishing its meal, become encysted upon the stalk of its victim.

It is obvious that the only way in which a case of heterogenesis could be proved would be by actually watching the transformation, and this no heterogenist has ever done; at the most, certain supposed intermediate stages between the extreme forms have been observed—say, between a Euglena and a thread-worm—and the rest of the process inferred. On the other hand, innumerable observations have been made on these and other organisms, the result being that each species investigated has been found to go through a definite series of changes in the course of its development, the ultimate result being invariably an organism resembling in all essential respects that which formed the starting-point of the observations : Euglenæ always giving rise to Euglenæ and nothing else, Bacteria to Bacteria and nothing else, and so on.

There are many cases which imperfect knowledge might class under heterogenesis, such as the origin of frogs from tadpoles or of jelly-fishes from polypes (Lesson XXIII. Fig. 53), but in these and many other cases the apparently anomalous transformations have been found to be part of the normal and invariable cycle of changes undergone by the organism in the course of its development; the frog always gives rise ultimately to a frog, the jelly-fish to a jelly-fish. If a frog at one time produced a tadpole, at another a trout, at another a worm : if jelly-fishes gave rise sometimes to polypes, sometimes to infusoria, sometimes to cuttle-fishes, and all without any regular sequence—*that* would be heterogenesis.

It is perhaps hardly necessary to caution the reader against the error that there is any connection between the theory of heterogenesis and that of organic evolution. It might be said—if, as naturalists tell us, dogs are descended from wolves and jackals and birds from reptiles, why should not, for instance, thread-worms spring from Euglenæ or Infusoria from Bacteria? To this it is sufficient to answer that the evolution of one form from another takes place by a series of slow, orderly, progressive changes going on through a long series of generations (see Lesson XIII.); whereas heterogenesis presupposes the casual occurrence of sudden transformations in any direction—*i.e.*, leading to either a less or a more highly organized form—and in the course of a single generation.

LESSON X

PARAMŒCIUM, STYLONYCHIA, AND OXYTRICHA

IT will have been noticed with regard to the simple unicellular organisms hitherto considered that all are not equally simple: that Protamœba (Fig. 2, p. 9) and Micrococcus (Fig. 15, p. 86) may be considered as the lowest of all, and that the others are raised above these forms in the scale of being in virtue of the possession of nucleus or contractile vacuole, or of flagella, or even, as in the case of Euglena (Fig. 5, p. 45), of a mouth or gullet.

Thus we may speak of any of the organisms already studied as relatively "high" or "low" with regard to the rest: the lowest or least differentiated forms being those which approach most nearly to the simplest conception of a living thing—a mere lump of protoplasm: the highest or most differentiated those in which the greatest complication of structure has been attained. It must be remembered, too, that this increase in structural complexity is always accompanied by some degree of division of physiological labour, or, in other words, that morphological and physiological differentiation go hand in hand.

We have now to consider certain organisms in which this differentiation has gone much further; which have, in fact, acquired many of the characteristics of the higher animals and plants while remaining unicellular. The study of several of these more or less highly differentiated though unicellular forms will occupy the next seven Lessons.

It was mentioned above that, in the earlier stages of the putrefaction of an organic infusion, bacteria only were found, and that later, monads made their appearance. Still later organisms much larger than monads are seen, generally of an ovoidal form, moving about very quickly, and seen by the use of a high power to be covered with innumerable fine cilia. These are called *ciliate Infusoria*, in contradistinction to monads, which are often known as *flagellate Infusoria*: many kinds are common in putrefying infusions, some occur in the intestines of the higher animals, while others are among the commonest inhabitants of both fresh and salt water. Five genera of these Infusoria will form the subjects of this and the four following Lessons.

A very common ciliate infusor is the beautiful "slipper animalcule," *Paramœcium aurelia*, which from its comparatively large size and from the ease with which all essential points of its organization can be made out is a very convenient and interesting object of study.

Compared with the majority of the organisms which have come under our notice it may fairly be considered as gigantic, being no less than $\frac{1}{5}-\frac{1}{4}$ mm. (200—260μ) in length : in fact it is just visible to the naked eye as a minute whitish speck.

Its form (Fig. 20 A) can be fairly well imitated by making out of clay or stiff dough an elongated cylinder rounded at one end and bluntly pointed at the other ; then giving the broader end a slight twist ; and finally making on the side

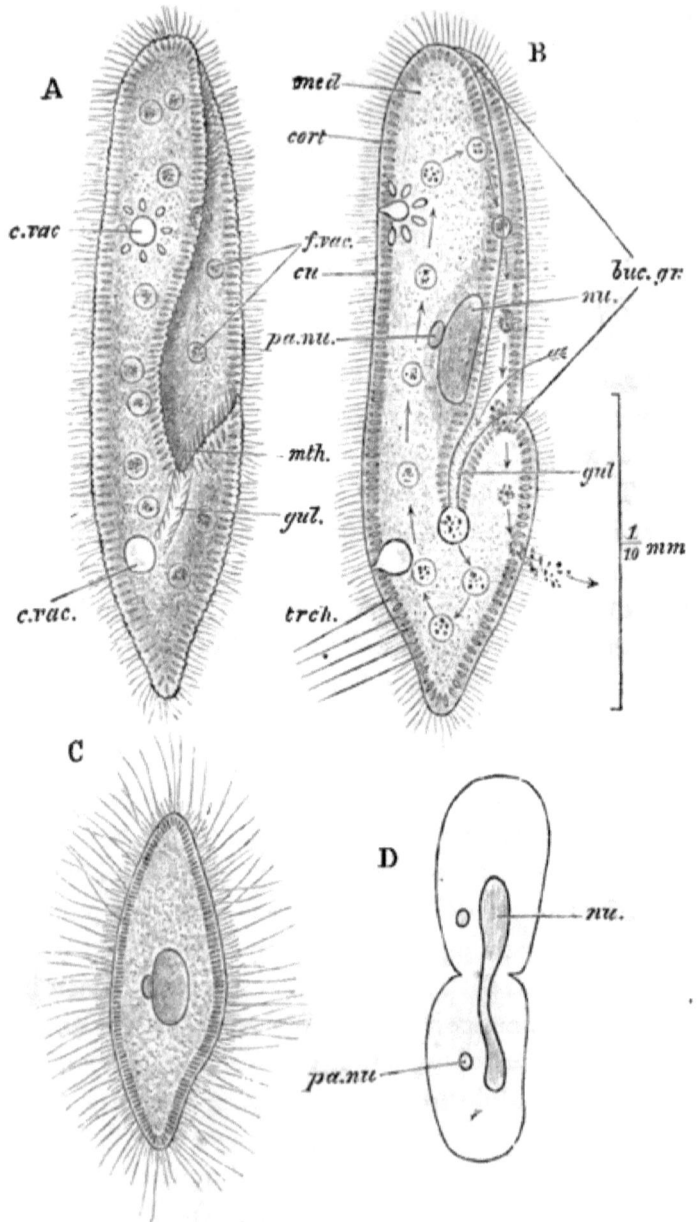

FIG. 20.—*Paramœcium aurelia.*
A, the living animal from the ventral aspect, showing the covering of cilia, the buccal groove (to the right) ending posteriorly in the mouth

(*mth*) and gullet (*gul*); several food vacuoles (*f. vac*), and the two contractile vacuoles (*c. vac*).

B, the same in optical section, showing cuticle (*cu*), cortex (*cort*), and medulla (*med*); buccal groove (*buc. gr*), mouth, and gullet (*gul*); numerous food vacuoles (*f. vac*) circulating in the direction indicated by the arrows, and containing particles of indigo, which are finally ejected at an anal spot; meganucleus (*nu*), micronucleus (*pa. nu*), and trichocysts, some of which (*trch*) are shown with their threads ejected.

The scale to the right of this figure applies to A and B.

C, a specimen killed with osmic acid, showing the ejection of trichocyst-threads, which project considerably beyond the cilia.

D, diagram of binary fission: the micronucleus (*pa. nu*) has already divided, the nucleus (*nu*) is in the act of dividing.

(D after Lankester.)

rendered somewhat concave by the twist a wide shallow groove beginning at the broad end and gradually narrowing to about the middle of the body, where it ends in a tolerably deep depression.

The grove is called the *buccal groove* (Fig. 20, A & B, *buc. gr*): at the narrow end is a small aperture the mouth (*mth*), which, like the mouth of Euglena (Fig. 5), leads into the soft internal protoplasm of the body. The surface of the creature on which the groove is placed is distinguished as the ventral surface, the opposite surface being upper or dorsal; the broad end is anterior, the narrow end posterior, the former being directed forwards as the animalcule swims. These descriptive terms being decided upon, it will be seen from Fig. 20 A, that the buccal groove begins on the left side of the body, and gradually curves over to the middle of the ventral surface.

As the animal swims its form is seen to be permanent, exhibiting no contractions of either an amœboid or a euglenoid nature. It is however distinctly flexible, often being bent in one or other direction when passing between obstacles such as entangled masses of weed. This permanence of contour is due to the presence of a tolerably firm though delicate cuticle (*cu*) which invests the whole surface.

The protoplasm thus enclosed by the cuticle is distinctly divisible into two portions—an external somewhat dense layer, the *cortical layer* or *cortex* (*cort*), and an internal more fluid material, the *medullary substance* or *medulla* (*med*). It will be remembered that a somewhat similar distinction of the protoplasm into two layers is exhibited by Amœba (p. 3), the ectosarc being distinguished from the endosarc simply by the absence of granules. In Paramœcium the distinction is a far more fundamental one : the cortex is radially striated and is comparatively firm and dense, while the medulla is granular and semi-fluid, as may be seen from the fact that food particles (*f. vac*, see below, p. 112,) move freely in it, whereas they never pass into the cortex. It has recently been found that the medulla has a reticular structure similar to that of the protoplasm of the ordinary animal cell (Fig. 9, p. 62), consisting of a delicate granular network the meshes of which are filled with a transparent material. In the cortex the meshes of the network are closer, and so form a comparatively dense substance. The cortex also exhibits a superficial oblique striation, forming what is called the *myophan layer*.

The mouth (*mth*) leads into a short funnel-like tube, the gullet (*gul*), which is lined by cuticle and passes through the cortex to end in the soft medulla, thus making a free communication between the latter and the external water.

The cilia with which the body is covered are of approximately equal size, quite short in relation to the entire animal, and arranged in longitudinal rows over the whole outer surface. They consist of prolongations of the cortex, and each passes through a minute perforation in the cuticle, They are in constant rhythmical movement, and are thereby distinguished from the flagella of Hæmatococcus, Euglena, &c., which exhibit more or less intermittent lashing movements (see p. 25, note, and p. 59). Their rapid motion and minute size make them somewhat difficult to see while the Paramœcium is alive and active, but after death they are very obvious, and look quite like a thick covering of fine silky hairs.

Near the middle of the body, in the cortex, is a large oval nucleus (B, *nu*), which is peculiar in taking on a uniform tint when stained, showing none of the distinction into chromatin and nuclear matrix which is so marked a feature in many of the nuclei we have studied (see especially Fig. 1, p. 2, and Fig. 9, p. 62). It has also a further peculiarity: against one side of it is a small oval structure (*pa. nu*) which is also deeply stained by magenta or carmine. This is the *micronucleus:* it is to be considered as a second, smaller nucleus, the larger body being distinguished as the *meganucleus*.

There are two contractile vacuoles (*c. vac*), one situated at about a third of the entire length from the anterior end of the body, the other at about the same distance from the posterior end: they occur in the cortex.

The action of the contractile vacuoles is very beautifully seen in a Paramœcium at rest: it is particularly striking in a specimen subjected to slight pressure under a cover glass, but is perfectly visible in one which has merely temporarily suspended its active swimming movements. It is then seen that during the *diastole*, or phase of expansion of each vacuole, a number—about six to ten—of delicate radiating, spindle-shaped spaces filled with fluid appear round it, like the rays of a star (upper vacuole in A & B): the vacuole itself contracts or performs its *systole*, completely disappearing from view, and immediately afterwards the radiating canals flow together and re-fill it, becoming themselves emptied and therefore invisible for an instant (lower vacuole in A & B) but rapidly appearing once more. There seems to be no doubt that the water taken in with the food is collected into these canals, emptied into the vacuole, and finally discharged into the surrounding medium.

The process of feeding can be very conveniently studied in Paramœcium by placing in the water some finely-divided carmine or indigo. When the creature comes into the neighbourhood of the coloured particles, the latter are swept about in various directions by the action of the cilia: some of these are however certain to be swept into the neighbourhood of the buccal groove and gullet, the cilia of which all work downwards, *i.e.* towards the inner end of the gullet. The grains of carmine are thus carried into the gullet, where for an instant they lie surrounded by the water of which it is full: then, instantaneously, probably by the contraction of the tube itself, the animalcule performs a sort of gulp, and the grains with an enveloping globule of water or food-vacuole are forced into the medullary protoplasm. This process is repeated again and again, so that in any well-nourished Paramœcium there are to be seen numerous globular spaces filled with water and containing particles of food—or in the present instance of carmine or indigo. At every gulp the newly formed food-vacuole pushes, as it were, its predecessor before it: contraction of the medullary protoplasm also takes place in a definite direction, and thus a circulation of food-vacuoles is produced, as indicated in Fig. 20, B, by arrows.

After circulating in this way for some time the water of the food-vacuoles is gradually absorbed, being ultimately excreted by the contractile vacuoles, so that the contained particles come to lie in the medulla itself (refer to figure). The circulation still continues, until finally the particles are brought to a spot situated about half-way between the mouth and the posterior end of the body: here if carefully watched they are seen to approach the surface and then to be suddenly ejected. The spot in question is therefore to be looked upon as a potential *anus*, or aperture for the egestion of fæces or undigested food-matters. It is a potential and not an actual anus, because it is not a true aperture but only a soft place in the cortex through which by the contractions of the medulla solid particles are easily forced.

Of course when Paramœcium ingests, as it usually does, not carmine but minute living organisms, the latter are digested as they circulate through the medullary protoplasm, and only the non-nutritious parts cast out at the anal spot. It has been found by experiment that this infusor can digest not only proteids but also starch and perhaps fats. The starch is probably converted into *dextrin*, a carbohydrate having the same formula ($C_6H_{10}O_5$) but soluble and diffusible. Oils or fats seem to be partly converted into fatty acids and glycerine. The nutrition of Paramœcium is therefore characteristically holozoic.

It was mentioned above (p. 108) that the cortex is radially striated in optical section. Careful examination with a very high power shows that this appearance is due to the presence in the cortex of minute spindle-shaped bodies (A and B, *trch*) closely arranged in a single layer and perpendicular to the surface. These are called *trichocysts*.

When a Paramœcium is killed, either by the addition of osmic acid or some other poisonous reagent or by simple pressure of the cover glass, it frequently assumes a remarkable appearance. Long delicate threads suddenly appear, projecting from its surface in all directions (C) and looking very much as if the cilia had suddenly protruded to many times their original length. But these filaments have really nothing to do with the cilia; they are contained under ordinary circumstances in the trichocysts, probably coiled up; and by the contraction of the cortex consequent upon any sudden irritation they are projected in the way indicated. In Fig. 20 B, a few trichocysts (*trch*) are shown in the exploded condition, *i.e.* with the threads protruded. Most likely these bodies are weapons of offence like the very similar structures (nematocysts) found in polypes (see Lesson XXII. Fig 51).

Paramœcium multiplies by simple fission, the division of the body being always preceded by the elongation and subsequent division of the mega- and micronucleus (Fig. 20, D). Division of the meganucleus is direct, that of the micronucleus indirect, *i.e.* takes place by karyokinesis.

Conjugation also occurs, usually after multiplication by fission has gone on for some time, but the details and the results of the process are very different from what are found to obtain in Heteromita (p. 62). Two Paramœcia come into contact by their ventral faces (Fig. 21, A) and the meganucleus (*mg. nu*) of each gradually breaks up into minute fragments (D—G) which are either absorbed into the protoplasm or ejected. At the same time the micronucleus (*mi. nu*) divides, by karyokinesis, and the process is repeated, the result being that each gamete contains four micronuclei (B). Two of these become absorbed and disappear, (C *mi. nu′*, *mi. nu″*) of the remaining two one is now distinguished as the *active pronucleus,* the other as the *stationary pronucleus.* Next, the active pronucleus of each gamete passes into the body of the other (C) and fuses with its stationary pronucleus (D): in this way each gamete contains a single nuclear body, the *conjugation-nucleus* (E), formed by the union of two similar pronuclei one of which is derived from another individual. It is this fusion of two nuclear bodies, one from each of the conjugating cells, which is the essential part of the whole

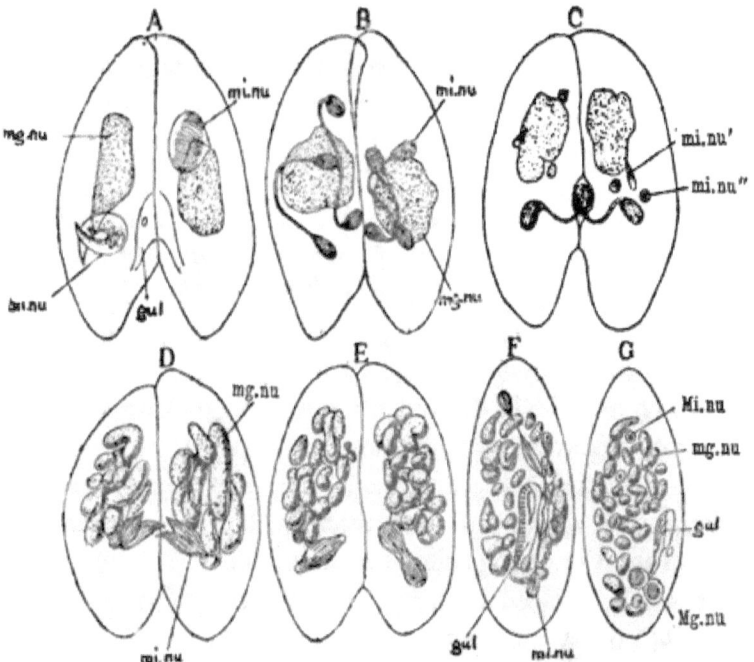

FIG. 21.—*Stages in the Conjugation of Paramæcium.*

A, Commencement of conjugation: the meganuclei (*mg. nu*) of the two gametes are almost unaltered: the micronuclei (*mi. nu*) are in an early stage of karyokinesis.

B, The micronuclei have divided twice, each gamete now containing four.

C, Two of the micronuclei (*mi. nu′*, *mi. nu″*) of each gamete are degenerating: of the remaining two one—the active pronucleus—is passing into the other gamete.

D, The active pronucleus of each gamete has passed into the other gamete and is conjugating with its stationary pronucleus. The meganucleus (*mg. nu*) has begun to break up.

E, Each gamete contains a single conjugation-nucleus formed by the union of its own stationary pronucleus with the active pronucleus of the other gamete. On the right side the conjugation-nucleus is beginning to divide.

F, Conjugation is over and only one of the separated gametes is shown. It contains the fragments of the meganucleus (dotted) and four nuclear bodies (*mi. nu*) produced by the division and re-division of the conjugation-nucleus.

G, Two of the products of division of the conjugation-nucleus (*Mg. nu*) are enlarging to form mega-nuclei, the other two (*Mi. nu*) are taking on the characters of micronuclei.

(After Hortwig.)

process. Soon after this the gametes separate from one another and begin once more to lead an independent existence; the conjugation nucleus of each undergoing a twice repeated process of division, the infusor thus acquiring four small nuclei (F). Two of these enlarge and take on the character of meganuclei (G, *Mg. nu*), the other two remaining unaltered and having the character of micronuclei (*Mi. nu*). Thus shortly after the completion of conjugation each individual contains two mega- and two micronuclei all derived from the conjugation-nucleus. Ordinary transverse fission now takes place, as described in the preceding paragraph, each of the two daughter cells having one mega- and one micronucleus, and thus the normal form of the species is re-acquired.

It will be noticed that, in the present instance, conjugation is not a process of multiplication: it has been ascertained that during the time two infusors are conjugating each might have produced several thousand offspring by continuing to undergo fission at the usual rate. The importance of the process lies in the exchange of nuclear material between the two conjugating individuals: without such exchange these organisms have been shown to undergo a gradual process of senile decay characterized by diminution in size and degeneration in structure.

Another ciliated infusor common in stagnant water and organic infusions is *Stylonychia mytilus*, an animalcule varying from $\frac{1}{11}$mm. to $\frac{1}{3}$mm.

Like Paramœcium it is often to be seen swimming rapidly in the fluid, but unlike that genus it frequently creeps about, almost like a wood-louse or a caterpillar, on the surface of the plants or other solid objects among which it lives. In correspondence with this, instead of being nearly cylindrical, it is flattened on one—the ventral—side, and is thus irregularly plano-convex in transverse section (Fig. 22, C).

It resembles Paramœcium in general structure (compare

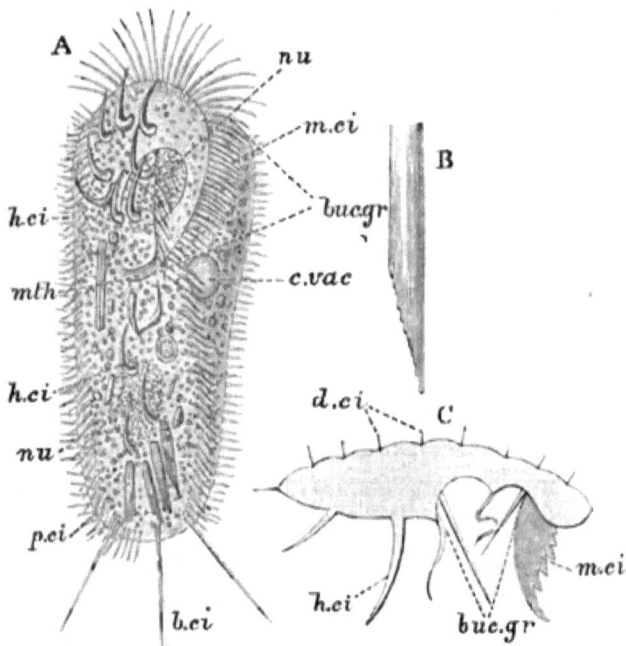

FIG. 22.—A, *Stylonychia mytilus*, ventral aspect, showing the buccal groove (*buc. gr.*) and mouth (*mth*), two nuclei (*nu*, *nu*), contractile vacuole (*c.vac*), and cilia differentiated into hook-like (*h. ci*), bristle-like (*b. ci*), plate-like (*p. ci*), and fan-like (*m. ci*) organs.

B, one of the plate-like cilia of the same (*p. ci* in A), showing its frayed extremity.

C, transverse section of *Gastrostyla*, a form allied to Stylonychia, showing buccal groove (*buc. gr.*), small dorsal cilia (*d. ci*), hook-like cilium (*h. ci*), and the various cilia of the buccal groove, including an expanded fan-like organ (*m. ci*). A and B after Claparède and Lachmann: C after Sterki.

Fig. 22, A, with Fig. 20, A); but owing to the absence of trichocysts the distinction between cortex and medulla is less obvious: moreover, it has two nuclei (*nu*, *nu*) and only one contractile vacuole (*c. vac*).

But it is in the character of its cilia that Stylonychia is most markedly distinguished from Paramœcium : these structures, instead of being all alike both in form and size, are modified in a very extraordinary way.

On the dorsal surface the cilia are represented only by very minute processes of the cortex (c, *d. ci.*) set in longitudinal grooves and exhibiting little movement. It seems probable that these are to be looked upon as *vestigial* or *rudimentary* cilia, *i.e.*, as the representatives of cilia which were of the ordinary character in the ancestors of Stylonychia, but which have undergone partial *atrophy*, or diminution beyond the limits of usefulness, in correspondence with the needs of an animalcule which has taken to creeping on its ventral surface, instead of swimming freely and so using all its cilia equally.

On the other hand, the cilia on the ventral surface have undergone a corresponding enlargement or *hypertrophy*. Near the anterior and posterior ends and about the middle are three groups of cilia of comparatively immense size, shaped either like hooks (*h. ci.*), or like flattened rods frayed at their ends (*p. ci*, and B). All these structures neither vibrate rhythmically like ordinary cilia nor perform lashing movements like flagella, but move at the base only like one-jointed legs. The movement is under the animal's control, so that it is able to creep about by the aid of these hooks and plates in much the same way as a caterpillar by means of its legs.

Notice that we have here a third form of contractility: in amœboid movement there is an irregular flowing of the protoplasm (pp. 4 and 10); in ciliary movement a flexion of a protoplasmic filament from side to side (p. 33); while in the present case we have sudden contractions taking place at irregular intervals. The movements of these locomotor hooks and plates are therefore very similar to the muscular contractions to which the movements of the higher animals are due: it cannot be said that definite muscles are present in Stylonychia, but the protoplasm in certain regions of the unicellular body is so modified as to be able to perform a sudden contraction in a definite direction. The nature of muscular contraction will be further discussed in the next Lesson (see p. 130).

The remainder of the ventral surface, with the exception of the buccal groove, is bare, but along each side of the margin is a row of large vibratile cilia, of which three at the posterior end are modified into long, stiff, bristle-like processes (A, *b. ci*).

There is also a special differentiation of the cilia of the buccal groove (*buc. gr.*). On its left side is a single row of very large and powerful cilia (A and C, *m. ci*) which are the chief organs for causing the food-current as well as the main swimming-organs: each has the form of a triangular fan-like plate (C, *m. ci*). On the right side of the buccal groove is a row of smaller but still large cilia of the ordinary form, and in the interior of the gullet a row of extremely delicate cilia which aid in forcing particles of food down the gullet into the medulla.

In Stylonychia and allied genera intermediate forms are found between these peculiar hooks, plates, bristles, and fans, and ordinary cilia; from which we may conclude that these diverse appendages are to be looked upon as highly modified or *differentiated* cilia. Probably they have been evolved in the course of time from ordinary cilia, and on the principle that the more complicated or specialized organisms are descended from simpler or more generalized forms (see Lesson XIII.), we may consider Stylonychia as the highly-specialized descendant of some uniformly-ciliated progenitor.

A third genus of ciliated Infusoria must be just referred to in concluding the present Lesson. We have seen how the nucleus of a Paramœcium which has just conjugated breaks up and apparently disappears (Fig. 21, K—O). In *Oxytricha*, a genus closely resembling Stylonychia, the two nuclei have been found to break up into a large number of minute granules (Fig. 23), which can be seen only after

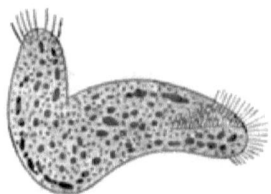

FIG. 23.—*Oxytricha flava*, killed and stained, showing the fragmentation of the nuclei. (After Gruber.)

careful staining and by the use of high magnifying powers. This process is called *fragmentation* of the nucleus; in other cases it goes even further, and the nucleus is reduced to an almost infinite number of chromatin granules only just visible under the highest powers. From this it seems very probable that organisms which, like Protamœba (p. 9) and Protomyxa (p. 49), appear non-nucleate, are actually provided with a nucleus in this pulverized condition, and that a nucleus in some form or other is an essential constituent of the cell.

LESSON XI

OPALINA

The large intestine of the common frog often contains numbers of ciliate Infusoria belonging to two or three genera. One of these parasitic animalcules, called *Opalina ranarum*, will now be described. It is easily obtained by killing a frog, opening the body, making an incision in the rectum, and spreading out a little of its blackish contents in a drop of water on a slide.

Opalina has a flattened body with an oval outline (Fig. 24, A, B), and full-sized specimens may be as much as one millimetre in length. The protoplasm is divided into cortex and medulla, and is covered with a cuticle, and the cilia are equal-sized and uniformly arranged in longitudinal rows over the whole surface (A).

On a first examination no nucleus is apparent, but after staining a large number of nuclei can be seen (B, *nu*), each being a globular body (C, 1), consisting of a nuclear matrix surrounded by a membrane and containing a coil or network of chromatin. These nuclei multiply within the body of the infusor, and in so doing pass through the various changes characteristic of karyokinesis or indirect nuclear division (compare Fig. 10, p. 64, with Fig. 23): the

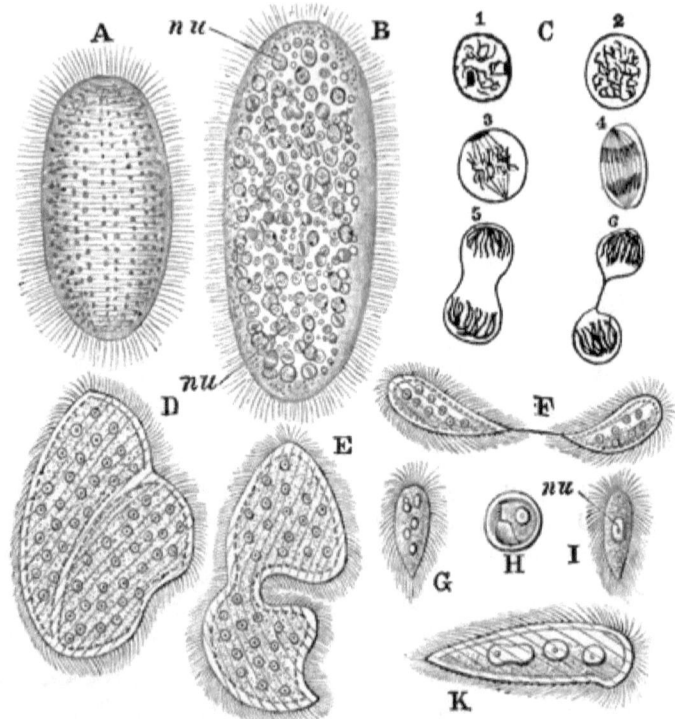

FIG. 24.—*Opalina ranarum.*
A, living specimen, surface view, showing longitudinal rows of cilia.
B, the same, stained, showing numerous nuclei (*nu*) in various stages of division.
C, 1—6, stages in nuclear division.
D, longitudinal fission.
E, transverse fission.
F, the same in a specimen reduced in size by repeated division.
G, final product of successive divisions.
H, encysted form.
I, uninucleate form produced from cyst.
K, the same after multiplication of the nucleus has begun.
(A—C, after Pfitzner; D—K, from Saville Kent after Zeller.)

chromatin breaks up (C, 2), a spindle is formed with the chromosomes across its equator (3), the chromosomes pass to the poles of the spindle (4, 5), and the nucleus becomes constricted (5), and finally divides into two (6).

The presence of numerous nuclei in Opalina is a fact worthy of special notice. The majority of the organisms we have studied are uninucleate as well as unicellular: the higher animals and plants we found (Lesson VI.) to consist of numerous cells each with a nucleus, so that they are multicellular and multinucleate: Opalina, on the other hand, is multinucleate but unicellular. An approach to this condition of things is furnished by Stylonychia, which is unicellular and binucleate (Fig. 24, A), but the only organisms we have yet studied in which numerous nuclei of the ordinary character occur in an undivided mass of protoplasm are the Mycetozoa (p. 52), and in them the multinucleate condition of the plasmodium is largely due to its being formed by the fusion of separate cells, while in Opalina it is due, as we shall see, to the repeated binary fission of an originally single nucleus.

There is no contractile vacuole, and no trace of either mouth or gullet, so that the ingestion of solid food is impossible. The creature lives, as already stated, in the intestine of the frog: it is therefore an *internal parasite*, or *endoparasite*, having the frog as its *host*. The intestine contains the partially-digested food of the frog, and it is by the absorption of this that the Opalina is nourished. Having no mouth, it feeds solely by imbibition: whether it performs any kind of digestive process itself is not certainly known, but the analogy of other mouthless parasites leads us to expect that it simply absorbs food ready digested by its host, upon which it is dependent for a constant supply of soluble and diffusible nutriment.

Thus Opalina, in virtue of its parasitic mode of life, is saved the performance of certain work—the work of digestion, that work being done for it by its host. This is the essence of internal parasitism: an organism exchanges a free life, burdened with the necessity of finding food for itself, for existence in the interior of another organism, on which, in one way or another, it levies blackmail.

Note the close analogy between the nutrition of an internal parasite like Opalina and the saprophytic nutrition of a monad (p. 39). In both the organism absorbs proteids rendered soluble and diffusible, in the one case by the digestive juices of the host, in the other by the action of putrefactive bacteria.

The reproduction of Opalina presents certain points of interest, largely connected with its peculiar mode of life. It is obvious that if the Opalinæ simply went on multiplying, by fission or otherwise, in the frog's intestine, the population would soon outgrow the means of subsistence: moreover, when the frog died there would be an end of the parasites. What is wanted in this as in other internal parasites is some mode of multiplication which shall serve as a *means of dispersal*, or in other words, enable the progeny of the parasite to find their way into the bodies of other hosts, and so start new colonies instead of remaining to impoverish the mother country.

Opalina multiplies by a somewhat peculiar process of binary fission: an animalcule divides in an oblique direction (Fig. 24, D), and then each half, instead of growing to the size of the parent cell, divides again transversely (E). The process is repeated again and again (F), the plane of division being alternately oblique and transverse, until finally small bodies are produced (G), about $\frac{1}{20}-\frac{1}{30}$ mm. in length, and containing two to four nuclei.

If the parent cell had divided simultaneously into a number of these little bodies the process would have been one of multiple fission: as it is it forms an interesting link between simple and multiple fission.

Opalina ranarum multiplies in this way in the spring—*i.e.* during the frog's breeding season. Each of the small products of division (G) becomes encysted (H), and in this passive condition is passed out with the frog's excrement, probably falling on to a water-weed or other aquatic object. Nothing further takes place unless the cyst is swallowed by a tadpole, as must frequently happen when these creatures, produced in immense numbers from the frogs' eggs, browse upon the water-weeds which form their chief food.

Taken into the tadpole's intestine, the cyst is burst or dissolved, and its contents emerge as a lanceolate mass of protoplasm (I), containing a single nucleus and covered with cilia. This, as it absorbs the digested food in the intestine of its host, grows, and at the same time its nucleus divides repeatedly (K) in the way already described, until by the time the animalcule has attained the maximum size it has also acquired the large number of nuclei characteristic of the genus.

Here, then, we have another interesting case of development (see p. 43): the organism begins life as a very small uninucleate mass of protoplasm, and as it increases in size increases also in complexity by the repeated binary fission of its nucleus.

LESSON XII

VORTICELLA AND ZOOTHAMNIUM

THE next organism we have to consider is a ciliated infusor even commoner than those described in the two previous lessons. It is hardly possible to examine the water of a pond with any care without finding in it, sometimes attached to weeds, sometimes to the legs of water-fleas, sometimes to the sticks and stones of the bottom, numbers of exquisitely beautiful little creatures, each like an inverted bell with a very long handle, or a wine-glass with a very long stem. These are the well-known "bell-animalcules;" the commonest among them belong to various species of the genus *Vorticella*.

The first thing that strikes one about Vorticella (Fig. 25, A) is the fact that it is permanently fixed, like a plant, the *proximal* or near end of the stalk being always firmly fixed to some aquatic object, while to the *distal* or far end the body proper of the animalcule is attached.

But in spite of its peculiar form it presents certain very obvious points of resemblance to Paramœcium, Stylonychia, and Opalina. The protoplasm is divided into cortex (Fig. 25, C, *cort*) and medulla (*med*), and is invested with a

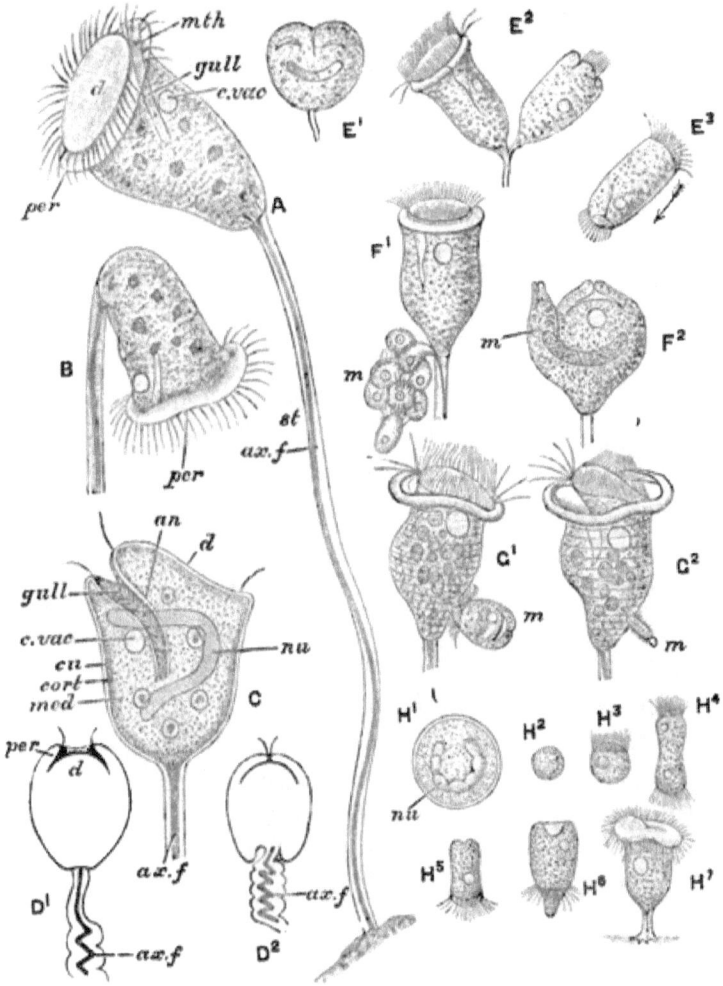

FIG. 25.—*Vorticella*.

A, living specimen fully expanded, showing stalk (*st*) with axial fibre (*ax. f.*), peristome (*per*), disc (*d*), mouth (*mth*), gullet (*gull*), and contractile vacuole.

B, the same, bent on its stalk and with the disc turned away from the observer.

C, optical section of the same, showing cuticle (*cu*), cortex (*cort*), medulla (*med*), nucleus (*nu*), gullet (*gull*), several food-vacuoles, and anus (*an*), as well as the structures shown in A.

D^1, a half-retracted and D^2 a fully-retracted specimen, showing the coiling of the stalk and overlapping of the disc by the peristome.

E^1, commencement of binary fission; E^2, completion of the process; E^3, the barrel-shaped product of division swimming freely in the direction indicated by the arrow.

F^1, a specimen dividing into a megazooid and several microzooids (m); F^2, division into one mega- and one microzooid.

G^1, G^2, two stages in conjugation showing the gradual absorption of the microgamete (m) into the megagamete.

H^1, multiple fission of encysted form, the nucleus dividing into numerous masses: H^2, spore formed by multiple fission; H^3—H^7, development of the spore; H^4 is undergoing binary fission.

(E—H after Saville Kent.)

delicate cuticle (*cu*). There is a single contractile vacuole (*c. vac*) the movements of which are very readily made out owing to the ease with which the attached organism is kept under observation. There is a meganucleus (*nu*) remarkable for its elongated band-like form, and having in its neighbourhood a small rounded micronucleus. Cilia are also present, but the way in which they are disposed is very peculiar and characteristic. To understand it we must study the form of the body a little more closely.

The conical body is attached by its apex or proximal end to the stalk: its base or distal end is expanded so as to form a thickened rim, the *peristome* (*per*), within which is a plate-like body elevated on one side, called the *disc* (*d*), and looking like the partly raised lid of a chalice. Between the raised side of the disc and the peristome is a depression, the mouth (*mth*), leading into a conical gullet (*gull*).

There is reason for thinking that the whole proximal region of Vorticella answers to the ventral surface of Paramœcium, and its distal surface with the peristome and disc to the dorsal surface of the free-swimming genus: the mouth is to the left in both.

A single row of cilia is disposed round the inner border of the peristome, and continued on the one hand down the gullet, and on the other round the elevated portion of the disc; the whole row of cilia thus takes a spiral direction. The rest of the body is completely bare of cilia.

The movements of the cilia produce a very curious optical illusion: as one watches a fully-expanded specimen it is hardly possible to believe that the peristome and disc are not actually revolving—a state of things which would imply that they were discontinuous from the rest of the body. As a matter of fact the appearance is due to the successive contraction of all the cilia in the same direction, and is analogous to that produced by a strong wind on a field of corn or long grass. The bending down of successive blades of grass produces a series of waves travelling across the field in the direction of the wind. If instead of a field we had a large circle of grass, and if this were acted upon by a cyclone, the wave would travel round the circle, which would then appear to revolve.

Naturally the movement of the circlet of cilia produces a small whirlpool in the neighbourhood of the Vorticella, as can be seen by introducing finely-powdered carmine into the water. It is through the agency of this whirlpool that food particles are swept into the mouth, surrounded, as in Paramœcium, by a globule of water: the food-vacuoles (*f. vac*) thus constituted circulate in the medullary protoplasm, and the non-nutritive parts are finally egested at an anal spot (*an*) situated near the base of the gullet.

The stalk (*st*) consists of a very delicate, transparent, outer substance, which is continuous with the cuticle of the body and contains a delicate *axial fibre* (*ax. f.*) running along it from end to end in a somewhat spiral direction. This fibre is a prolongation of the cortex of the body (c, *ax. f.*): under a very high power it appears granular or delicately striated, the striæ being continued into the cortex of the proximal part of the body.

K

A striking characteristic of Vorticella is its extreme irritability, *i.e.*, the readiness with which it responds to any external stimulus (see p. 10). The slightest jar of the microscope, the contact of some other organism, or even a current of water produced by some free-swimming form like Paramœcium, is felt directly by the bell-animalcule and is followed by an instantaneous change in the relative position of its parts. The stalk becomes coiled into a close spiral (D^1, D^2) so as to have a mere fraction of its original length, and the body from being bell-shaped becomes globular, the disc being withdrawn and the peristome closed over it (D^1, D^2).

The coiling of the stalk leads us to the consideration of the particular form of contractility called *muscular*, which we have already met with in Stylonychia (p. 116). It was mentioned above that while the stalk in its fully expanded condition is straight, the axial fibre is not straight, but forms a very open spiral, *i.e.*, it does not lie in the centre of the stalk but at any transverse section is nearer the surface at one spot than elsewhere, and this point as we ascend the stalk is directed successively to all points of the compass.

Now suppose that the axial fibre undergoes a sudden contraction, that is to say, a decrease in length accompanied by an increase in diameter, since as we have already seen (p. 10) there is no decrease in volume in protoplasmic contraction. There will naturally follow a corresponding shortening of the elastic cuticular substance which forms the outer layer of the stalk. If the axial fibre were entirely towards one side of the stalk, the result of the contraction would be a flexure of the stalk towards that side, but, as its direction is spiral, the stalk is bent successively in every direction, that is, is thrown into a close spiral coil.

The axial fibre is therefore a portion of the protoplasm

which possesses the property of contractility in a special degree; in which moreover contraction takes place in a definite direction—the direction of the length of the fibre—so that its inevitable result is to shorten the fibre and consequently to bring its two ends nearer together. This is the essential characteristic of a muscular contraction, and the axial fibre in the stalk of Vorticella is therefore to be looked upon as the first instance of a clearly differentiated *muscle* which has come under our notice.

There are some interesting features in the reproduction of Vorticella. It multiplies by binary fission, dividing through the long axis of the body (Fig. 25, E^1, E^2). Hence it is generally said that fission is longitudinal, not transverse, as in Paramœcium. But on the theory (p. 107) that the peristome and disc are dorsal and the attached end ventral, fission is really transverse in this case also.

It will be seen from the figures that the process takes place by a cleft appearing at the distal end (E^1), and gradually deepening until there are produced two complete and full-sized individuals upon a single stalk (E^2). This state of things does not last long: one of the two daughter-cells takes on a nearly cylindrical form, keeps its disc and peristome retracted, and acquires a new circlet of cilia near its proximal end (E^3): it then detaches itself from the stalk, which it leaves in the sole possession of its sister-cell, and swims about freely for a time in the direction indicated by the arrow. Sooner or later it settles down, becomes attached by its proximal end, loses its basal circlet of cilia, and develops a stalk, which ultimately attains the normal length.

The object of this arrangement is obvious. If when a Vorticella divided, the plane of fission extended down the stalk until two ordinary fixed forms were produced side by side, the constant repetition of the process would so increase the numbers of the species in a given spot that the food-supply would inevitably run short. This is prevented by one of the two sister-cells produced by fission leading a free existence long enough to enable it to emigrate and settle in a new locality, where the competition with its fellows will be less keen. The production of these free-swimming zooids is therefore a means of dispersal (see p. 122): contrivances having this object in view are a very general characteristic of fixed as of parasitic organisms.

Conjugation occasionally takes place, and presents certain peculiarities. A Vorticella divides either into two unequal halves (F^2) or into two equal halves, one of which divides again into from two to eight daughter-cells (F^1). There are thus produced from one to eight *microzooids* which resemble the barrel-shaped form (E^3) in all but size, and like it become detached and swim freely by means of a basal circlet of cilia. After swimming about for a time, one of these microzooids comes in contact with an ordinary form or *megazooid*, when it attaches itself to it near the proximal end (G^1), and undergoes gradual absorption (G^2), the mega- and microzooids becoming completely and permanently fused. As in Paramœcium, conjugation is followed by increased activity in feeding and dividing (p. 113).

Notice that in this case the conjugating bodies or gametes are not of equal size and similar characters, but one, which is conveniently distinguished as the *microgamete* (= microzooid) is relatively small and active, while the other or *megagamete* (= megazooid, or ordinary individual) is relatively large and passive. As we shall see in a later lesson, this differentiation of the gametes is precisely what we get in almost all organisms with two sexes: the microgamete being the male, the megagamete the female conjugating body (see Lesson XVI.).

The result of conjugation is strikingly different in the three cases already studied: in Heteromita (p. 41) the two gametes unite to form a zygote, a motionless body provided with a cell-wall, the protoplasm of which divides into spores: in Paramœcium (p. 113) no zygote is formed, conjugation being a mere temporary union: in Vorticella the zygote is an actively moving and feeding body, indistinguishable from an ordinary individual of the species.

Vorticella sometimes encysts itself (Fig. 25, H^1), and the nucleus of the encysted cell has been observed to break up into a number of separate masses, each doubtless surrounded by a layer of protoplasm. After a time the cyst bursts, and a number of small bodies or spores (H^2) emerge from it, each containing one of the products of division of the nucleus. These acquire a circlet of cilia (H^3), by means of which they swim freely, and they are sometimes found to multiply by simple fission (H^4). Finally, they settle down (H^5) by the end at which the cilia are situated, the attached end begins to elongate into a stalk (H^6), this increases in length, the basal circlet of cilia is lost, and a ciliated peristome and disc are formed at the free end (H^7). In this way the ordinary form is assumed by a process of development recalling what we found to occur in Heteromita (p. 42), but with an important difference: the free-swimming young of Vorticella (H^3), to which the spores formed by division of the encysted protoplasm give rise, differ strikingly in form and habits from the adult. This is expressed by saying that development is in this case accompanied by a *metamorphosis*, this word, literally meaning simply a change, being always used in biology to express a striking and fundamental difference in form and habit between the young and the adult; as, for instance, between the tadpole and the frog, or between the caterpillar and the butterfly. It is obvious that in the present instance metamorphosis is another means of ensuring dispersal.

In Vorticella, as we have seen, fission results not in the

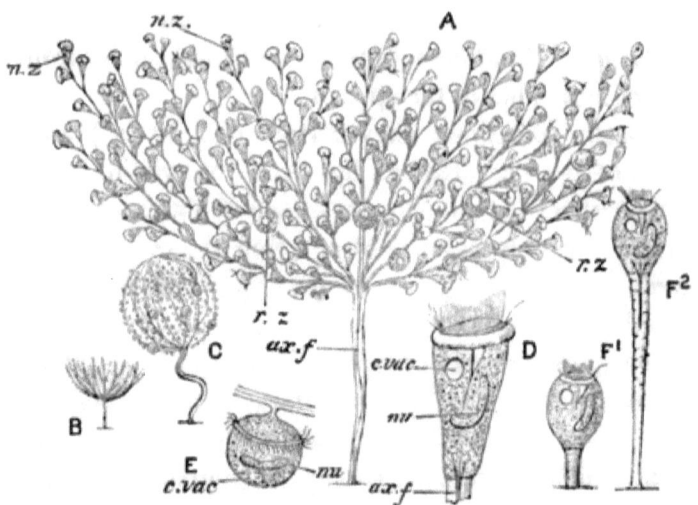

FIG. 26.—*Zoothamnium arbuscula.*

A, entire colony, magnified, showing nutritive (*n. z*) and reproductive (*r. z*) zooids; *ax. f* axial fibre of the stem.

B, the same, natural size.

C, the same, magnified, in the condition of retraction.

D, nutritive zooid, showing nucleus (*nu*), contractile vacuole (*c. vac*), gullet, and axial fibre (*ax. f*).

E, reproductive zooid, showing nucleus (*nu*) and contractile vacuole (*c. vac*), and absence of mouth and gullet.

F^1, F^2, two stages in the development of the reproductive zooid. (After Saville Kent.)

production of equal and similar daughter-cells, but of one stalked and one free-swimming form. It is however quite possible to conceive of a Vorticella-like organism in which the parent cell divides into two equal and similar products, each retaining its connection with the stalk. If this process were repeated again and again, and if, further, the plane of fission were extended downwards so as to include the distal end of the stalk, the result would be a branched, tree-like stem with a Vorticella-like body at the end of every branch.

As a matter of fact, this process takes place not in Vorticella itself, but in a nearly allied infusor, the beautiful *Zoothamnium*, a common genus found mostly in sea-water attached to weeds and other objects.

Zoothamnium arbuscula (Fig. 26, A) consists of a main stem attached by its proximal end and giving off at its distal end several branches, on each of which numerous shortly-stalked bell-animalcules are borne, like foxgloves or Canterbury-bells on their stem. The entire tree is about 1 cm. high, and so can be easily seen by the naked eye: it is shown of the natural size in Fig. 26, B.

We see, then, that Zoothamnium differs from all our previous types in being a *compound organism*. The entire "tree" is called a *colony* or *stock*, and each separate bell-animalcule borne thereon is an *individual* or *zooid*, morphologically equivalent to a single Vorticella or Paramœcium.

As in Vorticella, the stem consists of a cuticular sheath with an axial muscle-fibre ($ax.\ f$), which, at the distal end of the main stem, branches like the stem itself, a prolongation of it being traceable to each zooid (D). So that the muscular system is common to the whole colony, and any shock causes a general contraction, the tree-like structure assuming an almost globular form (C).

It will be noticed from the figure that all the zooids of the colony are not alike: the majority are bell-shaped and resemble Vorticellæ (A, $n.\ z$, and D), but here and there are found larger bodies (A, $r.\ z$, and E) of a globular form, without mouth, peristome, or disc, and with a basal circlet of cilia. The characteristic band-like nucleus (nu) and the contractile vacuole (*c. vac*) are found in both the bell-shaped and the globular zooids.

It is to these globular, mouthless zooids that the functions of reproducing the whole colony and of ensuring dispersal are assigned. They become detached, swim about freely for a time, then settle down, develop a stalk and mouth (F^1, F^2), and finally, by repeated fission, give rise to the adult, tree-like colony.

The Zoothamnium colony is thus *dimorphic*, bearing individuals of two kinds : *nutritive zooids*, which feed and add to the colony by fission but are unable to give rise to a new colony, and *reproductive zooids*, which do not feed while attached, but are capable, after a period of free existence, of developing a mouth and stalk, and finally producing a new colony. Dimorphism is a differentiation of the individuals of a colony, just as the formation of axial fibre, gullet, contractile vacuole, and cilia are cases of differentiation of the protoplasm of a single cell.

LESSON XIII

SPECIES AND THEIR ORIGIN—THE PRINCIPLES OF CLASSIFICATION

MORE than once in the course of the foregoing lessons we have had occasion to use the word *species*—for instance, in Lesson I. (p. 8) it was stated that there were different kinds or species of Amœbæ, distinguished by the characters of their pseudopods, the structure of their nuclei, &c.

We must now consider a little more in detail what we mean by a species, and, as in all matters of this sort, the study of concrete examples is the best aid to the formation of clear conceptions, we will take, by way of illustration, some of the various species of Zoothamnium.

The kind described in the previous lesson is called *Zoothamnium arbuscula*. As Fig. 26, A, shows, it consists of a tolerably stout main stem, from the distal end of which spring a number of slender branches diverging in a brush-like manner, and bearing on short secondary branchlets the separate individuals of the colony: these are of two kinds, bell-shaped nutritive zooids, and globular reproductive zooids, so that the colony is dimorphic.

Zoothamnium (or, for the sake of brevity, *Z.*) *alternans* (Fig. 27, A) is found also in sea-water, and differs markedly from Z. arbuscula in the general form of the colony. The main stem is continued to the extreme distal end of the colony and terminates in a zooid; from it branches are given off right and left, and on these the remaining zooids are borne. To use Mr. Saville Kent's comparison, Z. arbus-

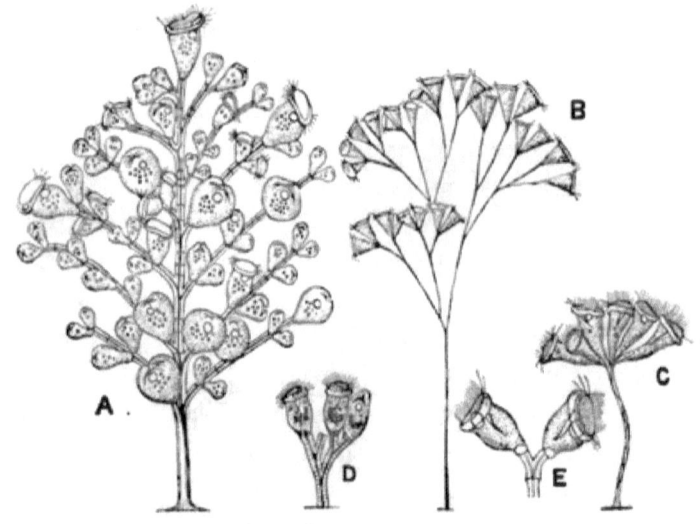

FIG. 27.—Species of Zoothamnium. A, *Z. alternans.* B, *Z. dichotomum.* C, *Z. simplex.* D, *Z. affine.* E, *Z. nutans.* (After Saville Kent.)

cula may be compared to a standard fruit tree, Z. alternans to an espalier. In this species also the colony is dimorphic.

Z. dichotomum (Fig. 27, B) is also dimorphic and presents a third mode of branching. The main stem divides into two, and each of the secondary branches does the same, so that a repeatedly forking stem is produced. The branching of this species is said to be *dichotomous,* while that of Z. alternans is *monopodial,* and that of Z. arbuscula *umbellate.*

Another mode of aggregation of the zooids is found in Z. *simplex* (Fig. 27, C) in which the stem is unbranched and bears at its distal end about six zooids in a cluster. The zooids are more elongated than in any of the preceding species, and there are no special reproductive individuals, so that the colony is *homomorphic.*

In *Z. affine* (Fig. 27, D) the stalk is dichotomous but is proportionally thicker than in the preceding species, and bears about four zooids, all alike. It is found in fresh water attached to insects and other aquatic animals.

The last species we shall consider is *Z. nutans* (Fig. 27, E), which is the simplest known, never bearing more than two zooids, and sometimes only one.

A glance at Figs. 26 and 27 will show that these six species agree with one another in the general form of the zooids, in the characters of the nucleus, contractile vacuole, &c., in the arrangement of the cilia, and in the fact that they are all compound organisms, consisting of two or more zooids attached to a common stem, the axial fibre of which branches with it, *i.e.*, is continuous throughout the colony.

On account of their possessing these important characters in common, the species described are placed in the single *genus* Zoothamnium, and the characters summarized in the preceding paragraph are called *generic characters.* On the other hand the points of difference between the various species, such as the forking of the stem in Z. dichotomum, the presence of only two zooids in Z. nutans, and so on, are called *specific characters.* Similarly the name *Zoothamnium*, which is common to all the species, is the *generic name*, while those which are applied only to a particular species, such as *arbuscula, simplex*, &c., are the *specific names.* As was mentioned in the first lesson (p. 8), this method of naming organisms is known as the Linnean system of binomial nomenclature.

It will be seen from the foregoing account that by a

species we understand an assemblage of individual organisms, whether simple or compound, which agree with one another in all but unessential points, such as the precise number of zooids in Zoothamnium, which may vary considerably in the same species, and come, therefore, within the limits of *individual variation*. Similarly, what we mean by a genus is a group of species agreeing with one another in the broad features of their organization, but differing in detail, the differences being constant.

A comparison of the six species described brings out several interesting relations between them. For instance, it is clear that Z. arbuscula and Z. alternans are far more complex *i.e.*, exhibit greater differentiation of the entire colony, than Z. simplex, or Z. nutans ; so that, within the limits of the one genus, we have comparatively low or generalized, and comparatively high or specialized species. Nevertheless, a little consideration will show that we cannot arrange the species in a single series, beginning with the lowest and ending with the highest, for, although we should have no hesitation in placing Z. nutans at the bottom of such a list, it would be impossible to say whether Z. affine was higher or lower than Z. simplex, or Z. arbuscula than Z. alternans.

It is, however, easy to arrange the species into groups according to some definite system. For instance, if we take the mode of branching as a criterion, Z. nutans, affine, and dichotomum will all be placed together as being dichotomous, and Z. simplex and arbuscula as being umbellate— the zooids of the one and the branches of the other all springing together from the top of the main stem : on this system Z. alternans will stand alone on account of its monopodial branching. Or, we may make two groups, one of dimorphic forms, including Z. arbuscula, alternans, and dichotomum, and another of homomorphic species, including Z. affine, simplex, and nutans. We have thus two very obvious ways of arranging or *classifying* the species of Zoothamnium, and the question arises—which of these, if either, is the right one? Is there any standard by which we can judge of the accuracy of a given classification of these or any other organisms, or does the whole thing depend upon the fancy of the classifier, like the arrangement of books in a library? In other words, are all possible classifications of living things more or less artificial, or is there such a thing as a *natural classification?*

Suppose we were to try and classify all the members of a given family—parents and grandparents, uncles and aunts, cousins, second cousins, and so on. Obviously there are a hundred ways in which it would be possible to arrange them—into dark and fair, tall and short, curly-haired and straight-haired and so on. But it is equally obvious that all these methods would be purely artificial, and that the only natural way, *i.e.*, the only way to show the real connection of he various members of the family with one another would be to classify them according to blood-relationship, in other words to let our classification take the form of a genealogical tree.

It may be said—what has this to do with the point under discussion, the classification of the species of Zoothamnium?

There are two theories which attempt to account for the existence of the innumerable species of living things which inhabit our earth: the theory of *creation* and the theory of *evolution.*

According to the theory of creation, all the individuals of every species existing at the present day—the tens of thousands of dogs, oak trees, amœbæ, and what not—are derived by a natural process of descent from a single individual, or from a pair of individuals, in each case precisely resembling, in all essential respects, their existing descendants, which came into existence by a process outside the ordinary course of nature and known as Creation. On this hypothesis the history of the genus Zoothamnium would be represented by the diagram (Fig. 28); each of the species being derived from a single individual which came into

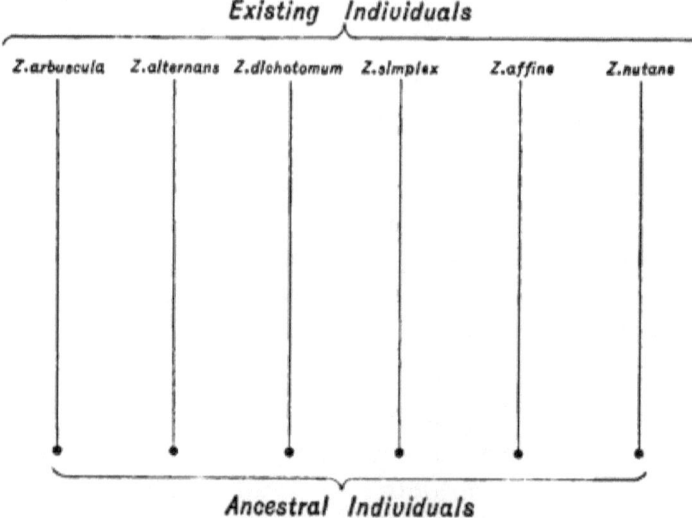

FIG. 28.—Diagram illustrating the origin of the species of Zoothamnium by creation.

existence, independently of the progenitors of all the other species, at some distant period of the earth's history.

Notice that on this theory the various species are no more actually *related* to one another than is either of them to Vorticella, or for the matter of that to Homo. The individuals of any one species are truly related since they all share a common descent, but there is no more relationship between the individuals of any two independently created species than between any two independently manufactured chairs or tables. The words affinity, relationship, &c., as applied to different species are, on the theory of Creation purely metaphorical, and mean nothing more than that a certain likeness or community of structure exists; just as we might say that an easy chair was more nearly related to a kitchen chair than either of them to a three-legged stool.

We see therefore that on the hypothesis of creation the varying degrees of likeness and unlikeness between the species receive no explanation, and that we get no absolute criterion of classification : we may arrange our organisms, as nearly as our knowledge allows, according to their resemblances and differences, but the relative importance of the characters relied on becomes a purely subjective matter.

According to the rival theory—that of Descent or Organic Evolution—every species existing at the present day is derived by a natural process of descent from some other species which lived at a former period of the world's history. If we could trace back from generation to generation the individuals of any existing species we should, on this hypothesis, find their characters gradually change, until finally a period was reached at which the differences were so considerable as to necessitate the placing of the ancestral forms in a different species from their descendants at the present day. And in the same way if we could trace back the species of any one genus, we should find them gradually approach one another in structure until they finally converged in a single species, differing from those now existing but standing to all in a true parental relation.

Let us illustrate this by reference to Zoothamnium. As a matter of fact we know nothing of the history of the genus, but the comprehension of what is meant by the evolution of species will be greatly faciltated by framing a working hypothesis.

Suppose that at some distant period of the world's history

there existed a Vorticella-like organism which we will call A (Fig. 29), having the general characters of a single stalked zooid of Zoothamnium (compare Fig. 26, F^2), and suppose that, of the numerous descendants of this form, represented by the lines diverging from A, there were some in which both the zooids formed by the longitudinal division of the body remained attached to the stalk instead of one of them swimming off as in Vorticella. The result—it matters

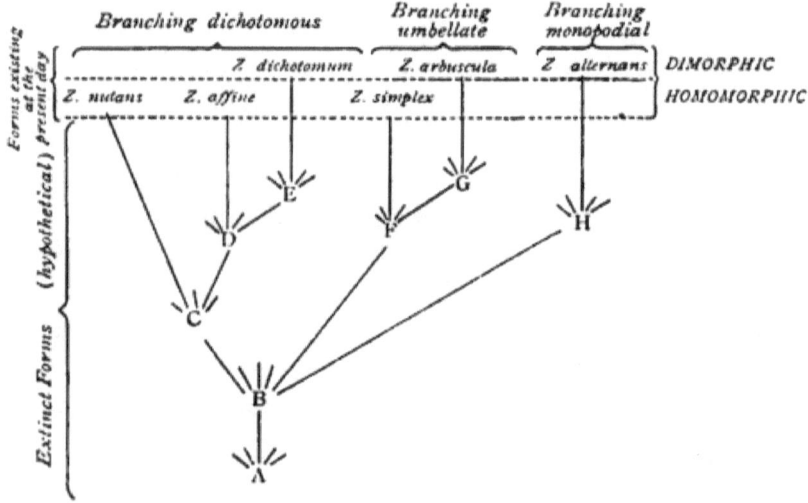

FIG. 29.—Diagram illustrating the origin of the species of Zoothamnium by evolution.

not for our present purpose how it may have been caused—would be a simple colonial organism consisting of two zooids attached to the end of a single undivided stalk. Let us call this form B.

Next let us imagine that in some of the descendants of B, represented as before by the diverging lines, the plane of division was continued downwards so as to include the distal end of the stalk: this would result in the production of a form (C) consisting of two zooids borne on a forked stem and resembling Z. nutans. If in some of the descendants of C this process were repeated, each of the two zooids again dividing into two fixed individuals and the division as before affecting the stem, we should get a species (D) consisting of four zooids on a dichotomous stem, like Z. affine. Let the same process continue from generation to generation, the colony becoming more and more complex; we should finally arrive at a species E, consisting of numerous zooids on a complicated dichotomously branching stem, and therefore resembling Z. dichotomum.

Let us further suppose that, in some of the descendants of our hypothetical form B, repeated binary fission took place without affecting the stem: the result would be a new form F, consisting of numerous zooids springing in a cluster from the end of the undivided stem, after the manner of Z. simplex. From this a more complicated umbellate form (G), like Z. arbuscula, may be supposed to have originated, and again starting from B with a different mode of branching a monopodial form (H) might have arisen.

Finally, let it be assumed that while some of the descendants of the forms C, D, and F became modified into more and more complex species, others survived to the present time with comparatively little change, forming the existing species nutans, affine, and simplex: and that, in the similarly surviving representatives of E, G, and H, a differentiation of the individual zooids took place resulting in the evolution of the dimorphic species dichotomum, arbuscula, and alternans.

It will be seen that, on this hypothesis, the relative likeness and unlikeness of the species of Zoothamnium are explained as the result of their descent with greater or less modification or *divergence of character* from the ancestral form A. And that we get an arrangement or classification in the form of a genealogical tree, which on the hypothesis is a strictly natural one, since it shows accurately the relationship of the various species to one another and to the parent stock. So that, on the theory of evolution, a natural classification of any given group of allied organisms is simply a genealogical tree, or as it is usually called, a *phylogeny*.

It must not be forgotten that the forms A, B, C, D, E, F, G, and H are purely hypothetical: their existence has been assumed in order to illustrate the doctrine of descent by a concrete example. The only way in which we could be perfectly sure of an absolutely natural classification of the species of Zoothamnium would be by obtaining specimens as far back as the distant period when the genus first came into existence; and this is out of the question, since minute soft-bodied organisms like these have no chance of being preserved in the fossil state.

It will be seen that the theory of evolution has the advantage over that of creation of offering a reasonable explanation of certain facts. First of all the varying degrees of likeness and unlikeness of the species are explained by their having branched off from one another at various periods: for instance, the greater similarity of structure between Z. affine and Z. dichotomum than between either of them and any other species is due to these two species having a common ancestor in D, whereas to connect either of them, say with Z. arbuscula, we have to go back to B. Then again the fact that all the species, however complex in their fully developed state, begin life as a simple zooid which by repeated branching gradually attains the adult complexity, is a result of the repetition by each organism, in the course of its single life, of the series of changes passed through by its ancestors in the course of ages. In other words *ontogeny*, or the evolution of the individual, is, in its main features, a recapitulation of *phylogeny* or the evolution of the race.

One other matter must be referred to in concluding the present lesson. It is obvious that the evolution of one species from another presupposes the occurrence of variations in the ancestral form. As a matter of fact such *individual variation* is of universal occurrence : it is a matter of common observation that no two leaves, shells, or human beings are precisely alike, and in our type genus Zoothamnium the number of zooids, their precise arrangement, the details of branching, &c., are all variables. This may be expressed by saying that *heredity*, according to which the offspring tends to resemble the parent in essentials, is modified by *variability*, according to which the offspring tends to differ from the parent in details. If from any cause an individual variation is perpetuated there is produced what is known as a *variety* of the species, and, according to the theory of the origin of species by evolution, such a variety may in course of time become a new species. Thus a variety is an incipient species, and a species is a (relatively) permanent variety.

It does not come within the scope of the present work to discuss either the causes of variability or those which determine the elevation of a variety to the rank of a species : both questions are far too complex to be adequately treated except at considerable length, and anything of the nature of a brief abstract could only be misleading. As a preliminary to the study of Darwin's *Origin of Species*, the student is recommended to read Romanes's *Evidences of Organic Evolution*, in which the doctrine of Descent is expounded as briefly as is consistent with clearness and accuracy.

LESSON XIV

FORAMINIFERA, RADIOLARIA, AND DIATOMS

In the four previous lessons we have learnt how a unicellular organism may attain very considerable complexity by a process of differentiation of its protoplasm. In the present lesson we shall consider briefly certain forms of life in which, while the protoplasm of the unicellular body undergoes comparatively little differentiation, an extraordinary variety and complexity of form is produced by the development of a *skeleton*, either in the shape of a hardened cell-wall or by the formation of hard parts within the protoplasm itself.

The name *Foraminifera* is given to an extensive group of organisms which are very common in the sea, some living near the surface, others at various depths. They vary in size from a sand-grain to a shilling. They consist of variously-shaped masses of protoplasm, containing nuclei, and produced into numerous pseudopods which are extremely long and delicate, and frequently unite with one another to form networks, as at × in Fig. 30. The cell-body of these organisms is therefore very simple, and may be compared to that of a multinucleate Amœba with fine radiating pseudopods.

But what gives the Foraminifera their special character is the fact that around the protoplasm is developed a cell-wall, sometimes membranous, but usually impregnated with calcium carbonate, and so forming a *shell*. In some cases, as in the genus *Rotalia* (Fig. 30), this is perforated by numerous small holes, through which the pseudopods are protruded, in others it has only one large aperture (Fig. 31),

FIG. 30.—A living Foraminifer (*Rotalia*), showing the fine radiating pseudopods passing through apertures in the chambered shell : at × several of them have united. (From Gegenbaur.)

through which the protoplasm protrudes, sending off its pseudopods and sometimes flowing over and covering the outer surface of the shell. Thus while in some cases the shell has just the relations of a cell-wall with one or more holes in it, in others it becomes an internal structure, being covered externally as well as filled internally by protoplasm.

The mode of growth of Foraminifera is largely determined by the hard and non-distensible character of the cell-wall, which when once formed is incapable of being enlarged. In he young condition they consist of a simple mass of protoplasm covered by a more or less globular shell, having at least one aperture. But in most cases as the cell-body grows, it protrudes through the aperture of the shell as a mass of protoplasm at first naked, but soon becoming covered by the secretion around it of a second compartment or chamber of the shell. The latter now consists of two

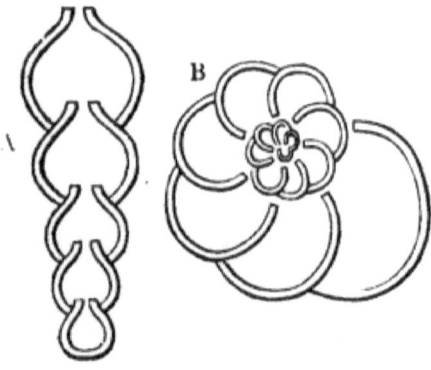

FIG. 31.—A, diagram of a Foraminifer in which new chambers are added in a straight line : the smallest first-formed chamber is below, the newest and largest is above and communicates with the exterior.

B, diagram of a Foraminifer in which the chambers are added in a flat spiral : the oldest and smallest chamber is in the centre, the newest and largest as before communicates with the exterior. (From Carpenter.)

chambers communicating with one another by a small aperture, and one of them—the last formed—communicating with the exterior. This process may go on almost indefinitely, the successive chambers always remaining in communication by small apertures through which continuity of the protoplasm is maintained, while the last formed chamber has a terminal aperture placing its protoplasm in free communication with the outer world.

The new chambers may be added in a straight line (Fig. 31, A) or in a gentle curve, or in a flat spiral (Fig. 31, B), or like the segments of a Nautilus shell, or more or less irregularly. In this way shells of great variety and beauty

FIG. 32.—Section of one of the more complicated Foraminifera (*Aveolina*), showing the numerous chambers containing protoplasm (dotted), separated by partitions of the shell (white). × 60. (From Gegenbaur after Carpenter.)

of form are produced, often resembling the shells of Mollusca, and sometimes attaining a marvellous degree of complexity (Fig. 32). The student should make a point of examining mounted slides of some of the principal genera and of consulting the plates in Carpenter's *Introduction to the Study of Foraminifera* (Ray Society, 1862), or in Brady's *Report on the Foraminifera of the " Challenger" Expedition*, in order to get some notion of the great amount of differentiation attained by the shells of these extremely simple organisms.

The *Radiolaria* form another group of marine animalcules, the numerous genera of which are, like the Foraminifera, amongst the most beautiful of microscopic objects. They also (Fig. 33) consist of a mass of protoplasm giving off numerous delicate pseudopods (*psd*) which usually have a radial direction and sometimes unite to form networks. In the centre of the protoplasmic cell-body one or more nuclei (*nu*) of unusual size and complex structure are found.

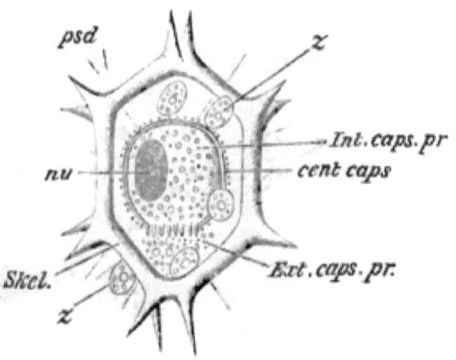

FIG. 33.—*Lithocircus annularis*, one of the Radiolaria, showing central capsule (*cent. caps.*), intra- and extra capsular protoplasm (*int. caps. pr.*, *ext. caps. pr.*), nucleus (*nu*), pseudopods (*psd*), silicious skeleton, (*skel*), and symbiotic cells of Zooxanthella (*z*). (After Bütschli.)

In the interior of the protoplasm, surrounding the nucleus, is a sort of shell, called the *central capsule* (*cent. caps.*), formed of a membranous material, and perforated by pores which place the inclosed or *intra-capsular* protoplasm (*int. caps. pr.*) in communication with the surrounding or *extra-capsular* protoplasm (*ext. caps. pr.*). But besides this simple membranous shell there is often developed, mainly in the extra-capsular protoplasm, a skeleton (*skel*) formed in the majority of cases of pure silica, and often of surpassing beauty and complexity. One very exquisite form is shown in Fig. 34 : it consists of three perforated concentric spheres connected by radiating spicules : the material of which it is composed resembles the clearest glass.

The student should examine mounted slides of the silicious shells of these organisms—sold under the name of *Polycystineæ*—and should consult the plates of Haeckel's *Die*

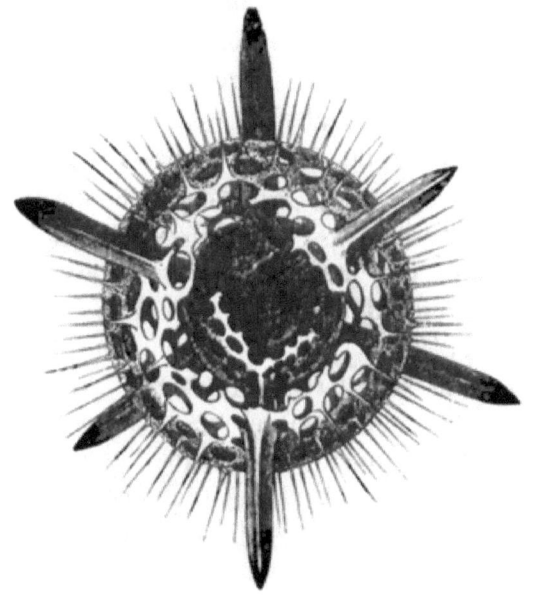

FIG. 34.—Skeleton of a Radiolarian (*Actinomma*), consisting of three concentric perforated spheres—the two outer partly broken away to show the inner—connected by radiating spicules. (From Gegenbaur after Haeckel.)

Radiolarien: he cannot fail to be struck with the complexity and variety attained by the skeletons of organisms which are themselves little more complex than Amœbæ.

Before leaving the Radiolaria, we must touch upon a matter of considerable interest connected with the physiology of the group. Imbedded usually in the extra-capsular protoplasm are found certain little rounded bodies of a yellow colour, often known as "yellow cells" (Fig. 33, z). Each consists of protoplasm surrounded by a cell-wall of cellulose, and coloured by chlorophyll, with which is associated a yellow pigment of similar character called *diatomin*.

For a long time these bodies were a complete puzzle to biologists, but it has now been conclusively proved that they are independent organisms resembling the resting condition of Hæmatococcus, and called *Zooxanthella nutricola*.

Thus an ordinary Radiolarian, such as Lithocircus (Fig. 33), consists of two quite distinct things, the Lithocircus in the strict sense of the word *plus* large numbers of Zooxanthellæ associated with it. The two organisms multiply quite independently of one another: indeed Zooxanthella has been observed to multiply by fission after the death of the associated Radiolarian.

This living together of two organisms is known as *Symbiosis*. It differs essentially from parasitism (see p. 121), in which one organism preys upon another, the host deriving no benefit but only harm from the presence of the parasite. In symbiosis, on the contrary, the two organisms are in a condition of mutually beneficial partnership. The carbon dioxide and nitrogenous waste given off by the Radiolarian serve as a constant food-supply to the Zooxanthella: at the same time the latter by decomposing the carbon dioxide provides the Radiolarian with a constant supply of oxygen, and at the same time with two important food-stuffs—starch and proteids, which, after solution, diffuse from the protoplasm of the Zooxanthella into that of the Radiolarian. The Radiolarian may therefore be said to keep the Zooxanthellæ constantly manured, while the Zooxanthellæ in return supply the Radiolarian with abundance of oxygen and of ready-digested food. It is as if a Hæmatococcus ingested by an Amœba retained its vitality instead of being digested : it would under these circumstances make use of the carbon dioxide and nitrogenous waste formed as products of katabolism by the Amœba, at the same time giving off oxygen and forming starch and proteids. The oxygen evolved would give an additional supply of this necessary gas to the Amœba, and the starch after conversion into sugar and the proteids after being rendered diffusible would in part diffuse through the cell-wall of the Hæmatococcus into the surrounding protoplasm of the Amœba, to which they would be a valuable food.

Thus, as it has been said, the relation between a Radiolarian and its associated yellow-cells are precisely those which obtain between the animal and vegetable kingdoms generally.

The *Diatomaceæ*, or *Diatoms*, as they are often called for the sake of brevity, are a group of minute organisms, included under a very large number of genera and species, and so common that there is hardly a pond or stream in which they do not occur in millions.

Diatoms vary almost indefinitely in form : they may be rod-shaped, triangular, circular, and so on. Their essential structure is, however, very uniform : the cell-body contains a nucleus (Fig. 35, A, *nu*) and vacuoles (*vac*), as well as two large chromatophores (*chr*) of a brown or yellow colour ; these are found to contain chlorophyll, the characteristic green tint of which is veiled, as in Zooxanthella, by diatomin. The cell is motile, executing curious, slow, jerky or gliding movements, the cause of which is still obscure.

The most interesting feature in the organization of diatoms is however the structure of the cell-wall : it consists of two parts or *valves* (B, C, *c. w*, *c. w'*), each provided with a rim or *girdle*, and so disposed that in the entire cell the girdle of one valve (*c. w*) fits over that of the other (*c. w'*) like the

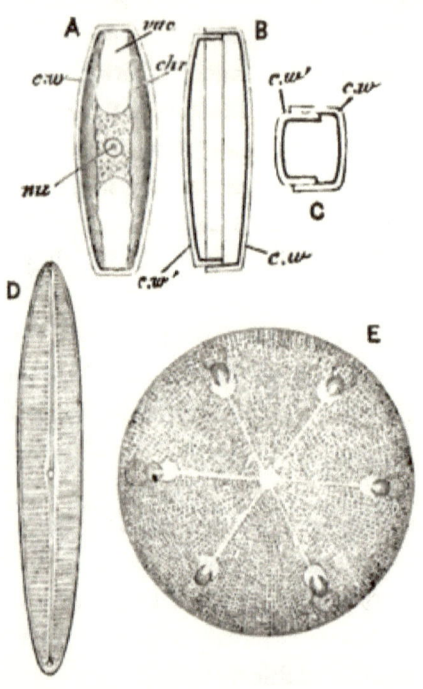

Fig. 35.—A, semi-diagrammatic view of a diatom from its flat face, showing cell-wall (*c. w*) and protoplasm with nucleus (*nu*), two vacuoles (*vac*), and two chromatophores (*chr*).

B, diagram of the shell of a diatom from the side, *i.e.*, turned on its long axis at right angles to A, showing the two valves (*c. w, c. w'*) with their overlapping girdles.

C, the same in transverse section.
D, surface view of the silicious shell of *Navicula truncata*.
E, surface view of the silicious shell of *Aulacodiscus sollittianus*.
(D, after Donkin ; E, after Norman.)

lid of a pill-box. The cell-wall is impregnated with silica, so that diatoms can be boiled in strong acid or exposed to the heat of a flame without losing their form : the protoplasm is of course destroyed, but the flinty cell-wall remains uninjured.

Moreover, the cell-walls of diatoms are remarkable for the beauty and complexity of their markings, which are in some cases so delicate that even now microscopists are not agreed as to the precise interpretation of the appearances shown by the highest powers of the microscope. Two species are shown in Fig. 35, D and E, but, in order to form some conception of the extraordinary variety in form and ornamentation, specimens of the mounted cell-walls should be examined and the plates of some illustrated work consulted. See especially Schmidt's *Atlas für Diatomaceenkunde* and the earlier volumes of the *Quarterly Journal of Microscopical Science*.

We see then that while Diatoms are in their essential structure as simple as Hæmatococcus, they have the power of extracting silica from the surrounding water, and of forming from it structures which rival in beauty of form and intricacy of pattern the best work of the metal-worker or the ivory-carver.

LESSON XV

MUCOR

THE five preceding lessons have shown us how complex a cell may become either by internal differentiation of its protoplasm, or by differentiation of its cell-wall. In this and the following lesson we shall see how a considerable degree of specialization may be attained by the elongation of cells into filaments.

Mucor is the scientific name of the common white or grey mould which every one is familiar with in the form of a cottony deposit on damp organic substances, such as leather, bread, jam, &c. For examination it is readily obtained by placing a piece of damp bread or some fresh horse-dung under an inverted tumbler or bell-jar so as to prevent evaporation and consequent drying. In the course of two or three days a number of delicate white filaments will be seen shooting out in all directions from the bread or manure; these are filaments of Mucor. The species which grows on bread is called *Mucor stolonifer*, that on horse-dung, *M. mucedo*.

The general structure and mode of growth of the mould can be readily made out with the naked eye. It first appears, as already stated, in the form of very fine white threads projecting from the surface of them ouldy substance; and these free filaments (Fig. 36, A, *a. hy*) can be easily

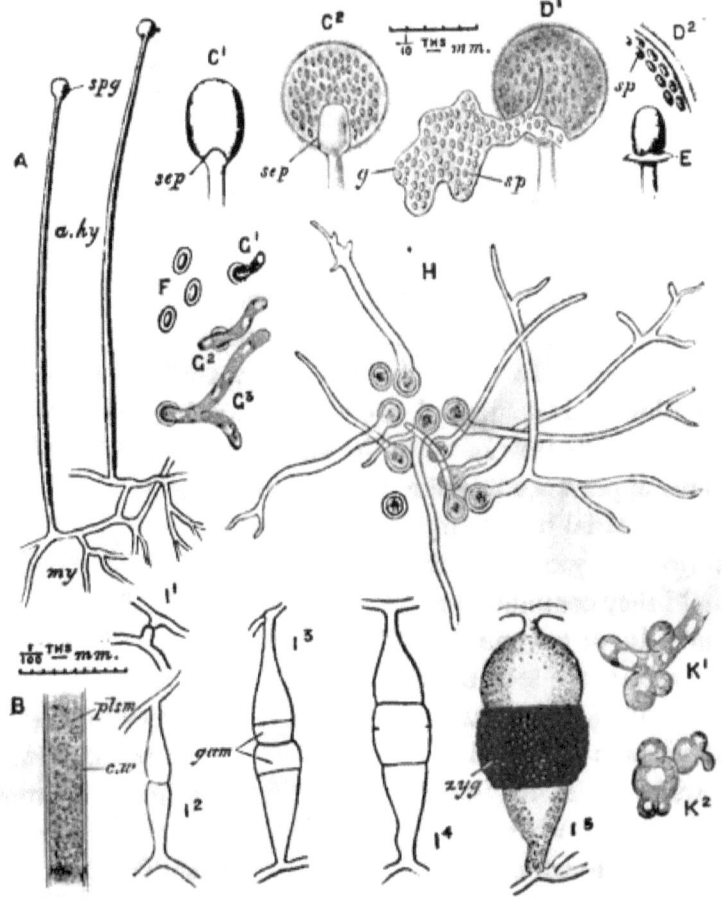

Fig. 36.—*Mucor.*

A, portion of mycelium of M. mucedo (*my*) with two aërial hyphæ (*a. hy*), each ending in a sporangium (*spg*).

B, small portion of an aërial hypha, highly magnified, showing protoplasm (*plsm*) and cell-wall (*c w*). The scale above applies to this figure only.

C¹, immature sporangium, showing septum (*sep*) and undivided protoplasm: C², mature sporangium in which the protoplasm has divided into spores; the septum (*sep*) has become very convex distally, forming the columella.

D¹, mature sporangium in the act of dehiscence, showing the spores (*sp*) surrounded by mucilage (*g*); D², small portion of the same, more highly magnified, showing spicules of calcium oxalate attached to wall.

E, a columella, left by complete dehiscence of a sporangium, showing the attachment of the latter as a black band.

The scale above C² applies to C¹ C², D¹, and E.

F, spores.

G^1, G^2, G^3, three stages in the germination of the spores.

H, a group of germinating spores forming a small mycelium.

I^1,—I^5, five stages in conjugation, showing two gametes (*gam*) uniting to form the zygote (*zyg*).

K^1, K^2, development of ferment cells from submerged hyphæ.

(A, C^2 D, E, F, G, and K, after Howes; I, after De Bary.)

ascertained to be connected with others (*my*) which form a network ramifying through the substance of the bread or horse-dung. This network is called a *mycelium*; the threads of which it is composed are *mycelial hyphæ*; and the filaments which grow out into the air and give the characteristic fluffy appearance to the growth are *aërial hyphæ*.

The aërial hyphæ are somewhat thicker than those which form the mycelium, and are at first of even diameter throughout: they continue to grow until they attain a length, in M. mucedo, of 6-8 cm. (two or three inches). As they grow their ends are seen to become dilated, so that each is terminated by a minute knob (A, *spg*): this increases in size and darkens in tint until it finally becomes dead black. In its earlier stages the knobs may be touched gently without injury, but when they have attained their full size the slightest touch causes them to burst and apparently to disappear—their actual fate being quite invisible to the naked eye. As we shall see, the black knobs contain *spores*, and are therefore called *sporangia* or spore-cases.

Examined under the microscope, a hypha is found to be a delicate more or less branched tube, with a clear transparent wall (B, *c. w*) and slightly granular contents (*plsm*): its free end tapers slightly (H), and the wall is somewhat thinner at the extremity than elsewhere. If a single hypha could be obtained whole and unbroken, its opposite end would be found to have much the same structure, and each of its branches would also be seen to end in the same way.

So that the mould consists of an interlacement of branched cylindrical filaments, each consisting of a granular substance completely covered by a kind of thin skin of some clear transparent material.

By the employment of the usual reagents, it can be ascertained that the granular substance is protoplasm, and the surrounding membrane cellulose. The protoplasm moreover contains vacuoles at irregular intervals and numerous small nuclei.

Thus a hypha of Mucor consists of precisely the same constituents as a yeast-cell—protoplasm, containing nuclei and vacuoles, surrounded by cellulose. Imagine a yeast cell to be pulled out—as one might pull out a sphere of clay or putty—until it assumed the form of a long narrow cylinder, and suppose it also to be pulled out laterally at intervals so as to form branches: there would be produced by such a process a very good imitation of a hypha of Mucor. We may therefore look upon a hypha as an elongated and branched cell, so that Mucor is, like Opalina, a multinucleate but unicellular organism. We shall see directly however that this is strictly true of the mould only in its young state.

As stated above, the aërial hyphæ are at first of even calibre, but gradually swell at their ends, forming sporangia. Under the microscope the distal end of an aërial hypha is found to dilate (Fig. 36, c^1): immediately below the dilatation the protoplasm divides at right angles to the long axis of the hypha, the protoplasm in the dilated portion thus becoming separated from the rest. Between the two a cellulose partition or *septum* (*sep*) is formed, as in the ordinary division of a plant cell (Fig. 11, p. 66). The portion thus separated is the rudiment of a sporangium.

Let us consider precisely what this process implies. Before it takes place the protoplasm is continuous throughout the whole organism, which is therefore comparable to the undivided plant-cell shown in Fig. 9, B. As in that case, the protoplasm divides into two and a new layer of cellulose is formed between the daughter-cells. Only whereas in the ordinary vegetable cell the products of division are of equal size (Fig 10, 1), in Mucor they are very unequal, one being the comparatively small sporangium, the other the rest of the hypha.

Thus a Mucor-plant with a single aërial hypha becomes, by the formation of a sporangium, *bicellular*: if, as is ordinarily the case, it bears numerous aërial hyphæ, each with its sporangium, it is *multicellular*.

Under unfavourable conditions of nutrition, septa frequently appear at more or less irregular intervals in the mycelial hyphæ: the organism is then very obviously multicellular, being formed of numerous cylindrical cells arranged end to end.

The sporangium continues to grow, and as it does so, the septum becomes more and more convex upwards, finally taking the form of a short, club-shaped projection, the *columella*, extending into the interior of the sporangium (c^2): at the same time the protoplasm of the sporangium undergoes multiple fission, becoming divided into numerous ovoid masses each of which surrounds itself with a cellulose coat and becomes a *spore* (D^1, D^2, *sp*). A certain amount of the protoplasm remains unused in the formation of spores, and is converted into a gelatinous material (*g*), which swells up in water.

The original cell-wall of the sporangium is left as an exceedingly delicate, brittle shell around the spores: minute needle-like crystals of calcium oxalate are deposited in it, and give it the appearance of being closely covered with short cilia (D^2).

In the ripe sporangium the slightest touch suffices to rupture the brittle wall and liberate the spores, which are dispersed by the swelling of the transparent intermediate substance. The aërial hypha is then left terminated by the columella (E), around the base of which is seen a narrow black ring indicating the place of attachment of the sporangium.

The spores (F) are clear, bright-looking, ovoidal bodies consisting of protoplasm containing a nucleus and sur-

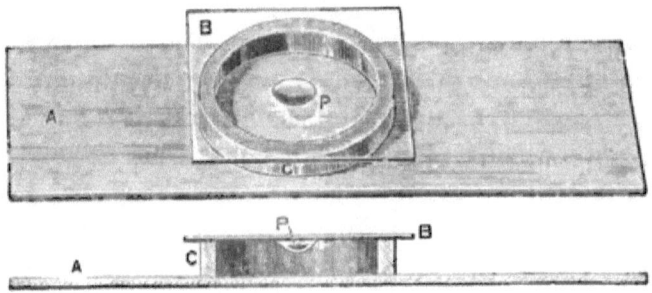

FIG. 37.—Moist chamber formed by cementing a ring of glass or metal (C) on an ordinary glass slide (A), and placing over it a cover-slip (B), on the under side of which is a hanging drop of nutrient fluid (P). The upper figure shows the apparatus in perspective, the lower in vertical section. (From Klein.)

rounded by a thick cell-wall. A spore is therefore an ordinary encysted cell, quite comparable to a yeast-cell.

The development of the spores is a very instructive process, and can be easily studied in the following way : A glass or metal ring (Fig. 37, C) is cemented to an ordinary microscopic slide (A) so as to form a shallow cylindrical chamber. The top of the ring is oiled, and on it is placed a cover glass (B), with a drop of Pasteur's solution on its under surface. Before placing the cover-glass in position a ripe sporangium of Mucor is touched with the point of a needle, which is then stirred round in the drop of Pasteur's solution, so as to sow it with spores. By this method the drop of nutrient fluid is prevented from evaporating, and the changes undergone by the spores can be watched by examination from time to time under a high power.

The first thing that happens to a spore under these conditions is that it increases in size by imbibition of fluid, and instead of appearing bright and clear becomes granular and develops one or more vacuoles. Its resemblance to a yeast-cell is now more striking than ever. Next the spore becomes bulged out in one or more places (G^1, Fig. 36), looking not unlike a budding Saccharomyces. The buds, however, instead of becoming detached increase in length until they become filaments of a diameter slightly less than that of the spore and somewhat bluntly pointed at the end (G^2). These filaments continue to grow, giving off as they do so side branches (G^3) which interlace with similar threads from adjacent spores (H). The filaments are obviously hyphæ, and the interlacement is a mycelium.

Thus the statement made in a previous paragraph (p. 161), that Mucor was comparable to a yeast-cell pulled out into a filament, is seen to be fully justified by the facts of development, which show that the branched hyphæ constituting the Mucor-plant are formed by the growth of spores each strictly comparable to a single Saccharomyces.

It will be noticed that the growth of the mycelium is centrifugal: each spore or group of spores serves as a centre from which hyphæ radiate in all directions (H), continuing to grow in a radial direction until, in place of one or more spores quite invisible to the naked eye, we have a white patch more or less circular in outline, and having the spores from which the growth proceeded in its centre. Owing to the centrifugal mode of growth the mycelium is always thicker at the centre than towards the circumference, since it is the older or more central portions of the hyphæ which have had most time to branch and become interlaced with one another.

Under certain circumstances a peculiar process of conjugation occurs in Mucor. Two adjacent hyphæ send out short branches (Fig. 36, 1^1), which come into contact with one another by their somewhat swollen free ends (1^2). In each a septum appears so as to shut off a separate terminal cell (1^3, *gam*) from the rest of the hypha. The opposed walls of the two cells then become absorbed (1^4) and their contents mingle, forming a single mass of protoplasm (1^5, *zyg*), the cell-wall of which becomes greatly thickened and divided into two layers, an inner delicate and transparent, and an outer dark in colour, of considerable thickness, and frequently ornamented with spines.

Obviously the swollen terminal cells (*gam*) of the short lateral hyphæ are gametes or conjugating bodies, and the large spore-like structure (*zyg*) resulting from their union is a zygote. The striking feature of the process is that the gametes are non-motile, save in so far as their growth towards one another is a mode of motion. In Heteromita both gametes are active and free-swimming (p. 41): in Vorticella one is free-swimming, the other fixed but still capable of active movement (p. 132); here both conjugating bodies exhibit only the slow movement in one direction due to growth.

There are equally important differences in the result of the process in the three cases. In Heteromita the protoplasm of the zygote breaks up almost immediately into spores; in Vorticella the zygote is active, and the result of conjugation is merely increased activity in feeding and fissive multiplication ; in Mucor the zygote remains inactive for a longer or shorter time, and then under favourable conditions germinates in much the same way as an ordinary spore, forming a mycelium from which sporangium-bearing aërial hyphæ arise. A *resting zygote* of this kind, formed by the conjugation of equal-sized gametes, is often distinguished as a *zygospore*.

Notice that differentiation of a very important kind is exhibited by Mucor. In accordance with its comparatively large size the function of reproduction is not performed by the whole organism, as in all previously studied types, but a certain portion of the protoplasm becomes shut off from the rest, and to it—as spore or gamete—the office of reproducing the entire organism is assigned. So that we have for the first time true *reproductive organs*, which may be of two kinds, asexual—the sporangia, and sexual—the gametes.[1]

In describing the reproduction of Amœba it was pointed out (p. 20) that as the entire organism divided into two daughter-cells, each of which began an independent life, an Amœba could not be said ever to die a natural death. The same thing is true of the other unicellular forms we have considered in the majority of which the entire organism produces by simple fission two new individuals.[2] But in Mucor the state of things is entirely altered. A compara-

---

[1] In Mucor no distinction can be drawn between the conjugating body (gamete) and the organ which produces it (gonad). See the description of the sexual process in Vaucheria (Lesson XVI.) and in Spirogyra (Lesson XIX.).

[2] An exception is formed by colonial forms such as Zoothamnium, in which life is carried on from generation to generation by the reproductive zooids only. In all probability the colony itself, like an annual plant, dies down after a longer or shorter time. Moreover the ciliate infusoria are found, as already stated (p. 116), to sink into decrepitude after multiplying by fission for a long series of generations.

tively small part of the organism is set apart for reproduction, and it is only the reproductive cells thus formed—spores or zygote—which carry on the life of the species the remainder of the organism, having exhausted the available food supply and produced the largest possible number of reproductive products, dies. That is, all vital manifestations such as nutrition cease, and decomposition sets in, the protoplasm becoming converted into progressively simpler compounds, the final stages being chiefly carbon dioxide, water, and ammonia.

Mucor is able to grow either in Pasteur's or in some similar nutrient solution, or on various organic matters such as bread, jam, manure, &c. In the latter cases it appears to perform some fermentative action, since food which has become "mouldy" is found to have experienced a definite change in appearance and flavour without actual putrefaction. When growing on decomposing organic matter, as it often does, the nutrition of Mucor is saprophytic, but in some instances, as when it grows on bread, it seems to approach very closely to the holozoic method. M. stolonifer is also known to send its hyphæ into the interior of ripe fruits, causing them to rot, and thus acting as a parasite. The parasitism in this case is, however, obviously not quite the same thing as that of Opalina (p. 121): the Mucor feeds not upon the ready digested food of its host but upon its actual living substance, which it digests by the action of its own ferments. Thus a parasitic fungus such as Mucor, unlike an endo-parasitic animal such as Opalina or a tapeworm, is no more exempted from the work of digestion than a dog or a sheep: the organism upon which it lives is to be looked upon rather as its prey than as its host.

It is a remarkable circumstance that, under certain conditions, Mucor is capable of exciting alcoholic fermentation in a saccharine solution. When the hyphæ are submerged in such a fluid they have been found to break up, forming rounded cells (Fig. 36, κ^1, κ^2), which not only resemble yeast-cells in appearance but are able like them to set up alcoholic fermentation.

The aërial hyphæ of Mucor exhibit in an interesting way what is known as heliotropism, *i.e.*, a tendency to turn towards the light. This is very marked if a growth of the fungus is placed in a room lighted from one side : the long aërial hyphæ all bend towards the window. This is due to the fact that growth is more rapid on the side of each hypha turned away from the light than on the more strongly illuminated aspect.

LESSON XVI

VAUCHERIA AND CAULERPA

STAGNANT ponds, puddles, and other pieces of still, fresh water usually contain a quantity of green scum which in the undisturbed condition shows no distinction of parts to the naked eye, but appears like a homogeneous slime full of bubbles if exposed to sunlight. If a little of the scum is spread out in a saucer of water, it is seen to be composed of great numbers of loosely interwoven green filaments.

There are many organisms which have this general naked-eye character, all of them belonging to the *Algæ*, a group of plants which includes most of the smaller fresh-water weeds, and the vast majority of sea-weeds. One of these filamentous Algæ, occurring in the form of dark-green, thickly-matted threads, is called *Vaucheria*. Besides occurring in water it is often found on the surface of moist soil, *e.g.*, on the pots in conservatories.

Examined microscopically the organism is found to consist of cylindrical filaments with rounded ends and occasionally branched (Fig. 38, A). Each filament has an outer covering of cellulose (B, *c.w*) within which is protoplasm containing a vacuole so large that the protoplasm has the

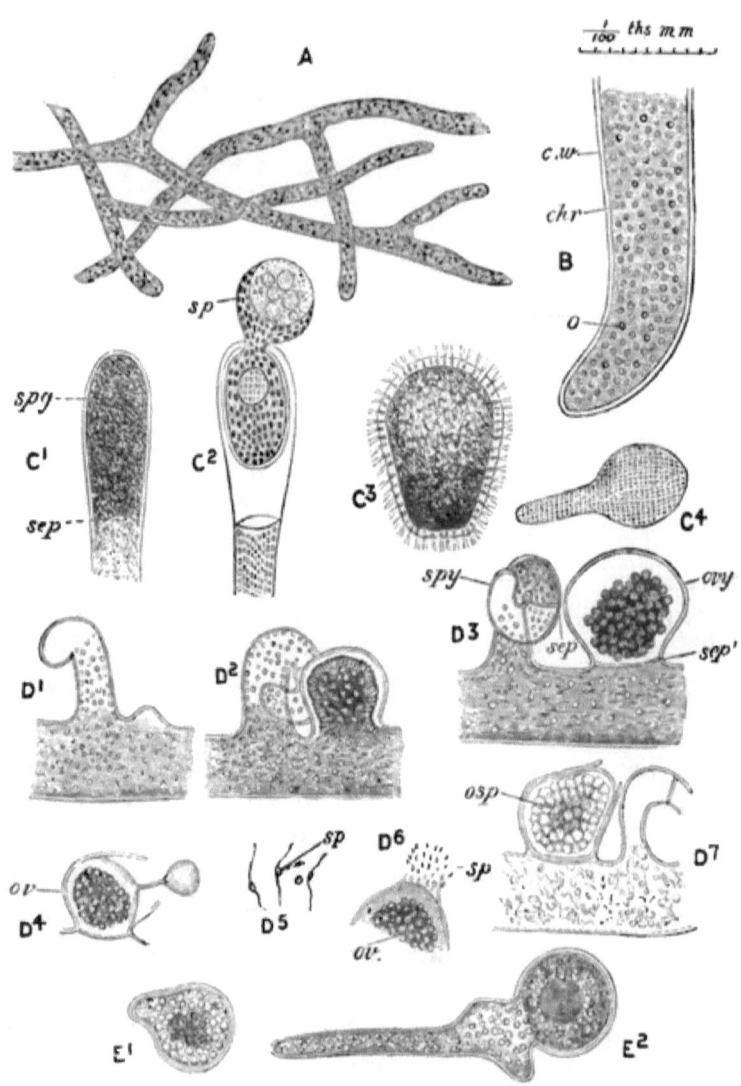

FIG. 38.—*Vaucheria*.

A, tangled filaments of the living plant, showing mode of branching.

B, extremity of a filament, showing cell-wall (*c. w*) and protoplasm with chromatophores (*chr*), and oil-drops (*o*). The scale above applies to this figure only.

C^1, immature sporangium (*spg*) separated from the filament by a septum; C^2, mature sporangium with the spore (*sp*) in the act of escaping; C^3, free-swimming spore, showing cilia, colourless ectoplasm containing nuclei, and endoplasm containing the green chromatophores; c^4, the same at the commencement of germination.

D^1, early, and D^2, later stages in the development of the gonads, the spermary to the left, the ovary to the right; D^3, the fully-formed spermary (*spy*) and ovary (*ovy*), each separated by a septum (*sep*) from the filament.

D^4, the ovary after dehiscence, showing the ovum (*ov*), with small detached portion of protoplasm; D^5, sperms; D^6, distal end of ripe ovary, showing sperms (*sp*) passing through the aperture towards the ovum (*ov*).

D^7, the gonads after fertilization, showing the oosperm (*osp*) still inclosed in the ovary and the dehisced spermary.

E^1, oosperm about to germinate: E^2, further stage in germination.

(c^1 and c^3, after Strasburger; c^2 and c^4, after Sachs; D and E, after Pringsheim.)

character of a membrane lining the cellulose coat. Numerous small nuclei occur in the protoplasm, as well as oil-globules (*o*), and small, close-set, ovoid chromatophores (*chr*) coloured with chlorophyll and containing starch.

Thus a Vaucheria-plant, like a Mucor-plant, is comparable to a single multinucleate cell, extended in one dimension of space so as to take on the form of a filament.

Various modes of asexual reproduction occur in different species of Vaucheria: of these we need only consider that which obtains in *V. sessilis*. In this species the end of a branch swells up (c^1) and becomes divided off by a septum (*sep*), forming a sporangium (*spg*) in principle like that of Mucor, but differing in shape. The protoplasm of the sporangium does not divide but separates itself from the wall, and takes on the form of a single naked ovoidal spore (c^3), formed of a colourless cortical layer containing numerous nuclei and giving off cilia arranged in pairs, and of an inner or medullary substance containing numerous chromatophores.

The wall of the sporangium splits at its distal end (c^2), and the contained spore (*sp*) escapes and swims freely in the water for some time by the vibration of its cilia (c^3). After a short active life it comes to rest, develops a cell-wall, and germinates (c^4), *i.e.*, gives out one or more processes which extend and take on the form of ordinary Vaucheria-filaments, so that in the present case, as in Mucor (p. 164), the development of the plant shows it to be a single immensely elongated multinucleate cell.

In its mode of sexual reproduction Vaucheria differs strikingly not only from Mucor, but from all the organisms we have hitherto studied.

The filaments are often found to bear small lateral processes arranged in pairs (D^1), and each consisting of a little bud growing from the filament and quite continuous with it. These are the rudiments of the sexual reproductive organs or *gonads*. The shorter of the two becomes swollen and rounded (D^2), and afterwards bluntly pointed (D^3, *ovy*) : its protoplasm becomes divided from that of the filament, and a septum (D^3, *sep'*) is formed between the two : the new cell thus constituted is the *ovary*.[1] The longer of the two buds undergoes further elongation and becomes bent upon itself (D^2), its distal portion is then divided off by a septum (D^3, *sep*) forming a separate cell (*spy*), the *spermary*.[2]

Further changes take place which are quite different in the two organs. At the bluntly-pointed distal end of the ovary the cell-wall becomes gelatinized and the protoplasm protrudes through it as a small prominence which divides off and is lost (D^4). The remainder of the protoplasm then separates from the wall of the ovary and becomes a naked cell, the *ovum*[3] or egg-cell (D^4, *ov*), which, by the gelatinization and subsequent disappearance of a portion of the

---

[1] Usually called the *oogonium*.
[2] Usually called the *antheridium*.
[3] Frequently called *oosphere*.

wall of the ovary, is in free contact with the surrounding water.

At the same time the protoplasm of the spermary undergoes multiple fission, becoming converted into numerous minute green bodies (D^5), each with two flagella, called *sperms*.[1] These are liberated by the rupture of the spermary (D^7) at its distal end, and swim freely in the water.

Some of the sperms make their way to an ovary, and, as it has been expressed, seem to grope about for the aperture, which they finally pass through (D^6), and are then seen moving actively in the space between the aperture and the colourless distal end of the ovum. One of them, and probably only one, then attaches itself to the ovum and becomes completely united with it, forming the *oosperm*,[2] a body which we must carefully distinguish from the ovum, since, while agreeing with the latter in form and size, it differs in having incorporated with it the substance of a sperm.

Almost immediately the oosperm (D^7, *osp*) surrounds itself with a cellulose wall, and numerous oil-globules are formed in its interior. It becomes detached from the ovary, and, after a period of rest, germinates (E^1, E^2) and forms a new Vaucheria plant.

It is obvious that the fusion of the sperm with the ovum is a process of conjugation in which the conjugating bodies differ strikingly in form and size, one—the *megagamete* or ovum—being large, stationary, and more or less amœboid : the other—the *microgamete* or sperm—small, active, and flagellate. In other words, we have a more obvious case of sexual differentiation than was found to occur in Vorticella,

[1] Often called *spermatozooids* or *antherozooids*.
[2] Often called *oospore*.

(p. 132): the large inactive egg-cell which furnishes by far the greater portion of the material of the oosperm is the female gamete; the small active sperm-cell, the function of which is probably (see Lesson XXIV.) to furnish additional nuclear material, is the male gamete.

Similarly the oosperm is evidently a zygote, but a zygote formed by the union of the highly differentiated gametes,

FIG. 39.—*Caulerpa scalpelliformis* (⅔ nat. size), showing the stem-like, root-like, and leaf-like portions of the unicellular plant. (After Harvey.)

ovum and sperm, just as a zygospore (p. 164) is one formed by the union of equal-sized gametes.

As we shall see, this form of conjugation—often distinguished as *fertilization*—occurs in a large proportion of flowerless plants, such as mosses and ferns (Lessons XXVIII. and XXIX.), as well as in all animals but the very lowest. From lowly water-weeds up to ferns and club mosses, and from sponges and polypes up to man, the process of sexual reproduction is essentially the same, consisting in the conjugation of a microgamete or sperm with a megagamete or ovum, a zygote, the oosperm or unicellular embryo, being produced, which afterwards develops into an independent plant or animal of the new generation. It is a truly remarkable circumstance that what we may consider as the highest form of the sexual process should make its appearance so low down in the scale of life.

The nutrition of Vaucheria is purely holophytic; its food consists of a watery solution of mineral salts and of carbon dioxide, the latter being split up, by the action of the chromatophores, into carbon and oxygen.

Mucor and Vaucheria are examples of unicellular plants which attain some complexity by elongation and branching. The maximum differentiation attainable in this way by a unicellular plant may be illustrated by a brief description of a sea-weed belonging to the genus *Caulerpa*.

Caulerpa (Fig. 39) is commonly found in rock-pools between tide-marks, and has the form of a creeping stem from which root-like fibres are given off downwards and branched leaf-like organs upwards. These "leaves" may attain a length of 30 cm. (1 ft.) or more. So that, on a superficial examination, Caulerpa appears to be as complex an organism as a moss (compare Fig. 39 with Fig. 82, A). But microscopical examination shows that the plant consists of a single continuous mass of vacuolated protoplasm, containing numerous nuclei and green chromatophores and covered by a continuous cell-wall. Large and complicated in form as it is, the whole plant is therefore nothing more than a single branched cell, or, as it may be expressed, a continuous mass of protoplasm in which no cellular structure has appeared.

LESSON XVII

THE DISTINCTIVE CHARACTERS OF ANIMALS AND PLANTS

HITHERTO the words "animal" and "plant" have been either avoided altogether or used incidentally without any attempt at definition. We are now however in a position to consider in some detail the precise meaning of the two words, since in the last half-dozen lessons we have been dealing with several organisms which can be assigned without hesitation to one or other of the two great groups of living things. No one would dream of calling Paramœcium and Stylonychia plants, or Mucor and Vaucheria animals, and we may therefore use these forms as a starting-point in an attempt to form a clear conception of what the names *plant* and *animal* really signify, and how far it is possible to place the lowly organisms described in the earlier lessons in either the vegetable or the animal kingdom.

Let us consider, first of all, the chief points of resemblance and of difference between the indubitable animal Paramœcium on the one hand, and the two indubitable plants Mucor and Vaucheria on the other.

In the first place, the essential constituents of all three organisms is protoplasm, in which are contained one or more nuclei. But in Paramœcium the protoplasm is invested only by a delicate cuticle interrupted at the mouth and anus, while in Mucor and Vaucheria the outer layer is formed by a firm, continuous covering of cellulose.

We thus have as the first morphological difference between our selected animal and vegetable organisms the absence of a cellulose cell-wall in the former and its presence in the latter. This is a fundamental distinction, and applies equally well to the higher forms. The constituent cells of plants are in nearly all cases covered with a cellulose coat (p. 60), while there is no case among the higher animals of cells being so invested.

Next, let us take a physiological character. In all three organisms there is constant waste of substance which has to be made good by the conversion of food material into protoplasm: in other words, constructive and destructive metabolism are continually being carried on. But when we come to the nature of the food and the mode of its reception, we meet at once with a very fundamental difference. In Paramœcium the food consists of living organisms taken whole into the interior of the body, and the digestion of this solid proteinaceous food is the necessary prelude to constructive metabolism. In Vaucheria the food consists of a watery solution of carbon dioxide and mineral salts—*i.e.*, it is liquid and inorganic, its nitrogen being in the form of nitrates or of simple ammonia compounds. Mucor, like Paramœcium, contains no chlorophyll, and is therefore unable to use carbon dioxide as a food: like Vaucheria, it is prevented by its continuous cellulose investment from ingesting solid food, and is dependent upon an aqueous solution. It takes its carbon in the form of sugar or some such compound, while it can make use of nitrogen either in the simple form of a nitrate or an ammonia salt, or in the complex form of proteids or peptones.

In this case also our selected organisms agree with animals and plants generally. Animals, with the exception of some internal parasites, ingest solid food, and they must all have their nitrogen supplied in the form of proteids, being unable to build up their protoplasm from simpler compounds. Plants take their food in the form of a watery solution; those which possess chlorophyll take their carbon in the form of carbon dioxide and their nitrogen in that of a nitrate or ammonia salt: those devoid of chlorophyll cannot, except in the case of some bacteria, make use of carbon dioxide as a food, and are able to obtain nitrogen either from simple salts or from proteids. Chlorophyll-less plants are therefore nourished partly like green plants, partly like animals.

This difference in the character of the food is connected with a morphological difference. Animals have, as a rule, an ingestive aperture or mouth, and some kind of digestive cavity, either permanent (stomach) or temporary (food-vacuole). In plants neither of these structures exists.

Another difference which was referred to at length in an early lesson (p. 32), is not strictly one between plants and animals, but between organisms with and organisms without chlorophyll. It is that in green plants the nutritive processes result in deoxidation, more oxygen being given out than is taken in; while in animals and not-green plants the precise contrary is the case.

There is also a difference in the method of excretion. In Paramœcium there is a special structure, the contractile vacuole, which collects the superfluous water taken in with the food and expels it, doubtless along with nitrogenous and other waste matters. In Vaucheria and Mucor there is no contractile vacuole, and excretion is simply performed by diffusion from the general surface of the organism into the surrounding medium.

This character also is of some general importance. The large majority of animals possess a special organ of excretion, plants have nothing of the kind.

Another difference has to do with the general form of the organism. Paramœcium has a certain definite and constant shape, and when once formed produces no new parts. Vaucheria and Mucor are constantly forming new branches, so that their shape is always changing and their growth can never be said to be complete.

Finally, we have what is perhaps the most obvious and striking distinction of all. Paramœcium possesses in a conspicuous degree the power of automatic movement; in both Mucor and Vaucheria the organism, as a whole, exhibits no automatism but only the slow movements of growth. The spores and sperms of Vaucheria are, however, actively motile.

Thus, taking Paramœcium as a type of animals, and Mucor and Vaucheria as types of plants, we may frame the following definitions :—

Animals are organisms of fixed and definite form, in which the cell-body is not covered with a cellulose wall. They ingest solid proteinaceous food, their nutritive processes result in oxidation, they have a definite organ of excretion, and are capable of automatic movement.

Plants are organisms of constantly varying form in which the cell-body is surrounded by a cellulose wall; they cannot ingest solid food, but are nourished by a watery solution of nutrient materials. If chlorophyll is present the carbon dioxide of the air serves as a source of carbon, nitrogen is obtained from simple salts, and the nutritive processes result in deoxidation; if chlorophyll is absent carbon is obtained from sugar or some similar compound, nitrogen either from simple salts or from proteids, and the process of nutrition is one of oxidation. There is no special excretory organ, and, except in the case of certain reproductive bodies, there is usually no locomotion.

Let us now apply these definitions to the simple forms described in the first eight lessons, and see how far they will help us in placing those organism in one or other of the two "kingdoms" into which living things are divided.

Amœba has a cell-wall, probably nitrogenous, in the resting condition : it ingests solid proteids, its nutrition being therefore holozoic : it has a contractile vacuole : and it performs amœboid movements. It may therefore be safely considered as an animal.

Hæmatococcus has a cellulose wall : it contains chlorophyll and its nutrition is purely holophytic : a contractile vacuole is present in H. lacustris but absent in H. pluvialis : and its movements are ciliary.

Euglena has a cellulose wall in the encysted state : in virtue of its chlorophyll it is nourished by the absorption of carbon dioxide and mineral salts, but it can also ingest solid food through a special mouth and gullet : it has a contractile vacuole, and performs both euglenoid and ciliary movements.

In both these organisms we evidently have conflicting characters : the cellulose wall and holophytic nutrition would place them both among plants, while from the contractile vacuole and active movements of both genera and from the holozoic nutrition of Euglena we should group them with animals. That the difficulty is by no means easily overcome may be seen from the fact that both genera are claimed at the present day both by zoologists and by botanists. For instance, Prof. Huxley considers Hæmatococcus as a plant, and expresses doubts about Euglena; Mr. Saville Kent ranks Hæmatococcus as a plant and Euglena as an animal; Prof. Sachs and Mr. Thiselton Dyer place both genera in the vegetable kingdom; while Profs. Ray Lankester and Bütschli group them both among animals.

In Heteromita the only cell-wall is the delicate cuticle which in the zygote is firm enough to hold the spores up to the moment of their escape: food is taken exclusively by absorption and nutrition is wholly saprophytic: there is a contractile vacuole, and the movements are ciliary.

Here again the characters are conflicting: the probable absence of cellulose, the contractile vacuole and the cilia all have an "animal" look, but the mode of nutrition is that of a fungus.

In Protomyxa there is a decided preponderance of animal characteristics—ingestion or living prey, and both amœboid and ciliary movements. There is no chlorophyll, and the composition of the cell-wall is not known.

In the Mycetozoa, the life-history of which so closely resembles that of Protomyxa, the cyst in the resting stage consists of cellulose, and so does the cell-wall of the spore: nutrition is holozoic, a contractile vacuole is present in the flagellulæ, and both amœboid and ciliary movements are performed. Here again we have a puzzling combination of animal and vegetable characters, and as a consequence we find these organisms included among plants—under the name of *Myxomycetes* or "slime-fungi"—by Sachs and Goebel, while De Bary, Bütschli, and Ray Lankester place them in the animal kingdom.

In Saccharomyces there is a clear preponderance of vegetable characters. The cell-wall consists of cellulose, nutrition takes place by absorption and proteids are not essential, there is no contractile vacuole, and no motile phase.

Lastly, in the Bacteria the cell-wall is composed of cellulose, nutrition is usually saprophytic, there is no contractile vacuole, and the movements are ciliary. So that in all the characters named, save in the presence of cellulose and the absence of a contractile vacuole, the Bacteria agree with Heteromita, yet they are universally—except by Prof. Claus —placed among plants, while Heteromita is as constantly included among animals.

We see then that while it is quite easy to divide the higher organisms into the two distinct groups of plants and animals, any such separation is by no means easy in the case of the lowest forms of life. It was in recognition of this fact that Haeckel proposed, many years ago, to institute a third "kingdom," called *Protista*, to include all unicellular organisms. Although open to many objections in practice, there is a great deal to be said for the proposal. From the strictly scientific point of view it is quite as justifiable to make three subdivisions of living things as two : the line between animals and plants is quite as arbitrary as that between protists and plants or between protists and animals, and no more so : the chief objection to the change is that it doubles the difficulties by making two artificial boundaries instead of one.

The important point for the student to recognize is that these boundaries *are* artificial, and that there are no scientific frontiers in Nature. As in the liquefaction of gases there is a "critical point" at which the substance under experiment is neither gaseous nor liquid : as in a mountainous country it is impossible to say where mountain ends and valley begins: as in the development of an animal it is futile to argue about the exact period when, for instance, the egg becomes a tadpole or the tadpole a frog: so in the case under discussion. The distinction between the higher plants and animals is perfectly sharp and obvious, but when the two groups are traced downwards they are found gradually to merge, as it were, into an assemblage of organisms which partake of the characters of both kingdoms, and cannot without a certain violence be either included in or excluded from either. Where any given "protist" has to be classified the case must be decided on its individual merits: the organism must be compared in detail with all those which resemble it closely in structure, physiology, and life-history: and then a balance must be struck and the doubtful form placed in the kingdom with which it has, on the whole, most points in common.

It will no doubt occur to the reader that, on the theory of evolution, we may account for the fact of the animal and vegetable kingdoms being related to one another like two trees united at the root, by the hypothesis that the earliest organisms were protists, and that from them animals and plants were evolved along divergent lines of descent. And in this connection the fact that some bacteria—the simplest organisms known and devoid of chlorophyll—may flourish in solutions wholly devoid of organic matter, is very significant.

LESSON XVIII

PENICILLIUM AND AGARICUS

ONE of the commonest and most familiar of the lower organisms is the "green mould" which so quickly covers with a thick sage-green growth any organic substances exposed to damp, such as paste, jam, cheese, leather, &c. This mould is a plant belonging, like Mucor, to the group of Fungi, and is called *Penicillium glaucum*.

Examined with the naked eye a growth of Penicillium is seen to have a powdery appearance, and if the finger is passed over it a quantity of extremely fine dust of a sage-green colour comes away. This dust consists, as we shall see, of the spores of Penicillium. The best way to study the plant is to sow some of the spores in a saucer of Pasteur's solution by drawing a needle or brush over a growth of the mould and stirring it round in the fluid.

It is as well to study the naked-eye appearances first. If the quantity of spores taken is not too large and they are sufficiently well diffused through the fluid, little or no trace of them will be apparent to the naked eye. After a few days, however, extremely small white dots appear on the surface of the fluid; these increase in size and are seen, especially by the aid of a hand-magnifier, to consist of little discs, circular or nearly so in outline, and distinctly thicker in the centre than towards the edge : they float on the fluid so that their upper surfaces are dry. Each of these patches is a young Penicillium-growth, formed, as will be seen hereafter, by the germination of a group of spores.

As the growths are examined day by day they are found to increase steadily in size, and as they do so to become thicker and thicker in the middle : their growth is evidently centrifugal. The thicker central portion acquires a fluffy appearance, and, by the time the growth has attained a diameter of about 4 or 5 mm., a further conspicuous change takes place : the centre of the patch acquires a pale blue tint, the circumference still remaining pure white. When the diameter has increased to about 6–10 mm. the colour of the centre gradually changes to dull sage-green : around this is a ring of light blue, and finally an outer circle of white. In all probability some of the growths, several of which will most likely occur in the saucer, will by this time be found to have come together by their edges: they then become completely interwoven, their original boundaries remaining evident for some time by their white tint. Sooner or later, however, the white is replaced by blue and the blue by sage-green, until the whole surface of the fluid is covered by a single growth of a uniform green colour.

Even when they are not more than 2–3 mm. in diameter the growths are strong enough to be lifted up from the fluid, and are easily seen under a low power to be formed of a tough, felt-like substance, the *mycelium*, Fig. 40, A (*my*), from the upper surface of which delicate threads, the *aërial hyphæ* (*a. hy.*), grow vertically upwards into the air, while from its lower surface similar but shorter threads, the *submerged hyphæ* (*s. hy.*), hang vertically downwards into the fluid.

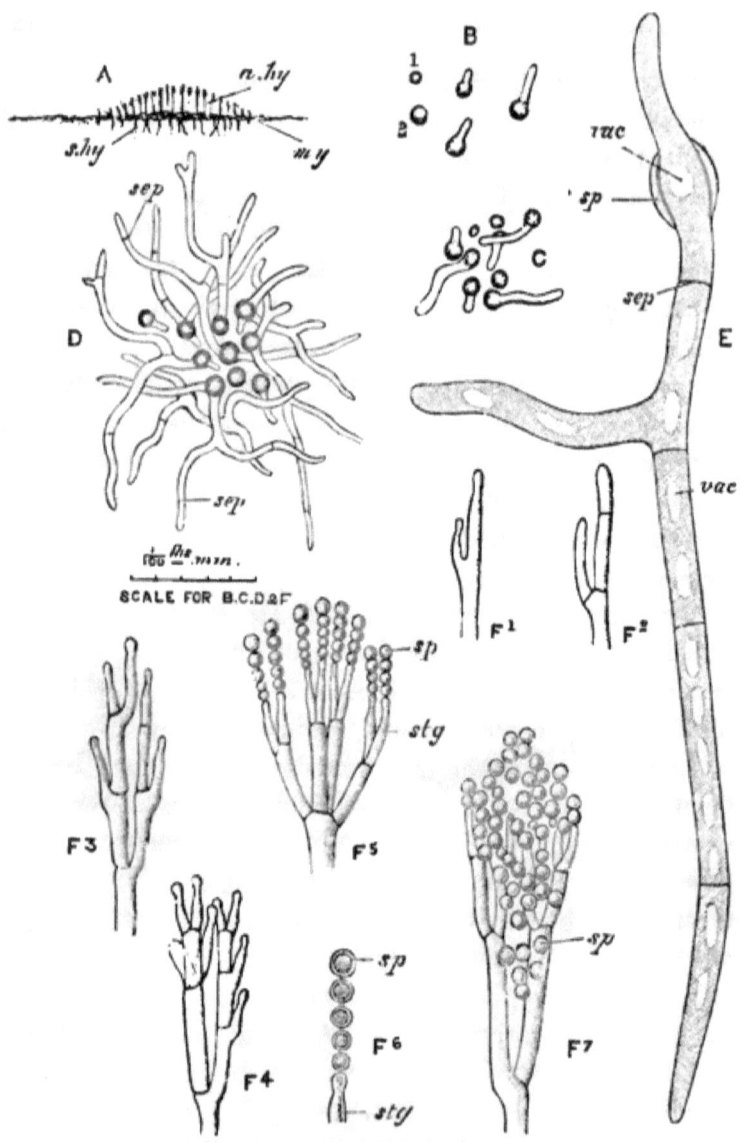

FIG. 40.—*Penicillium glaucum.*

A, Diagrammatic vertical section of a young growth (× 5), showing mycelium (*my*), submerged hyphæ (*s. hy*), and aërial hyphæ (*a. hy*).

B, group of spores: 1, before commencement of germination; 2, after imbibition of fluid: the remaining three have begun to germinate.

C, very young mycelium formed by a small group of germinating spores.

D, more advanced mycelium: the hyphæ have increased in length and begun to branch, and septa (*sep*) have appeared.

E, germinating spore (*sp*) very highly magnified, sending out one short and one long hypha, the latter with a short lateral branch and several septa (*sep*). Both spore and hyphæ contain vacuoles (*vac*) in their protoplasm.

F^1-F^4, development of the spore-bearing brushes by repeated branching of an aërial hypha: the short terminal branches or sterigmata are already being constricted to form spores.

F^5, a fully-developed brush with a row of spores developed from each sterigma (*stg*).

F^6, a single sterigma (*stg*) with its spores (*sp*).

F^7, an over-ripe brush in which the structure is obscured by spores which have dropped from the sterigmata.

B–D, F^1-F^5, and $F^7 \times 150$: $F^6 \times 200$: E $\times 500$.

As long as the growths are white or blue in colour no powder can be detached by touching the aërial hyphæ, showing that the spores are not yet fully formed, but as soon as the permanent green hue is attained the slightest touch is sufficient to detach large quantities of spores.

A bit of the felt-like mycelium is easily teased out or torn asunder with two needles, and is then found, like actual felt, to be formed of a close interlacement of delicate threads (D). These are the *mycelial hyphæ*: they are regularly cylindrical, about $\frac{1}{100}$ mm. in diameter, frequently branched, and differ in an important particular from the somewhat similar hyphæ of Mucor (p. 161). The protoplasm is not continuous, but is interrupted at regular intervals by transverse partitions or *septa* (D, E, *sep*). In other words, a hypha of Penicillium is normally what a hypha of Mucor becomes under unfavourable conditions (p. 162), *multicellular*, the septa dividing it into separate portions, each of which is morphologically comparable to a single yeast cell.

Penicillium shows therefore a very important advance in structure over the organisms hitherto considered. While in these latter the entire organism is a single cell; in Penicillium it is a *cell-aggregate*—an accumulation of numerous cells all in organic connection with one another. As the cells are arranged in a single longitudinal series, Penicillium is an example of a *linear aggregate*.

Each cell is surrounded, as already described, by a wall of cellulose: its protoplasm is more or less vacuolated (E, *vac*), sometimes so much so as to form a mere thin layer within the cell-wall, the whole interior of the cell being occupied by one large vacuole. Recently, by staining with logwood, numerous nuclei have been found, so that the Penicillium cell, like an Oxytricha (p. 120), or a filament of Mucor or Vaucheria, is multinucleate.

The submerged hyphæ have the same structure, but it is easier to find their actual ends than those of the mycelial hyphæ. The free extremity tapers to a blunt point where the cellulose wall is thinner than elsewhere (see E).

The aërial hyphæ from the youngest (white) part of a growth consist of unbranched filaments, but taken from a part which is just beginning to turn blue they are found to have a very characteristic appearance (F^1—F^4). Each sends off from its distal or upper end a larger or smaller number of branches which remain short and grow parallel to one another: the primary branches (F^1, F^2) form secondary ones (F^3), and the secondary tertiary (F^4), so that the hypha finally assumes the appearance of a little brush or pencil, or more accurately of a minute cactus with thick-set forking branches. The ultimate or distal branches are short cells called *sterigmata* (F^5, *stg*).

Next, the ends of the sterigmata become constricted, exactly as if a thread were tied round them and gradually tightened (F^1, F^6), the result being to separate the distal end of the sterigma as a globular daughter-cell, in very much the same way as a bud is separated in Saccharomyces (p. 72).

In this way a *spore* is produced. The process is repeated, the end of the sterigma is constricted again and a new spore formed, the old one being pushed further onwards. By a continual repetition of the same process a longitudinal row of spores is formed (F^5, F^6), of which the proximal or lower one is the youngest, the distal or upper one the oldest. The spores grow for some time after their formation, and are therefore found to become larger and larger in passing from the proximal to the distal end of the chain (F^6). Sooner or later they lose their connection with each other, become detached, and fall, covering the whole growth with a fine dust which readily adheres to all parts owing to the somewhat sticky character of the spores. In this stage it is by no means easy to make out the structure of the brushes, since they are quite obscured by the number of spores adhering to them (F^7).

It is at the period of complete formation of the spores that the growth turns green. The colour is not due to the presence of chlorophyll. Under a high power the spores appear quite colourless, whereas a cell of the same size coloured with chlorophyll would appear bright green.

The germination of the spores can be readily studied by sowing them in a drop of Pasteur's solution in a moist chamber (Fig. 37, p. 163). The spores, several of which usually adhere together, are at first clear and bright (B^1): soon they swell considerably, and the protoplasm becomes granular and vacuolated (B^2): in this stage they are hardly distinguishable from yeast-cells (compare Fig. 13, p. 71). Then one or more buds spring from each and elongate into hyphæ (B, C), just as in Mucor. But the difference between the two moulds is soon apparent: by the time a hypha has grown to a length equal to about six or eight times its own diameter, the protoplasm in it divides transversely and a cellulose septum is formed (D, E. *sep*) dividing the young hypha into two cells (compare Fig. 36, H, p. 159). The distal cell then elongates and divides again, and in this way the hyphæ are, almost from the first, divided into cells of approximately equal length.

The mode of growth of the distal or *apical cell* of a hypha is probably as follows. The free end tapers slightly (E) and the cellulose wall thins out as it approaches the apex. The protoplasm performing constructive more rapidly than destructive metabolism increases in volume, and its tendency is to grow in all directions: as, however, the cellulose membrane surrounding it is thinner at the apex than elsewhere, it naturally, on the principle of least resistance, extends in that direction, thus increasing the length of the cell without adding to its thickness. Thus the growth of a hypha of Penicillium is *apical*, *i.e.* takes place only at the distal end, the cells once formed ceasing to grow. Thus also the oldest cells are those nearest the original spore from which the hypha sprang, the youngest those furthest removed from it.

<small>A process which has been described as sexual, sometimes, but apparently very rarely, occurs in Penicillium, and is said to consist essentially in the conjugation of two gametes having the form of twisted hyphæ, and the subsequent development of spores in the resulting branched zygote. But as the details of the process are complicated and its sexual character is doubtful, it is considered best to do no more than call attention to it. The student is referred to Brefeld's original account of the process in the *Quarterly Journal of Microscopical Science*, vol. xv., p. 342. The so-called sexual reproduction of the closely-allied *Eurotium* is described in Huxley and Martin's *Elementary Biology* (new edition), p. 419, and figured in Howes's *Atlas of Elementary Biology*, pl. xix., figs. xxvi and xxvii.</small>

The nutrition of Penicillium is essentially like that of Mucor (p. 167). But, as it has been remarked, " it is often content with the poorest food which would be too bad for higher fungi. It lives in the human ear; it does not shun cast-off clothes, damp boots, or dried-up ink. Sometimes it contents itself with a solution of sugar with a very little [nitrogenous] organic matter, at other times it appears as if it preferred the purest solution of a salt with only a trace of organic matter. It will even tolerate the hurtful influence of poisonous solutions of copper and arsenious acid." It flourishes best in a solution of peptones and sugar.

This eclecticism in matters of diet is one obvious explanation of the universal occurrence of Penicillium ; another is the extraordinary vitality of the spores. They will germinate at any temperature between 1·5° and 43° C., the optimum being about 22° C. They are not killed by a dry heat of 108° C., and some will even survive a temperature of 120°. And lastly, they will germinate after being kept for two years.

We have seen that the form of a Penicillium growth is irregular, and is determined by the surface on which it grows. There are, however, certain fungi which are quite constant and determinate both in form and size, and are yet found on analysis to be formed exclusively of interlaced hyphæ, that is, to belong to the type of linear aggregates. Among the most striking of these are the mushrooms and toadstools.

A mushroom (*Agaricus*) consists of a stout vertical stalk (Fig. 41, A, *st*), on the upper or distal end of which is borne an umbrella-like disc or *pileus* (*p*). The lower or proximal end of the stalk is in connection with an underground mycelium (*my*), from which it springs.

On the underside of the pileus are numerous radiating vertical plates or *lamellæ* (*l*) extending a part or the whole of the distance from the circumference of the pileus to the stalk. In the common edible mushroom (*Agaricus cam-*

pestris) these lamellæ are pink in young specimens, and afterwards become dark brown.

The mushroom is too tough to be readily teased out like

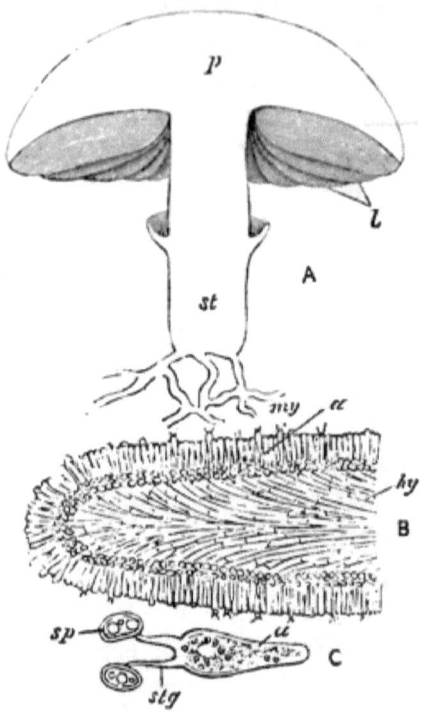

FIG. 41.—*Agaricus campestris.*

A, Diagrammatic vertical section, showing the stalk (*st*) springing from a mycelium (*my*), and expanding into the pileus (*p*), on the under side of which are the radiating lamellæ (*l*).

B, transverse vertical section of a lamella, showing the hyphæ (*hy*) turning outwards to form the layer of club-shaped cells (*a*) from which the sterigmata spring.

C, one of the club-shaped cells (*a*), highly magnified, showing its two sterigmata (*stg*), each bearing a spore (*sp*).

(B and C after Sachs.)

the mycelium of Penicillium, and its structure is best investigated by cutting thin sections of various parts and examining them under a high power.

Such sections show the whole mushroom to be composed of immense numbers of closely interwoven, branched hyphæ (B) divided by numerous septa into cells. In the stalk the hyphæ take a longitudinal direction; in the pileus they turn outwards, passing from the centre to the circumference, and finally send branches downwards to form the lamellæ. Frequently the hyphæ are so closely packed as to be hardly distinguishable one from another.

At the surfaces of the lamellæ the hyphæ turn outwards, so that their ends are perpendicular to the free surfaces of those plates. Their terminal cells become dilated or club-shaped (B, C, *a*), and give off two small branches or sterigmata (C, *stg*), the ends of which swell up and become constricted off as spores (*sp*). These fall on the ground and germinate, forming a mycelium from which more or fewer mushrooms are in due course produced.

Thus in point of structure a mushroom bears much the same relation to Penicillium as Caulerpa (p. 175) bears to Vaucheria. Caulerpa shows the extreme development of which a single branched cell is capable, the mushroom how complicated in structure and definite in form a simple linear aggregate may become.

LESSON XIX

SPIROGYRA

AMONGST the numerous weeds which form a green scum in stagnant ponds and slowly-flowing streams, one, called *Spirogyra*, is perhaps the commonest. It is recognised at once under a low power by the long delicate green filaments of which it is composed being marked with a regular green spiral band.

Examined under the microscope the filaments are seen to be, like the hyphæ of Penicillium, linear aggregates, that is, to be composed of a single row of cells arranged end to end. But in Penicillium the hyphæ are frequently branched, and it is always possible in an entire hypha to distinguish the slightly tapering distal end from the proximal end which springs either from another hypha or from a spore. In Spirogyra the filaments do not branch, and there is no distinction between their opposite ends.

The cells of which the filaments are composed (Fig. 42, A) are cylindrical, covered with a cellulose cell-wall (*c. w*), and separated from adjacent cells by septa (*sep*) of the same substance. The protoplasmic cell-body presents certain characteristic peculiarities.

It has been noticed in more than one instance that in the

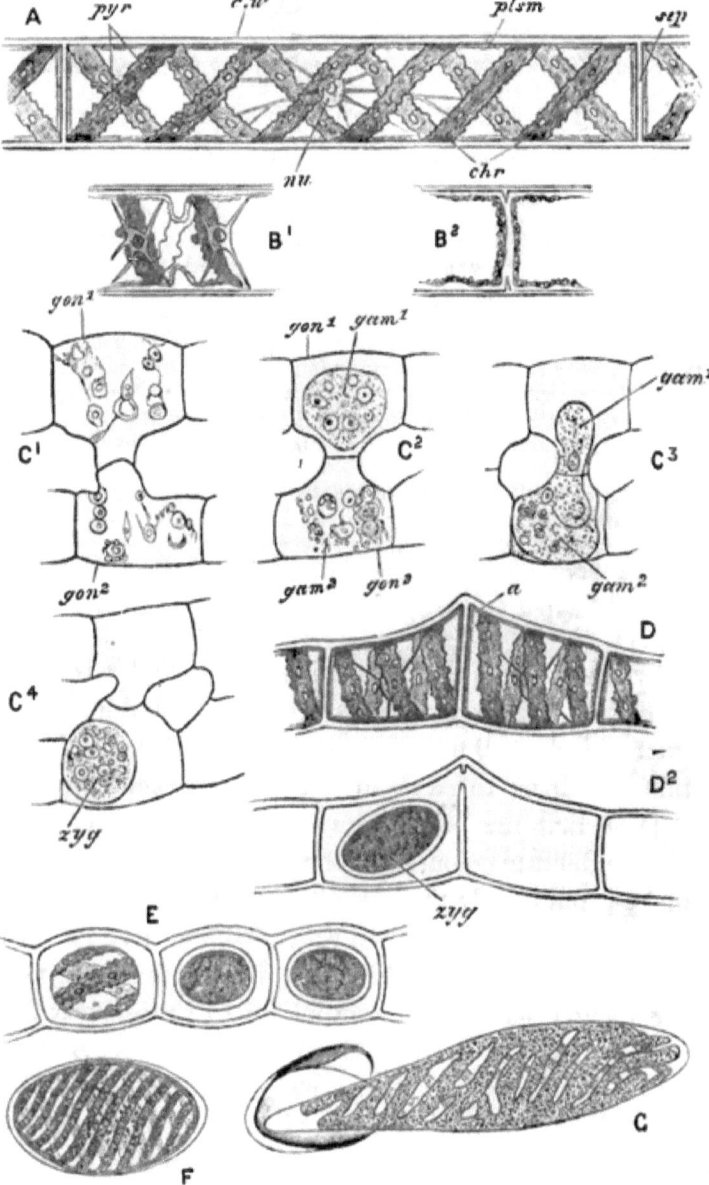

FIG. 42.—*Spirogyra*.

A, small portion of a living filament, showing a single cell, with cell-wall (*c. w*), septa (*sep*) separating it from adjacent cells, peripheral layer of protoplasm (*plsm*) connected by threads with a central mass containing the nucleus (nu), two spiral chromatophores (chr), and pyrenoids (pyr).

B^1, B^2, middle portion of a cell, showing two stages in binary fission.

C, four stages in diœcious conjugation: in C^1 the gonads (gon^1, gon^2) are connected by short processes of their adjacent sides: in C^2 the active or male gamete (gam^1) has separated from the wall of the gonad (gon^1) preparatory to passing across the connecting bridge to the stationary or female gamate (gam^2) which has not yet separated from its containing gonad (gon^2): in C^3 the female gamete (gam^2) has undergone separation, and the male gamete (gam^1) is in the act of conjugating with it: in C^4 the two have united to form a zygote (zyg) lying in the female gonad.

D, two stages in monœcious conjugation: in D^1 the adjacent cells (gonads) have sent out conjugating processes (a): in D^2 conjugation is complete, the male gamete having passed through the aperture between the conjugating processes and united with the female gamete to form the zygote (zyg).

E, parthenogenetic formation of zygotes.

F, fully developed zygote (zygospore).

G, early stage in the germination of the zygote.

(B after Sachs: C after Strasburger: F and G from Sachs after Pringsheim.)

larger cells of plants the development of vacuoles is so extensive that the protoplasm is reduced to a thin layer in contact with the cell-wall (see pp. 169 and 188). This state of things is carried to excess in Spirogyra: the central vacuole is so large that the protoplasm (A, *plsm*) has the character of a mere delicate colourless membrane within the cell-wall: to make it out clearly the specimen should be treated with a fluid of greater density than water, such as a 10 per cent. solution of sodium chloride, which by absorbing the water in the vacuole causes the protoplasm to shrink away from the cell-wall and so brings it clearly into view. It is to this layer of protoplasm that the name *primordial utricle* is applied by botanists, but the student should remember that a primordial utricle is not a special constituent of those cells in which it occurs, but is merely the protoplasm of a vegetable cell in which the vacuole is inordinately large.

The protoplasm of the cell of Spirogyra is not, however,

confined to the primordial utricle; towards the centre of the vacuole is a small irregular mass of protoplasm connected to the peripheral layer by extremely delicate protoplasmic strands. Imbedded in this central mass is the nucleus (nu), which has the form of a biconvex lens and contains a distinct nucleolus.

The chromatophores differ from anything we have yet considered, having the form of green spiral bands (chr), of which each cell may contain one (D^1) or two coiled in opposite directions (A). Imbedded in the chromatophores are numerous pyrenoids (pyr, see p. 27), to which the strands of protoplasm proceeding from the central nucleus-containing mass can be traced.

The process of growth in Spirogyra is brought about by the binary fission of its constituent cells. It takes place under ordinary circumstances during the night (11—12 P.M.), but by keeping the plant cold all night may be delayed until morning.

The nucleus divides by the complicated process (karyokinesis) already described in general terms (p. 67), so that two nuclei are found at equal distances from the centre of the cell. The cell-body with its chromatophores then begins to divide across the centre (B^1), the process commencing near the cell-wall and gradually proceeding inwards: as it goes on cellulose is secreted between the halves of the dividing protoplasm so that a ring of cellulose is formed lying transversely across the middle of the cell, and in continuity externally with the wall (B^2). The ring is at first very narrow, but as the annular furrow across the dividing cell-body deepens, so the ring increases in width, until by the time the protoplasm has divided it has become a complete partition separating the newly-formed daughter-cells from one another.

Any of the cells of a Spirogyra-filament may divide in this way, so that the filament grows by the intercalation of new cells between the old ones. This is an example of *interstitial growth*. Note its difference from the *apical growth* which was found to take place in Penicillium (p. 190), a difference which explains the fact mentioned above (p. 194) that there is no distinction between the two ends of a filament of Spirogyra, while in Penicillium the proximal and distal ends can always be distinguished in a complete hypha.

The sexual reproduction of Spirogyra is interesting, as being intermediate between the very different processes which were found to obtain in Mucor (p. 165) and in Vaucheria (p. 172).

In summer or autumn adjoining filaments become arranged parallel to one another and the opposite cells of each send out short rounded processes which meet (Fig. 43, c^1), and finally become united by the absorption of the adjacent walls, thus forming a free communication between the two connected cells or *gonads* (gon^1, gon^2). As several pairs of cells on the same two filaments unite simultaneously a ladder-like appearance is produced.

The protoplasmic cell-bodies (c^2, gam^1, gam^2) of the two gonads become rounded off and form *gametes* or conjugating bodies (see p. 166, note [1]): it is observable that this process of separation from the wall of the gonad always takes place earlier in one gamete (c^2, gam^1) than in the other (c^2, c^3, gam^2). Then the gamete which is ready first (gam^1) passes through the connecting canal (c^3) and conjugates with the other (gam^2), forming a *zygote* (c^4, zyg) which soon surrounds itself with a thick cell-wall. It has been ascertained that the nuclei of the gametes unite to form the single nucleus of the zygote.

Thus, as in Mucor, the gametes are similar and equal-sized, and the result of the process is a resting zygote or zygospore. But while in Mucor each gamete meets the other half way, so that there is absolutely no sexual differentiation, in Spirogyra, as in Vaucheria, one gamete remains passive, and conjugation is effected by the activity of the other. So that we have here the very simplest case of sexual differentiation: the gametes, although of equal size and similar appearance, are divisible into an active or male cell, corresponding with the sperm of Vaucheria, and a passive or female cell corresponding with the ovum. It will be seen that in Spirogyra the whole of the protoplasm of each gonad is used up in the formation of a single gamete, whereas in Vaucheria, while this is the case with the ovary, numerous gametes (sperms) are formed from the protoplasm of the spermary.

In some forms of Spirogyra conjugation takes place not between opposite cells of distinct filaments, but between adjacent cells of the same filament. Each of the gonads sends out a short process (D^1, a) which abuts against a corresponding process from the adjoining cell: the two processes are placed in communication with one another by a small aperture (D^2) through which the male gamete makes its way in order to conjugate with the female gamete and form a zygote (zyg).

In the ordinary ladder-like method of conjugation the conjugating filaments appear to be of opposite sexes, one producing only male, the other only female gametes: the plant in this case is said to be *diœcious*, *i.e.*, has the sexes lodged in distinct individuals, and conjugation is a process of *cross-fertilization*. But in the method described in the preceding paragraph the individual filaments are *monœcious*, *i.e.*, produce both male and female cells, and conjugation is a process of *self-fertilization*.

Sometimes filaments are found in which the protoplasm of certain cells separates from the wall, and surrounds itself with a thick coat of cellulose forming a body which is quite indistinguishable from a zygote (E). There seems to be some doubt as to whether such cells ever germinate, but they have all the appearance of female cells which for some reason have developed into zygote-like bodies without fertilization. Such development from an unfertilized female gamete, although it has not been proved in Spirogyra is known to occur in many cases, and is distinguished as *parthenogenesis*.

When the zygote is fully developed (F) its cell wall is divided into three layers, the middle one undergoing a peculiar change which renders it waterproof: at the same time the starch in its protoplasm is replaced by oil. In this condition it undergoes a long period of rest, its structure enabling it to offer great resistance to drought, frost, &c. Finally it germinates: the two outer coats are ruptured, and the protoplasm covered by the inner coat protrudes as a club-shaped process (G) which gradually takes on the form of an ordinary Spirogyra filament, dividing as it does so into numerous cells.

Thus in the present case, as in Penicillium and the mushroom, the multicellular adult organism is originally unicellular.

The nutrition of Spirogyra is purely holophytic: like Hæmatococcus and Vaucheria it lives upon the carbon dioxide and mineral salts dissolved in the surrounding water. Like these organisms also it decomposes carbon dioxide and forms starch only under the influence of sunlight.

LESSON XX

MONOSTROMA, ULVA, LAMINARIA, &C.

IT was pointed out in a previous lesson (p. 193) that the highest and most complicated fungi, such as the mushrooms, are found on analysis to be built up of linear aggregates of cells—to consist of hyphæ so interwoven as to form structures often of considerable size and of definite and regular form.

This is not the case with the Algæ or lower green plants—the group to which Vaucheria, Caulerpa, Spirogyra, the diatoms, and in the view of some authors Hæmatococcus and Euglena, belong. These agree with fungi in the fact that the lowest among them (*e.g.* Zooxanthella) are unicellular, and others (*e.g.* Spirogyra) simple linear aggregates, but the higher forms, such as the majority of sea-weeds, have as it were gone beyond the fungi in point of structure and attained a distinctly higher stage of morphological differentiation. This will be made clear by a study of three typical genera.

Amongst the immense variety of seaweeds found in rock-pools between high and low water-marks are several kinds having the form of flat irregular expansions, of a bright green colour and very transparent. One of these is the genus *Monostroma*, of which M. bullosum is a fresh-water species.

Examined microscopically the plant (Fig. 43) is found to consist of a single layer of close-set, green cells, the cell-walls of which are in close approximation, so that the cell-bodies appear as if embedded in a continuous layer of transparent cellulose. Thus Monostroma, like Spirogyra, is only one cell thick (B), but unlike that genus it is not one but many

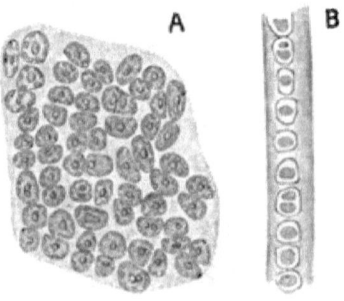

FIG. 43.—*Monostroma.*

A, surface view of M. bullosum, showing the cells embedded in a common layer of cellulose: many of them are in various stages of division.

B, vertical section of M. laceratum, showing the arrangement of the cells in a single layer.

(A after Reinke: B after Cooke.)

cells broad. In other words, instead of being a linear it is a *superficial aggregate.*

To use a geometrical analogy :—a unicellular organism like Hæmatococcus may be compared to a point; a linear aggregate like Penicillium or Spirogyra to a line; a superficial aggregate like Monostroma to a plane.

Growth takes place by the binary fission of the cells (A), but here again there is a marked and important difference from Spirogyra. In the latter the plane of division is always at right angles to the long axis of the filament, so that growth takes place in one dimension of space only, namely in length. In Monostroma the plane of division may be inclined in any direction provided it is perpendicular to the surface of the plant, so that growth goes on in two dimensions of space, namely in length and breadth.

Another of the flat, leaf-like, green sea-weeds is the very common genus *Ulva*, sometimes called "sea-lettuce." It consists of irregular, more or less lobed expansions with crinkled edges, and under the microscope closely resembles Monostroma, with one important difference: it is formed not of one but of two layers of cells, and is therefore not a superficial but a *solid aggregate*. To return to the geometrical analogy used above it is to be compared not to a plane but to a solid body.

As in Monostroma growth takes place by the binary fission of the cells. But these divide not only along variously inclined planes at right angles to the surface of the plant but also along a plane parallel to the surface, so that growth takes place in all three dimensions of space—in length, breadth, and thickness.

Ulva may be looked upon as the simplest example of a solid aggregate : the largest and most complicated sea-weeds are the great olive-brown forms known as "tangles" or "kelp," so common at low water-mark. They belong to various genera, of which the commonest British form is *Laminaria*.

Laminaria (Fig. 44, A) consists of a cylindrical stem, which may be as much as two metres (6 ft.) in length and 5 or 6 cm. in diameter: its proximal end is fastened to the rocks by a branched, root-like structure, while distally it expands into a great, flat, irregularly-cleft, leaf-like body, which may be as much as 2–3 metres long and 70–80 cm. wide.

Other genera of tangles attain even greater dimensions. A common New Zealand genus, *Lessonia* (Fig. 44, B) is a gigantic tree-like weed, the trunk of which is sometimes more than three metres (9–10 ft.) long, and as thick as a

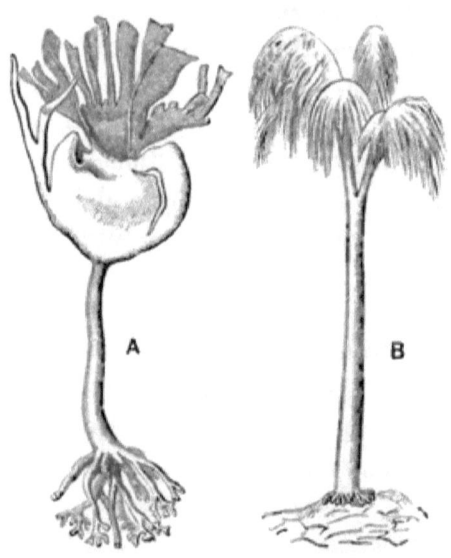

FIG. 44.—A, *Laminaria claustoni*, a young plant, showing stem with branched root-like organ of attachment, and deeply-cleft leaves (about $\frac{1}{8}$th natural size).

B, *Lessonia fuscescens*, showing tree-like form (about $\frac{1}{30}$th natural size).

(A after Sachs: B after Le Maout and Decaisne.)

man's thigh, while the graceful *Macrocystis*, another southern genus, is believed to attain a length of over 200 metres (700 ft.), and is known to grow as much as $5\frac{1}{2}$ metres (over 18 ft.) in six months.

But in spite of their immense size these olive sea-weeds are comparatively simple solid aggregates of cells. Examined with the naked eye the difference between them and a tree or shrub is quite obvious : when cut across they are seen to consist of a nearly homogeneous substance of the consistency of soft gristle, neither bark, wood, nor pith being distinguishable. Under the microscope, however, the cells of which they are composed are seen to vary considerably in form and size, some of them even assuming the characters of what we shall learn in our studies of the higher plants (Lesson XXIX) to distinguish as sieve-tubes.

LESSON XXI

NITELLA

IN the linear, superficial, and solid aggregates discussed in the three previous lessons, the organism was seen to be composed of cells which in most cases differed but little from one another, all complications of structure being due to a continued repetition of the process of cell-multiplication accompanied, except in Laminaria and its allies, by little or no cell-differentiation. In the present lesson we shall make a detailed study of a solid aggregate in which the constituent cells differ very considerably from one another in form and size.

Nitella (Fig. 45, A) is a not uncommon fresh-water weed, found in ponds and water-races, and distinguished at once from such low Algæ as Vaucheria and Spirogyra by its external resemblance to one of the higher plants, since it presents structures which may be distinguished as stem, branches, leaves, &c.

A Nitella plant consists of a slender cylindrical stem, some 15–20 cm. and upwards in length, but not more than about ½ mm. in diameter. The proximal end is loosely rooted to the mud at the bottom of the stream or pond by delicate root filaments or *rhizoids* (A, *rh*) : the distal end is

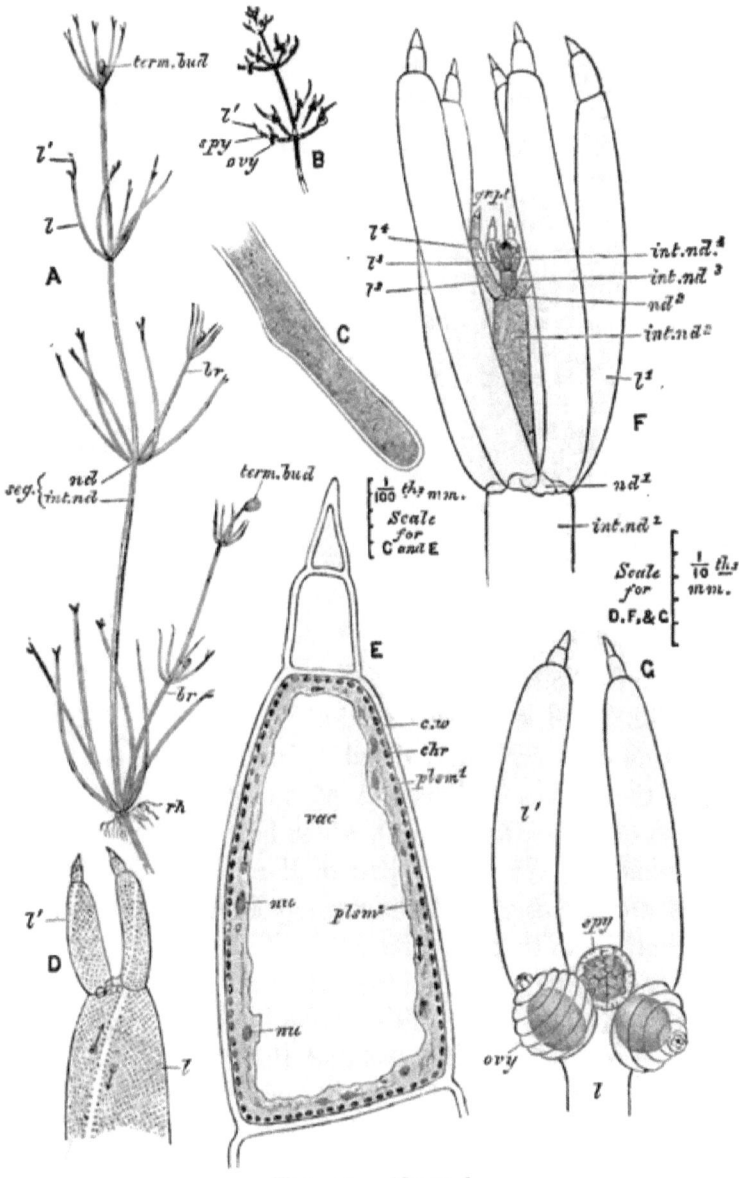

FIG. 45.—*Nitella.*[1]

A, the entire plant (nat. size), showing the segmented stem, each seg-

---

[1] This and the following figures are taken from a New Zealand species closely allied to, if not identical with, the British *N. flexilis.*

ment (*seg*) consisting of a proximal internode (*int. nd*) and distal node (*nd*) : the leaves (*l*) arranged in whorls and ending in leaflets (*l'*) : the rhizoids (*rh*) : and two branches (*br*), each springing from the axil of a leaf and ending, like the main stem, in a terminal bud (*term. bud*).

B, distal end of a shoot with gonads attached to the leaves : *ovy*, the ovaries ; *spy*, the spermaries.

C, distal end of a rhizoid.

D, distal end of a leaf (*l*) with two leaflets (*l'*), showing the chromatophores and the white line. The arrows indicate the direction of rotation of the protoplasm.

E, distal end of a leaflet, showing the general structure of a typical cell of Nitella in optical section : *c. w*, the cell-wall ; $plsm^1$, the quiescent outer layer of protoplasm containing chromatophores (*chr*) ; $plsm^2$, the inner layer, rotating in the direction indicated by the arrows, and containing nuclei (*nu*) ; *vac*, the large vacuole.

F, terminal bud, partly dissected, showing the nodes (*nd*), internodes (*int. nd*), and leaf-whorls (*l*), numbered from 1 to 4, starting from the proximal end ; *gr. pt*, growing point.

G, distal end of a leaf (*l*) with two leaflets (*l'*), at the base of which are attached a spermary (*spy*) and two ovaries (*ovy*).

free. Springing from it at intervals are circlets or *whorls* of delicate, pointed *leaves* (*l*).

Owing to the regular arrangement of the leaves the stem is divisible into successive sections or *segments* (*seg*), each consisting of a very short distal division or *node* (*nd*) from which the leaves spring, and of an elongated proximal division or *internode* (*int. nd*), which bears no leaves.

Throughout the greater part of the stem the whorls of leaves are disposed at approximately equal distances from one another, so that the internodes are of equal length, but towards the distal end the internodes become rapidly shorter and the whorls consequently closer together, until, at the actual distal end, a whorl is found the leaves of which, instead of spreading outwards like the rest, are curled upwards so that their points are in contact. In this way is formed the terminal bud (*term. bud*), by which the uninjured stem is always terminated distally.

The angle between the stem and a leaf, above (distad of) the attachment of the latter, is called the *axil* of the leaf.

There is frequently found springing from the axil of one of the leaves in a whorl a *branch* or *shoot* (*br*) which repeats the structure of the main stem, *i.e.* consists of an axis from which spring whorls of leaves, the whole ending in a terminal bud. The axis or stem of a shoot is called a *secondary axis*, the main stem of the plant being the *primary axis*. It is important to notice that both primary and secondary axes always end in terminal buds, and thus differ from the leaves which have pointed extremities.

The rhizoids or root-filaments (*rh*) arise, like the leaves and branches, exclusively from nodes.

In the autumn the more distal leaves present a peculiar appearance, owing to the development on them of the *gonads* or sexual reproductive organs (Fig. 45, B and G): of these the *spermaries* (antheridia) look very like minute oranges, being globular structures (*spy*) of a bright orange colour: the *ovaries* (oogonia) are flask-shaped bodies (*ovy*) of a yellowish brown colour when immature, but turning black after the fertilization of the ova.

Examined under the microscope each internode is found to consist of a single gigantic cell (F, *int. nd²*) often as much as 3 or 4 cm. long in the older parts of the plant. A node on the other hand is composed of a transverse plate of small cells (*nd¹*) separating the two adjacent internodes from one another. The leaves consist each of an elongated proximal cell like an internode (D, *l*; F, *l¹*), then of a few small cells having the character of a node, and finally of two or three leaflets (D, G, *l'*), each consisting usually of three cells, the distal one of which is small and pointed.

Thus the Nitella plant is a solid aggregate in which the cells have a very definite and characteristic arrangement.

The details of structure of a single cell are readily made

out by examining a leaflet under a high power. The cell is surrounded by a wall of cellulose (E, *c.w*) of considerable thickness. Within this is a layer of protoplasm (primordial utricle, p. 196), enclosing a large central vacuole (*vac*), and clearly divisible into two layers, an outer (*plsm*^{1}) in immediate contact with the cell-wall, and an inner (*plsm*^{2}) bounding the vacuole.

In the outer layer of protoplasm are the chromatophores or chlorophyll-corpuscles (*chr*) to which the green colour of the plant is due. They are ovoidal bodies, about $\frac{1}{100}$ mm. long, and arranged in obliquely longitudinal rows (D). On opposite sides of the cylindrical cell are two narrow oblique bands devoid of chromatophores and consequently colourless (D). The chromatophores contain minute starch grains.

The inner layer of protoplasm contains no chlorophyll corpuscles, but only irregular, colourless granules, many of which are nuclei (E, *nu :* see below, p. 213). If the temperature is not too low this layer is seen to be in active rotating movement, streaming up one side of the cell and down the other (E), the boundary between the upward and downward currents being marked by the colourless bands just mentioned, along which no movement takes place (D). This *rotation* of protoplasm is a form of contractility very common in vegetable cells in which, owing to the confining cell-wall, no freer movement is possible.

The numerous nuclei (E, *nu*) are rod-like and often curved : they can be seen to advantage only after staining (Fig. 46). Lying as they do in the inner layer of protoplasm, they are carried round in the rotating stream.

In the general description of the plant it was mentioned that the stem ended distally in a terminal bud (Fig. 45, A, *term. bud*) formed of a whorl of leaves with their apices curved towards one another. If these leaves (F, *l*^{1}) are dissected away, the node from which they spring (nd^1) is found to give rise distally to a very short internode (*int. nd^2*), above which is a node (nd^2) giving rise to a whorl of very small leaves (l^2), also curved inwards so as to form a bud. Within these is found another segment consisting of a still smaller internode (*int. nd^3*) and node, bearing a whorl of extremely small leaves (l^3), and within these again a segment so small that its parts (*int. nd^4, l^4*) are only visible under the microscope. The minute blunt projections (l^4), which are the leaves of this whorl, surround a blunt, hemispherical projection (*gr. pt*), the actual distal extremity of the plant—the *growing point* or *punctum vegetationis*.

The structure of the growing point and the mode of growth of the whole plant is readily made out by examining vertical sections of the terminal bud in numerous specimens (Fig. 46).

The growing point is formed of a single cell, the apical cell (A, *ap. c*), approximately hemispherical in form and about $\frac{1}{20}$ mm. in diameter. Its cell-wall is thick, and its cell-body formed of dense granular protoplasm containing a large rounded nucleus (*nu*) but no vacuole.

In the living plant the apical cell is continually undergoing binary fission. It divides along a horizontal plane, *i.e.*, a plane parallel to its base, into two cells, the upper (distal) of which is the new apical cell (B, *ap. c*), while the lower is now distinguished as the *sub-apical* or *segmental cell* (*s. ap. c*). The sub-apical cell divides again horizontally, forming two cells, the uppermost of which (C, nd^4) almost immediately becomes divided by vertical planes into several cells (D, nd^4); the lower (C, D, *int. nd^4*) remains undivided.

The sub-apical cell is the rudiment of an entire segment; the uppermost of the two cells into which it divides is the rudiment of a node, the lower of an internode. The future

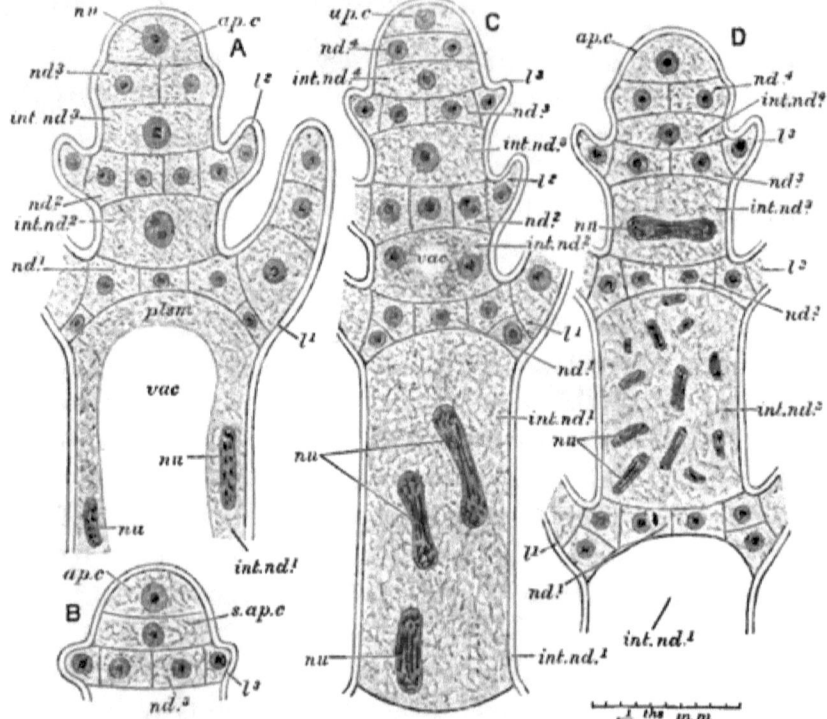

FIG. 46.—*Nitella*: Vertical sections of the growing point at four successive stages. The nodes (*nd*), internodes (*int. nd*), and leaf-whorls (*l*) are all numbered in order from the proximal to the distal end of the bud, the numbers corresponding in all the figures. The proximal segment (*int. nd¹*, *nd¹*, *l¹*) in these figures corresponds with the third segment (*int. nd³*, *l³*) shown in Fig. 46, F.

A, the apical cell (*ap, c*) is succeeded by a very rudimentary node (*nd³*) without leaves: *int. nd¹* is in vertical section, showing the protoplasm (*plsm*), vacuole (*vac*), and two nuclei (*nu*).

B, the apical cell has divided transversely, forming a new apical cell (*ap. c*) and a sub-apical cell (*s. ap. c*): the leaves (*l³*) of *nd³*) have appeared.

C, the sub-apical cell has divided transversely into the proximally-situated internode (*int. nd⁴*) and the distally-situated node (*nd⁴*) of a new segment; in the node the nucleus has divided preparatory to cell-division. The previously formed segments have increased in size: *int. nd²* has developed a vacuole (*vac*), and its nucleus has divided (comp. *int. nd²* in A): *int. nd¹* is shown in surface view with three dividing nuclei (*nu*).

D. *nd⁴* has divided vertically, forming a transverse plate of cells, and is now as far advanced as *nd³* in A: the nucleus of *int. nd³* is in the act of dividing, while *int. nd²*, shown in surface view, now contains numerous nuclei, some of them in the act of dividing.

fate of the two is shown at once by the node dividing into a horizontal plate of cells while the internode remains unicellular.

Soon the cells of the new node begin to send out short blunt processes arranged in a whorl: these increase in size, undergo division, and form leaves (A—D, l^2, l^3).

These processes are continually being repeated; the apical cell is constantly producing new sub-apical cells, the sub-apical cells dividing each into a nodal and an internodal cell; and the nodal cell dividing into a horizontal plate of cells and giving off leaves, while the internodal cell remains undivided.

The special characters of the fully-formed parts of the plant are due to the unequal growth of the new cells. The nodal cells soon cease to grow and undergo but little alteration (comp. nd^1 and nd^3), whereas the internodes increase immensely in length, being quite 3,000 times as long when full-grown as when first separated from the sub-apical cell. The leaves also, at first mere blunt projections (A, l^2), soon increase sufficiently in length to arch over the growing point and so form the characteristic terminal bud: gradually they open out and assume the normal position, their successors of the next younger whorl having in the meantime developed sufficiently to take their place as protectors of the growing point.

The multinucleate condition of the adult internodes is also a result of gradual change. In its young condition an internodal cell has a single rounded nucleus (A, *int. nd^2, int. nd^3*), but by the time it is about as long as broad the nucleus has begun to divide (D, *int. nd^3*; C, *int. nd^2*), and when the length of the cell is equal to about twice its breadth, the nucleus has broken up into numerous fragments (C, *int. nd^1*, D, *int. nd^2*), many of them still in active division. This repeated fission of the nucleus reminds us of what was found to occur in Opalina (p. 119).

Thus the growth of Nitella like that of Penicillium (p. 188), is apical: new cells arise only in the terminal bud, and, after the first formation of nodes, internodes, and leaves, the only change undergone by these parts is an increase in size accompanied by a limited differentiation of character.

A shoot arises by one of the cells in a node sending off a projection distad of a leaf, *i.e.*, in an axil: the process separates from the parent cell and takes on the characters of an apical cell of the main stem, the structure of which is in this way exactly repeated by the shoot.

The leaves, unlike the branches, are strictly limited in growth. At a very early period the apical cell of a leaf becomes pointed and thick-walled (Fig. 45, E), and after this no increase in the number of cells takes place.

The rhizoids also arise exclusively from nodal cells: they consist of long filaments (Fig. 45, C), not unlike Mucor-hyphæ, but occasionally divided by oblique septa into linear aggregates of cells, and increase in length by apical growth.

The structure of the gonads is peculiar and somewhat complicated.

As we have seen, the spermary (Fig. 45, G, *spy*) is a globular, orange-coloured body attached to a leaf by a short stalk. Its wall is formed of eight pieces or *shields*, which fit against one another by toothed edges, so that the entire spermary may be compared to an orange in which an equatorial incision and two meridional incisions at right angles to one another have been made through the rind dividing it into eight triangular pieces. Strictly speaking, however, only the four distal shields are triangular: the four proximal ones have each its lower angle truncated by the insertion of the stalk, so that they are actually four-sided.

Each shield (Fig. 47, A and B, *sh*) is a single concavo-convex cell having on its inner surface numerous orange-coloured chromatophores: owing to the disposition of these on the inner surface only, the spermary appears to have a

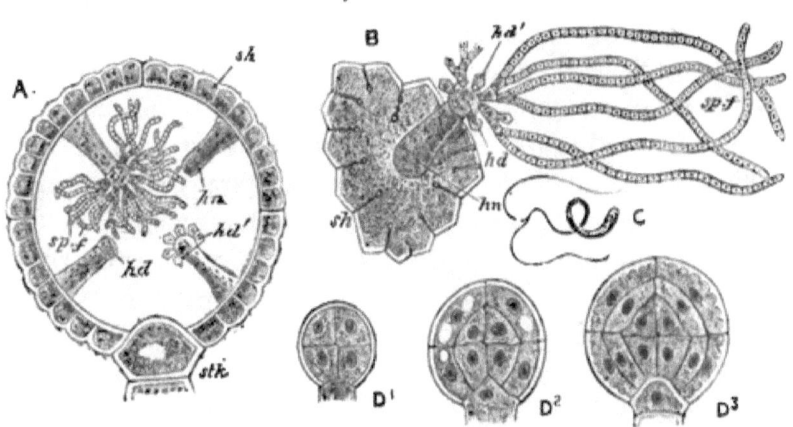

FIG. 47.—A, diagrammatic vertical section of the spermary of Nitella, showing the stalk (*stk*), four of the eight shields (*sh*), each bearing on its inner face a handle (*hn*), to which is attached a head-cell (*hd*): each head cell bears six secondary head-cells (*hd'*), to each of which four spermatic filaments (*sp. f.*) are attached.

B, one of the proximal shields (*sh*), with handle (*hn*), head-cell (*hd*), secondary head-cells (*hd'*), and spermatic filaments (*sp. f.*).

C, a single sperm.

D^1, D^2, D^3, three stages in the development of the spermary.

(C, after Howes.)

colourless transparent outer layer—like an orange inclosed in a close-fitting glass case.

Attached to the middle of the inner surface of each shield is a cylindrical cell, the *handle* (*hn*), which extends towards the centre of the spermary, and, like the shield itself, contains orange chromatophores. Each of the eight handles bears a colourless *head-cell* (*hd'*), to which six *secondary head* cells (*hd'*) are attached, and each of these latter bears four delicate coiled filaments (*sp. f.*) divided by septa into small cells arranged end to end, and thus not unlike the hyphæ of a fungus. There are therefore nearly two hundred of these *spermatic filaments* in each spermary, coiled up in its interior like a tangled mass of white cotton.

The cells of which the filaments are composed have at first the ordinary character, but as the spermary arrives at maturity there is produced in each a single sperm (C), having the form of a spirally-coiled thread, thicker at one end than the other, and bearing at its thin end two long flagella. In all probability the sperm proper, *i.e.*, the spirally-coiled body, is formed from the nucleus of the cell, the flagella from its protoplasm. As each of the 200 spermatic filaments consists of from 100 to 200 cells, a single spermary gives rise to between 20,000 and 40,000 sperms.

When the sperms are formed the shields separate from one another and the spermatic filaments protrude between them like cotton from a pod: the sperms then escape from the containing cells and swim freely in the water.

The ovary (Fig. 45, G, *ovy*, and Fig. 48 A) is ovoidal in form, attached to the leaf by a short stalk (*stk*), and terminated distally by a little chimney-like elevation or *crown* (*cr*). It is marked externally by spiral grooves which can be traced into the crown, and in young specimens its interior is readily seen to be occupied by a large opaque mass (*ov*). Sections show that this central body is the *ovum*, a large cell very rich in starch: it is connected with the unicellular stalk by a small cell (*nd*) from which spring five spirally-arranged cells (*sp. c.*): these coil round the ovum and their free ends —each divided by septa into two small cells—project at the distal end of the organ and form the crown, enclosing a narrow canal which places the distal end of the ovum in free communication with the surrounding water.

We saw how the various parts of the fully formed plant—nodal, and internodal cells, leaves, and rhizoids—were all formed by the modification of similar cells produced in the apical bud. It is interesting to find that the same is true of the diverse parts of the reproductive organs.

The spermary arises as a single stalked globular cell which

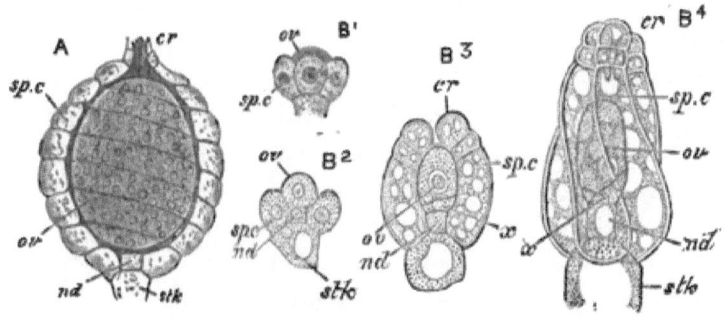

FIG. 48.—A, vertical section of the ovary of Nitella, showing the stalk (*stk*), small node (*nd*) from which spring the five spirally-twisted cells (*sp. c*), each ending in one of the two-celled sections of the crown (*cr*). The ovum contains starch grains, and is represented as transparent, the spiral cells being seen through it.

B^1, surface view, and B^2, section of a very young ovary: B^3, later stage in vertical section: B^4, still later stage, surface view, with the ovum seen through the transparent spiral cells. Letters as in A, except *x*, small cells formed by division from the base of the ovum. (B^2-B^4 after Sachs.)

becomes divided into eight octants (Fig. 47, D^1). Each of these then divides tangentially (*i.e.* parallel to the surface of the sphere) into two cells (D^2), the inner of which divides again (D^3) so that each octant is now composed of three cells. Of these the outermost forms the shield, the middle, the handle, and the inner the head-cell: from the latter the secondary head-cells and spermatic filaments are produced by budding. The entire spermary appears to be a modified leaflet.

The ovary also arises as a single cell, but soon divides and becomes differentiated into an axial row of three cells (Fig. 48, B^2, *ov*, *nd*, *stk*) surrounded by five others (*sp. c*) which arise as buds from the middle cell of the axial row (*nd*) and are at first knob-like and upright (B^1). The uppermost or distal cell of the axial row becomes the ovum (B^3, B^4, *ov*), the others the stalk (*stk*) and intermediate cells (*nd*, *x*) : the five surrounding cells elongate, and as they do so acquire a spiral twist which becomes closer and closer as growth proceeds (compare B^1, B^4, and Fig. 45, G, *ovy*). At the same time the distal end of each develops two septa (B^3) and, projecting beyond the level of the ovum, forms with its fellows the chimney or crown (*cr*) of the ovary. There is every reason to believe that the entire ovary is a highly-modified shoot : the stalk representing an internode, the cell *nd* a node, the spiral cells leaves, and the ovum an apical cell.

Thus while the ciliate Infusoria and Caulerpa furnish examples of cell-differentiation without cell-multiplication, and Spirogyra of cell-multiplication without cell-differentiation, Nitella is a simple example of an organism in which complexity is obtained by the two processes going on hand in hand. It is a solid aggregate, the constituent cells of which are so arranged as to produce a well-defined external form, while some of them undergo a more or less striking differentiation according to the position they have to occupy, and the function they have to perform.

Impregnation takes place in the same manner as in Vaucheria (p. 173). A sperm makes its way down the canal in the chimney-like crown of cells terminating the ovary, and conjugates with the ovum converting it into an oosperm.

After impregnation the ovary, with the contained oosperm, becomes detached and falls to the bottom, where, after a

FIG. 49.—Pro-embryo of Chara, showing the ovary (*ovy*) from the oosperm in which the pro-embryo has sprung: the two nodes (*nd*), apical cell (*ap. c*), rhizoids (*rh*), and leaves (*l*) of the pro-embryo: and the rudiment of the leafy plant ending in the characteristic terminal bud (*term. bud*). (After Howes, slightly altered.)

period of rest, it germinates. The process of germination does not appear to be known in Nitella, but has been followed in detail in the closely allied genus *Chara*.

The oosperm sends out a filament which consists at first of a single row of cells (Fig. 49) giving out a root-fibre (*rh*) at its proximal end. Soon two nodes (*nd*) are formed on the filament, or *pro-embryo*, from the lower of which rhizoids (*rh*) proceed, while the upper gives rise to a few leaves (*l*), not arranged in a whorl, and to a small process which is at first unicellular, but, behaving like an apical cell of Nitella, soon becomes a terminal bud (*term. bud*) and grows into the ordinary leafy plant.

This is an instance of what is known as *alternation of generations*. The Chara—and presumably the Nitella—plant gives rise by a sexual process to a pro-embryo which in turn produces, by an asexual process of budding, the Chara (or Nitella) plant. No case is known of the pro-embryo directly producing a pro-embryo or the leafy-plant a leafy-plant. In order to complete the cycle of existence or life-history of the species two generations which alternate with one another are required : a sexual generation or *gamobium*, which reproduces by the conjugation of gametes (ovum and sperm), and an asexual generation or *agamobium*, which reproduces by budding.

LESSON XXII

HYDRA

WE have seen that with plants, both Fungi and Algæ, the next stage of morphological differentiation after the simple cell is the linear aggregate. Among animals there are no forms known to exist in this stage, but coming immediately above the highest unicellular animals, such as the ciliate Infusoria, we have true solid aggregates. The characters of one of the simplest of these and the fundamental way in which it differs from the plants described in the two previous lessons will be made clear by a study of one of the little organisms known as "fresh-water polypes" and placed under the genus *Hydra*.

Although far from uncommon in pond-water, Hydra is not always easy to find, being rarely abundant and by no means conspicuous. In looking for it the best plan is to fill either a clear glass bottle or beaker or a white saucer with weeds and water from a pond and to let it remain undisturbed for a few minutes. If the gathering is successful there will be seen adhering to the sides of the glass, the bottom of the saucer, or the weeds, little white, tawny, or green bodies, about as thick as fine sewing cotton, and 2—6 mm. in length. They adhere pretty firmly by one end, and examin-

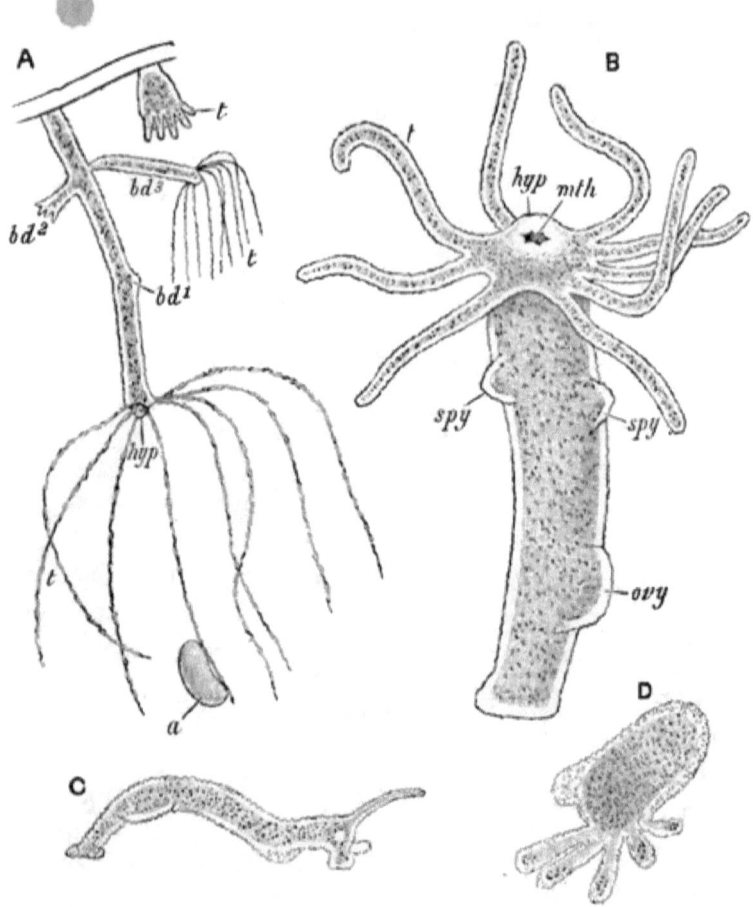

FIG. 50.—*Hydra*.

A, Two living specimens of *H. viridis* attached to a bit of weed. The larger specimen is fully expanded, and shows the elongated body ending distally in the hypostome (*hyp*), surrounded by tentacles (*t*), and three buds (*bd¹*, *bd²*, *bd³*) in different stages of development: a small water-flea (*a*) has been captured by one tentacle. The smaller specimen (to the right and above) is in a state of complete retraction, the tentacles (*t*) appearing like papillæ.

B, *H. fusca*, showing the mouth (*mth*) at the end of the hypostome (*hyp*), the circlet of tentacles (*t*), two spermaries (*spy*), and an ovary (*ovy*).

C, a Hydra creeping on a flat surface by looping movements.

D, a specimen crawling on its tentacles.

(C and D after W. Marshall.)

ation with a pocket lens shows that from the free extremity a number of very delicate filaments, barely visible to the naked eye, are given off.

Under the low power of a compound microscope, a Hydra (Fig. 50, B) is seen to have a cylindrical body attached by a flattened base to a weed or other aquatic object, and bearing at its opposite or distal end a conical structure, the *hypostome* (*hyp*), at the apex of which is a circular aperture, the mouth (*mth.*). At the junction of the hypostome with the body proper are given off from six to eight long delicate *tentacles* (*t*) arranged in a circlet or whorl. A longitudinal section shows that the body is hollow, containing a spacious cavity, the *enteron* (Fig. 51, A, *ent. cav*), which communicates with the surrounding water by the mouth. The tentacles are also hollow, their cavities communicating with the enteron.

There are three kinds of Hydra commonly found : one, *H. vulgaris*, is colourless or nearly so ; another, *H. fusca*, is of a pinkish-yellow or brown colour; the third, *H. viridis*, is bright green. In the two latter it is quite evident, even under a low power, that the colour is in the inner parts of the body-wall, the outside of which is formed by a transparent colourless layer (Fig. 50, A, B).

It is quite easy to keep a Hydra under observation on the stage of the microscope for a considerable time by placing it in a watch-glass or shallow "cell" with weeds, &c., and in this way its habits can be very profitably studied.

It will be noticed, in the first place, that its form is continually changing. At one time (Fig. 50, A, left-hand figure) it extends itself until its length is fully fifteen times its diameter and the tentacles appear like long delicate filaments : at another time (right-hand figure) it contracts itself into an almost globular mass, the tentacles then appearing like little blunt knobs.

Besides these movements of contraction and expansion, Hydra is able to move slowly from place to place. This it usually does after the manner of a looping caterpillar (Fig. 50, C): the body is bent round until the distal end touches the surface; then the base is detached and moved nearer the distal end, which is again moved forward, and so on. It has also been observed to crawl like a cuttle fish (D) by means of its tentacles, the body being kept nearly vertical.

It is also possible to watch a Hydra feed. It is a very voracious creature, and to see it catch and devour its prey is a curious and interesting sight. In the water in which it lives are always to be found numbers of "water-fleas," minute animals from about a millimetre downwards in length, belonging to the class *Crustacea*, a group which includes lobsters, crabs, shrimps, &c.

Water-fleas swim very rapidly, and occasionally one may be seen to come in contact with a Hydra's tentacle. Instantly its hitherto active movements stop dead, and it remains adhering in an apparently mysterious manner to the tentacle. If the Hydra is not hungry it usually liberates its prey after a time, and the water-flea may then be seen to drop through the water like a stone for a short distance, but finally to expand its limbs and swim off. If however the Hydra has not eaten recently it gradually contracts the tentacle until the prey is brought near the mouth, the other tentacles being also used to aid in the process. The water-flea is thus forced against the apex of the hypostome, the mouth expands widely and seizes it, and it is finally passed down into the digestive cavity. Hydræ can often be seen with their bodies bulged out in one or more places by recently swallowed water-fleas.

The precise structure of Hydra is best made out by cutting

it into a series of extremely thin sections and examining them under a high power. The appearance presented by a vertical section through the long axis of the body is shown in Fig. 51.

The whole animal is seen to be built up of cells, each consisting of protoplasm with a large nucleus (P, *nu*), and with or without vacuoles. As in the case of most animal cells, there is no cell-wall. Hydra is therefore a solid aggregate: but the way in which its constituent cells are arranged is highly characteristic and distinguishes it at once from a plant.

The essential feature in the arrangement of the cells is that they are disposed in two layers round the central digestive cavity or enteron (A, *ent. cav*) and the cavities of tentacles (*ent. cav'*). So that the wall of the body is formed throughout of an outer layer of cells, the *ectoderm* (*ect*), and of an inner layer, the *endoderm* (*end*), which bounds the enteric cavity. Between the two layers is a delicate transparent membrane, the *mesoglæa*, or *supporting lamella* (*msgl*). A transverse section shows that the cells in both layers are arranged radially (B).

Thus Hydra is a two-layered or *diploblastic* animal, and may be compared to a chimney built of two layers of radially arranged bricks with a space between the layers filled with mortar or concrete.

Accurate examination of thin sections, and of specimens teased out or torn into minute fragments with needles, shows that the structure is really much more complicated than the foregoing brief description would indicate.

The ectoderm cells are of two kinds. The first and most obvious (B, *ect* and C), are large cells of a conical form, the bases of the cones being external, their apices internal. Spaces

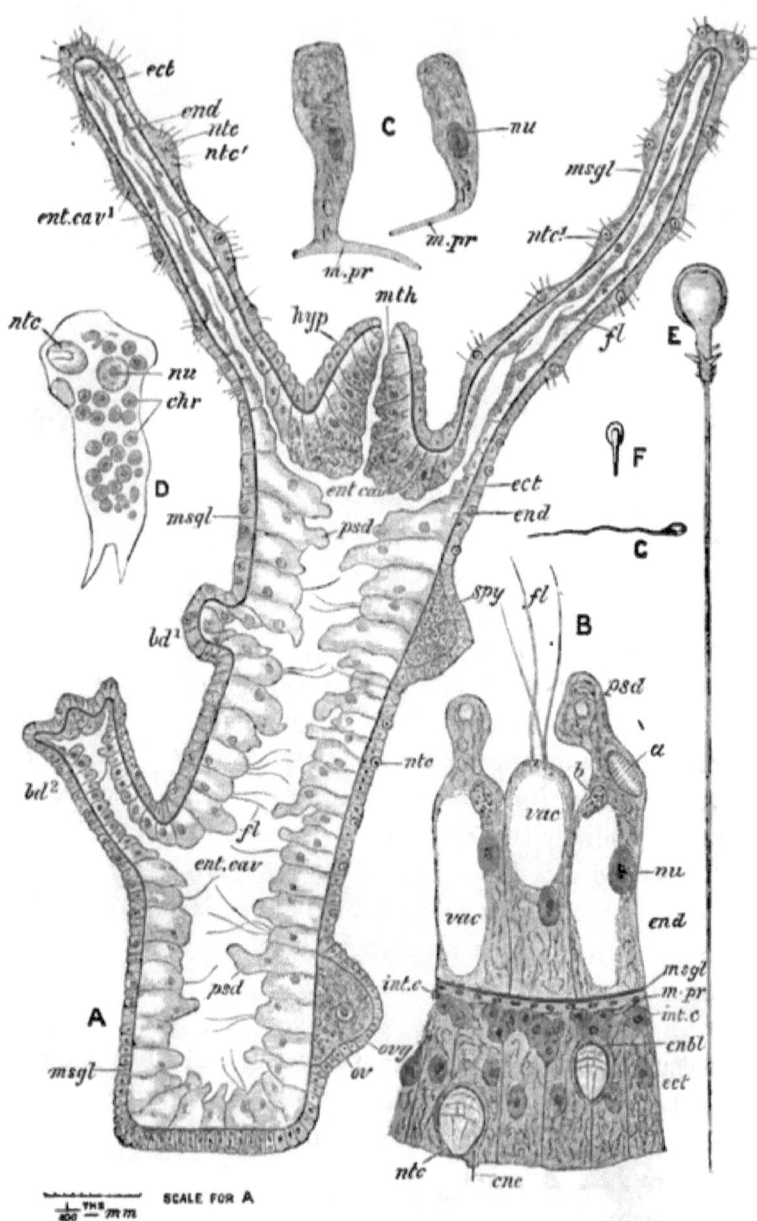

FIG. 51.—*Hydra*.

A, Vertical section of the entire animal, showing the body-wall composed of ectoderm (*ect*) and endoderm (*end*), enclosing an enteric cavit-

(*ent. cav*), which, as well as the two layers, is continued (*ent. cav'*) into the tentacles, and opens externally by the mouth (*mth*) at the apex of the hypostome (*hyp*). Between the ectoderm and endoderm is the mesoglœa (*msgl*), represented by a black line. In the ectoderm are seen large (*ntc*) and small (*ntc'*) nematocysts: some of the endoderm cells are putting out pseudopods (*psd*), others flagella (*fl*). Two buds (*bd^1, bd^2*) in different stages of development are shown on the left side, and on the right a spermary (*spy*) and an ovary (*ovy*) containing a single ovum (*ov*).

B, portion of a transverse section more highly magnified, showing the large ectoderm cells (*ect*) and interstitial cells (*int. c*): two cnidoblasts (*cnbl*) enclosing nematocysts (*ntc*), and one of them produced into a cnidocil (*cnc*): the layer of muscle-processes (*m. pr*) cut across just external to the mesoglœa (*msgl*): endoderm cells (*end*) with large vacuoles and nuclei (*nu*), pseudopods (*psd*), and flagella (*fl*). The endoderm cell to the right has ingested a diatom (*a*), and all enclose minute black granules.

C, two of the large ectoderm cells, showing nucleus (*nu*) and muscle-process (*m. pr*).

D, an endoderm cell of H. viridis, showing nucleus (*nu*), numerous chromatophores (*chr*), and an ingested nematocyst (*ntc*).

E, one of the larger nematocysts with extruded thread barbed at the base.

F, one of the smaller nematocysts.

G, a single sperm.

(D after Lankester: F and G after Howes.)

are necessarily left between their inner or narrow ends, and these are filled up with the second kind of cells (*int. c*), small rounded bodies which lie closely packed between their larger companions and are distinguished as *interstitial cells*.

The inner ends of the large ectoderm cells are continued into narrow, pointed prolongations (c, *m. pr*), placed at right angles to the cells themselves and parallel to the long axis of the body. There is thus a layer of these longitudinally-arranged *muscle-processes* lying immediately external to the mesoglœa (B, *m. pr*). They appear to possess, like the axial fibre of Vorticella (p. 129), a high degree of contractility, the almost instantaneous shortening of the body being due, in great measure at least, to their rapid and simultaneous contraction. It is probably correct to say that, while the ectoderm cells are both contractile and irritable, a special

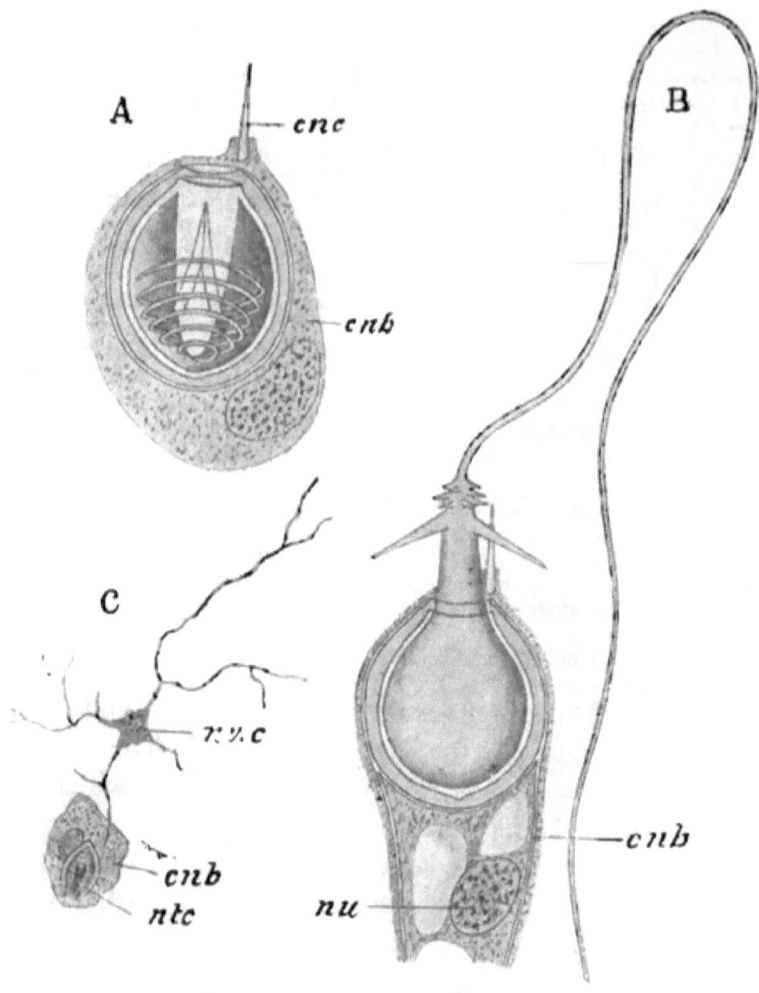

FIG. 52.—*Hydra*.

A, A nematocyst contained in its cnidoblast (*cnb*), showing the coiled filament and the cnidocil (*cnc*).

B, The same after extrusion of the thread, showing the larger and smaller barbs at the base of the thread. *nu*, the nucleus of the cnidoblast.

C, A cnidoblast, with its contained nematocyst, connected with one of the processes of a nerve-cell (*nv. c*).

(After Schneider.)

degree of contractility is assigned to the muscle-processes while the cells themselves are eminently irritable, the slightest stimulus applied to them being followed by an immediate contraction of the whole body.

Imbedded in some of the large ectoderm cells are found clear, oval sacs (A and B, *ntc*), with very well defined walls, and called *nematocysts*. Both in the living specimen and in sections they ordinarily present the appearance shown in Fig. 51, B, *ntc*, and Fig. 52 A, but are frequently met with in the condition shown in Fig. 51 E, and Fig. 52 B, that is, with a short conical tube protruding from the mouth of the sac, armed near its distal end with three recurved barbs, besides several similar processes of smaller size, and giving rise distally to a long, delicate, flexible filament.

Accurate examination of the nematocysts shows that the structure of these curious bodies is as follows:—each consists of a tough sac (Fig. 52, A), one end of which is turned in as a hollow pouch: the free end of the latter is continued into a hollow coiled filament, and from its inner surface project the barbs. The whole space between the wall of the sac and the contained pouch and thread is tensely filled with fluid. When pressure is brought to bear on the outside of the sac the whole apparatus goes off like a harpoon-gun (B), the compression of the fluid forcing out first the barbed pouch and then the filament, until finally both are turned inside out.

It is by means of the nematocysts—the resemblance of which to the trichocysts of Paramœcium (p. 113) should be noted—that the Hydra is enabled to paralyze its prey. Probably some specific poison is formed and ejected into the wound with the thread: in the larger members of the group to which Hydra belongs, such as jelly-fishes, the nematocysts produce an effect on the human skin quite like the sting of a nettle.

The nematocysts are formed in special interstitial cells called *cnidoblasts* (Fig. 51, D, *cnbl* and Fig. 52), and are thus in the first instance at a distance from the surface. But the cnidoblasts migrate outwards, and so come to lie quite superficially either in or between the large ectoderm cells. On its free surface the cnidoblast is produced into a delicate pointed process, the *cnidocil* or "trigger-hair" (*cnc*). In all probability the slightest touch of the cnidocil causes contraction of the cnidoblast, and the nematocyst thus compressed instantly explodes.

Nematocysts are found in the distal part of the body, but are absent from the foot or proximal end, where also there are no interstitial cells. They are especially abundant in the tentacles, on the knob-like elevations of which—due to little heaps of interstitial cells—they are found in great numbers. Amongst these occur small nematocysts with short threads and devoid of barbs (Fig. 51, A, *ntc'* and F).

There are sometimes found in connection with the cnidoblast small irregular cells with large nuclei: they are called *nerve-cells* (Fig. 52, C, *nv. c*), and constitute a rudimentary *nervous system*, the nature of which will be more conveniently discussed in the next lesson (p. 244).

The ectoderm cells of the foot differ from those of the rest of the body in being very granular (Fig. 51 A). The granules are probably the material of the adhesive substance by which the Hydra fixes itself, and are to be looked upon as products of destructive metabolism: *i.e.* as being formed by conversion of the protoplasm in something the same way as starch-granules (p. 33). This process of formation in a cell of a definite product which accumulates and is finally discharged at the free surface of the cell is called *secretion*, and the cell performing the function is known as a *gland cell*.

The endoderm consists for the most part of large cells which exceed in size those of the ectoderm, and are remarkable for containing one or more vacuoles, sometimes so large as to reduce the protoplasm to a thin superficial layer containing the nucleus (Fig. 51, A and B, *end*). Then again, their form is extremely variable, their free or inner ends undergoing continual changes of form. This can be easily made out by cutting transverse sections of a living Hydra, when the endoderm cells are seen to send out long blunt pseudopods (*psd*) into the digestive cavity, and now and then to withdraw the pseudopods and send out from one to three long delicate flagella (*fl*). Thus the endoderm cells of Hydra illustrate in a very instructive manner the essential similarity of flagella and pseudopods already referred to (p. 51). In the hypostome the endoderm is thrown into longitudinal folds, so as to allow of the dilatation of the mouth in swallowing.

Amongst the ordinary endoderm-cells are found long narrow cells of an extremely granular character. They are specially abundant in the distal part of the body, beneath the origins of the tentacles, and in the hypostome, but are absent in the tentacles and in the foot. There is no doubt that they are gland-cells, their secretion being a fluid used to aid in the digestion of the food.

In Hydra viridis the endoderm-cells (D) contain chromatophores (*chr*) coloured green by chlorophyll, which performs the same function as in plants, so that in this species holozoic is supplemented by holophytic nutrition. There is reason for believing that the chromatophores are to be regarded as symbiotic algæ, like those found in connection with Radiolaria (p. 154). In H. fusca bodies resembling these chromatophores are present, but are of an orange or brown colour, and devoid of chlorophyll. Brown and black granules occurring in the cells (B) seem to be due in part to the degeneration of the chromatophores, and in part to be products of excretion.

Muscle-processes exist in connection with the endoderm cells, and they are said to take a transverse or circular direction, *i.e.*, at right angles to the similar processes of the ectoderm cells.

When a water-flea or other minute organism is swallowed by a Hydra, it undergoes a gradual process of disintegration. The process is begun by a solution of the soft parts due to the action of a digestive fluid secreted by the gland-cells of the endoderm : it is apparently completed by the endoderm cells seizing minute particles with their pseudopods and engulfing them quite after the manner of Amœbæ. It is often found that the protrusion of pseudopods during digestion results in the almost complete obliteration of the enteric cavity.

It would seem therefore that in Hydra the process of digestion or solution of the food is to some extent at least *intra-cellular*, *i.e.*, takes place in the interior of the cells themselves, as in Amœba or Paramœcium : it is however mainly *extra-cellular* or *enteric*, *i.e.*, is performed in a special digestive cavity lined by cells.

The ectoderm cells do not take in food directly, but are nourished entirely by diffusion from the endoderm. Thus the two layers have different functions : the ectoderm is protective and sensory; it forms the external covering of the animal, and receives impressions from without : the endoderm, removed from direct communication with the outer world, performs a nutrient function, its cells alone having the power of digesting food.

The essential difference between digestion and assimilation is here plainly seen : all the cells of Hydra assimilate, all are constantly undergoing waste, and all must therefore form new protoplasm to make good the loss. But it is the endoderm cells alone which can make use of raw or undigested food : the ectoderm has to depend upon various products of digestion received by osmosis from the endoderm.

It will be evident from the preceding description that Hydra is comparable to a colony of Amœbæ in which particular functions are made over to particular individuals—just as in a civilized community the functions of baking and butchering are assigned to certain members of the community, and not performed by all. Hydra is therefore an example of *individuation :* morphologically it is equivalent to an indefinite number of unicellular organisms : but, these acting in concert, some taking one duty and some another, form, physiologically speaking, not a colony of largely independent units, but a single multicellular individual.

Like so many of the organisms which have come under our notice, Hydra has two distinct methods of reproduction, asexual and sexual.

Asexual multiplication takes place by a process of budding. A little knob appears on the body (Fig. 50, A, bd^1), and is found by sections to arise from a group of ectoderm cells ; soon however it takes on the character of a hollow outpushing of the wall containing a prolongation of the enteron, and made up of ectoderm, mesoglœa, and endoderm. (Fig. 51, A, bd^1). In the course of a few hours this prominence enlarges greatly, and near its distal end six or eight hollow buds appear arranged in a whorl (Fig. 50, A, bd^2; Fig. 51,

A, bd^2). These enlarge and take on the characters of tentacles: a mouth is formed at the distal end of the bud, which thus acquires the character of a small Hydra (Fig. 50, A, bd^3). Finally the bud becomes constricted at its base, separates from the parent, and begins an independent existence. Sometimes, however, several buds are produced at one time, and each of these buds again before becoming detached: in this way temporary colonies are formed. But the buds always separate sooner or later, although they frequently begin to feed while still attached.

It is a curious circumstance that Hydra can also be multiplied by artificial division: the experiment has been tried of cutting the living animal into pieces, each of which was found to grow into a perfect individual.

As in Vaucheria and Nitella, the sexual organs or gonads are of two kinds, spermaries and ovaries. Both are found in the same individual, Hydra being, like the plants just mentioned, *hermaphrodite* or *monœcious*.

The spermaries (Fig. 50, B, and Fig. 51, A, *spy*) are white conical elevations situated near the distal end of the body: as a rule not more than one or two are present at the same time, but there may be as many as twenty. They are perfectly colourless, even in the green and brown species, being obviously formed of ectoderm alone.

In the immature condition the spermary consists of a little heap of interstitial cells covered by an investment of somewhat flattened cells formed by a modification of the ordinary large cells of the ectoderm. When mature each of the small internal cells becomes converted into a sperm (Fig. 51, c), consisting of a small ovoid head formed from the nucleus of the cell, and of a long vibratile tail formed from its protoplasm. By the rupture of the investing cells or wall of the spermary the sperms are liberated and swim freely in the water.

The ovaries (Fig. 50, B, and Fig. 51, A, *ovy*) are found near the proximal end of the body, and vary in number from one to eight. When ripe an ovary is larger than a spermary, and of a hemispherical form. It begins, like the spermary, as an aggregation of interstitial cells, so that in their earlier stages the sex of the gonads is indeterminate. But while

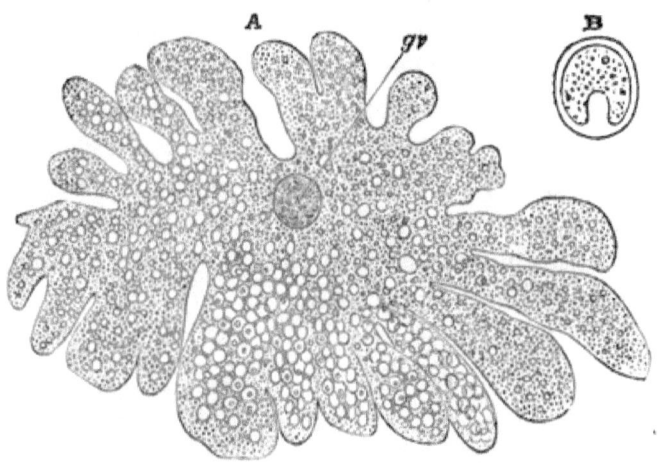

Fig. 53.—A, Ovum of *Hydra viridis*, showing pseudopods, nucleus (*gv*), and numerous chromatophores and yolk spheres.
B, a single yolk sphere. (From Balfour after Kleinenberg.)

in the spermary each cell is converted into a sperm, in the ovary one cell soon begins to grow faster than the rest, becomes amœboid in form (Fig. 51, A, *ov*, and Fig. 53, A), sending out pseudopods amongst its companions and ingesting the fragments into which they become broken up, thus continually increasing in size at their expense. Ultimately the ovary comes to consist only of this single amœboid *ovum*, and of a layer of superficial cells forming a capsule for it. As the ovum grows *yolk-spheres* (Fig. 53), small rounded masses of proteid material, are formed in it, and in Hydra viridis it also acquires green chromatophores.

When the ovary is ripe the ovum draws in its pseudopods and takes on a spherical form: the investing layer then bursts so as to lay bare the ovum and allow of the free access to it of the sperms. One of the latter conjugates with the ovum, producing an *oosperm* or unicellular embryo.

The oosperm divides into a number of cells, the outermost of which become changed into a hard shell or capsule. The embryo, thus protected, falls to the bottom of the water, and after a period of rest develops into a Hydra. As, however, there are certain abnormal features about the development of this genus which cannot well be understood by the beginner, it will not be described in detail, but the very important series of changes by which the oosperm of a multicellular animal becomes converted into the adult will be considered in the next lesson.

LESSON XXIII

HYDROID POLYPES :—BOUGAINVILLEA, DIPHYES, AND PORPITA

It was stated in the previous lesson (p. 234) that in a budding Hydra the buds do not always become detached at once, but may themselves bud while still in connection with the parent, temporary colonies being thus produced.

Suppose this state of things to continue indefinitely: the result would be a tree-like colony or compound organism consisting of a stem with numerous branchlets each ending in a Hydra-like zooid. Such a colony would bear much the same relation to Hydra as Zoothamnium bears to Vorticella (see p. 134).

As a matter of fact this is precisely what happens in a great number of animals allied to Hydra and known by the name of *Hydroid polypes*.

Every one is familiar with the common Sertularians of the sea-coast, often mistaken for sea-weeds : they are delicate, much-branched, semi-transparent structures of a horny consistency, the branches beset with little cups, from each of which, during life, a Hydra-like body is protruded.

A very convenient genus for our purpose is *Bougainvillea*, a hydroid polype found as little tufts a few centimetres long attached to rocks and other submarine objects. Fig. 54, A

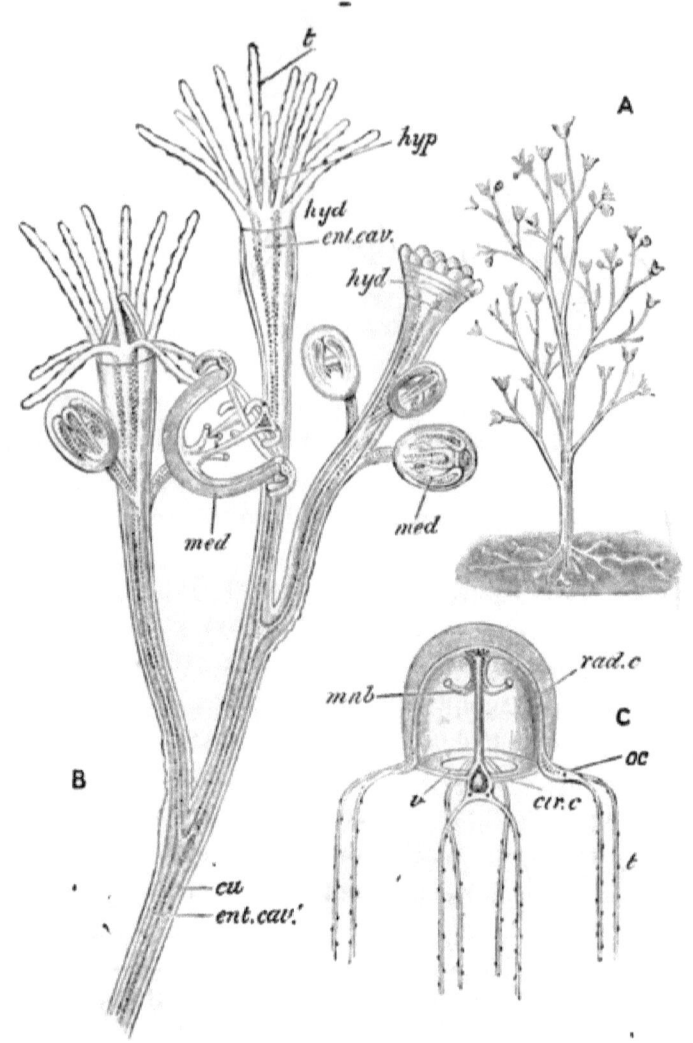

FIG. 54.—*Bougainvillea ramosa.*

A, a complete living colony of the natural size, showing the branched stem and root-like organ of attachment.

B, a portion of the same magnified, showing the branched stem bearing hydranths (*hyd*) and medusæ (*med*), one of the latter nearly mature, the others undeveloped: each hydranth has a circlet of tentacles (*t*) surrounding a hypostome (*hyp*), and contains an enteric cavity (*ent. cav*) continuous with a narrow canal (*ent. cav'*) in the stem. The stem is covered by a cuticle (*cu*).

C, a medusa after liberation from the colony, showing the bell with tentacles (*t*), velum (*v*), manubrium (*mnb*), radial (*rad. c*) and circular (*cir. c*) canals, and eye-spots (*oc*). (After Allman.)

shows a colony of the natural size, B a part of it magnified: it consists of a much-branched stem of a yellowish colour attached by root-like fibres to the support. The branches terminate in little Hydra-like bodies called hydranths (B, *hyd*), each with a hypostome (*hyp*) and circlet of tentacles (*t*). Lateral branchlets bear bell-shaped structures or medusæ (*med*): these will be considered presently.

Sections show that the hydranths have just the structure of a Hydra, consisting of a double layer of cells—ectoderm and endoderm—separated by a supporting lamella or mesoglœa and enclosing a digestive cavity (*ent. cav.*) which opens externally by a mouth placed at the summit of the hypostome.

The stem is formed of the same layers and contains a cavity (*ent. cav'*) continuous with those of the hydranths, and thus the structure of a hydroid polype is, so far, simply that of a Hydra in which the process of budding has gone on to an indefinite extent and without separation of the buds.

There is however an additional layer added in the stem for protective and strengthening purposes. It is evident that a colony of the size shown in Fig. 54, A would, if formed only of soft ectodermal and endodermal cells, be so weak as to be hardly able to bear its own weight even in water. To remedy this a layer of transparent, yellowish substance of horny consistency, called the *cuticle*, is developed outside the ectoderm of the stem, extending on to the branches and only stopping at the bases of the hydranths and medusæ. It is this layer which, when the organism dies and decays, is left as a semi-transparent branched structure resembling the living colony in all but the absence of hydranths and medusæ. The cuticle is therefore a supporting organ or skeleton, not like our own bones formed in the interior of the body (*endoskeleton*), but like the shell of a crab or lobster lying altogether outside the soft parts (*exoskeleton*).

As to the mode of formation of the cuticle:—we saw that many organisms, such as Amœba and Hæmatococcus, form, on entering into the resting condition, a cyst or cell-wall, by secreting or separating from the surface of their protoplasm a succession of layers either of cellulose or of a transparent horn-like substance. But Amœba and Hæmatococcus are unicellular, and are therefore free to form this protective layer at all parts of their surface. The ectoderm cells of Bougainvillea on the other hand are in close contact with their neighbours on all sides and with the mesoglœa at their inner ends, so that it is not surprising to find the secretion of skeletal substance taking place only at their outer ends. As the process takes place simultaneously in adjacent cells, the result is a continuous layer common to the whole ectoderm instead of a capsule to each individual cell. It is to an exoskeletal structure formed in this way, *i.e.* by the secretion of successive layers from the free faces of adjacent cells, that the name cuticle is in strictness applied in multicellular organisms.

The medusæ (B, *med.* and C), mentioned above as occurring on lateral branches of the colony, are found in various stages of development, the younger ones having a nearly globular shape, while when fully formed each resembles a bell attached by its handle to one of the branches of the colony and having a clapper in its interior. When quite mature the medusæ become detached and swim off as little jelly-fishes (C).

The structure of medusa must now be described in some detail. The *bell* (C) is formed of a gelatinous substance (Fig. 55, D, *msgl*) covered on both its inner and outer surfaces by a thin layer of delicate cells (*ect*): The clapper-like organ or *manubrium* (Fig. 54, C and Fig. 55 D and D', *mnb*) is formed of two layers of cells, precisely resembling the ectoderm and endoderm of Hydra, and separated by a thin mesoglœa; it is hollow, its cavity (Fig. 55, D, *ent. cav*) opening below, *i.e.* at its distal or free end, by a rounded aperture, the *mouth* (*mth*), used by the medusa for the ingestion of food. At its upper (attached or proximal) end the cavity of the manubrium is continued into four narrow, *radial canals* (Fig. 54, C, *rad. c*, and Fig. 55, D and D' *rad*) which extend through the gelatinous substance of the bell at equal distances from one another, like four meridians, and finally open into a *circular canal* (*cir. c*) which runs round the edge of the bell. The whole system of canals is lined by a layer of cells (Fig. 55, D and D', *end*) continuous with the inner layer or endoderm of the manubrium; and extending from one canal to another in the gelatinous substance of the bell, is a delicate sheet of cells, the *endoderm-lamella* (D', *end. la*).

From the edge of the bell four pairs of tentacles (Fig. 54, C and Fig. 55, D, *t*) are given off, one pair corresponding to each radial canal, and close to the base of each tentacle is a little speck of pigment (Fig. 54, *oc*), the *ocellus* or eye-spot. Lastly, the margin of the bell is continued inwards into a narrow circular shelf, the *velum* (*v*).

At first sight there appears to be very little resemblance between a medusa and a hydranth, but it is really quite easy to derive the one form from the other.

Suppose a short hydranth or Hydra-like body with four tentacles (Fig. 55, A, A') to have the region from which the tentacles spring pulled out so as to form a hollow, transversely extended disc (B). Next, suppose this disc to become bent into the form of a cup with its concavity towards the

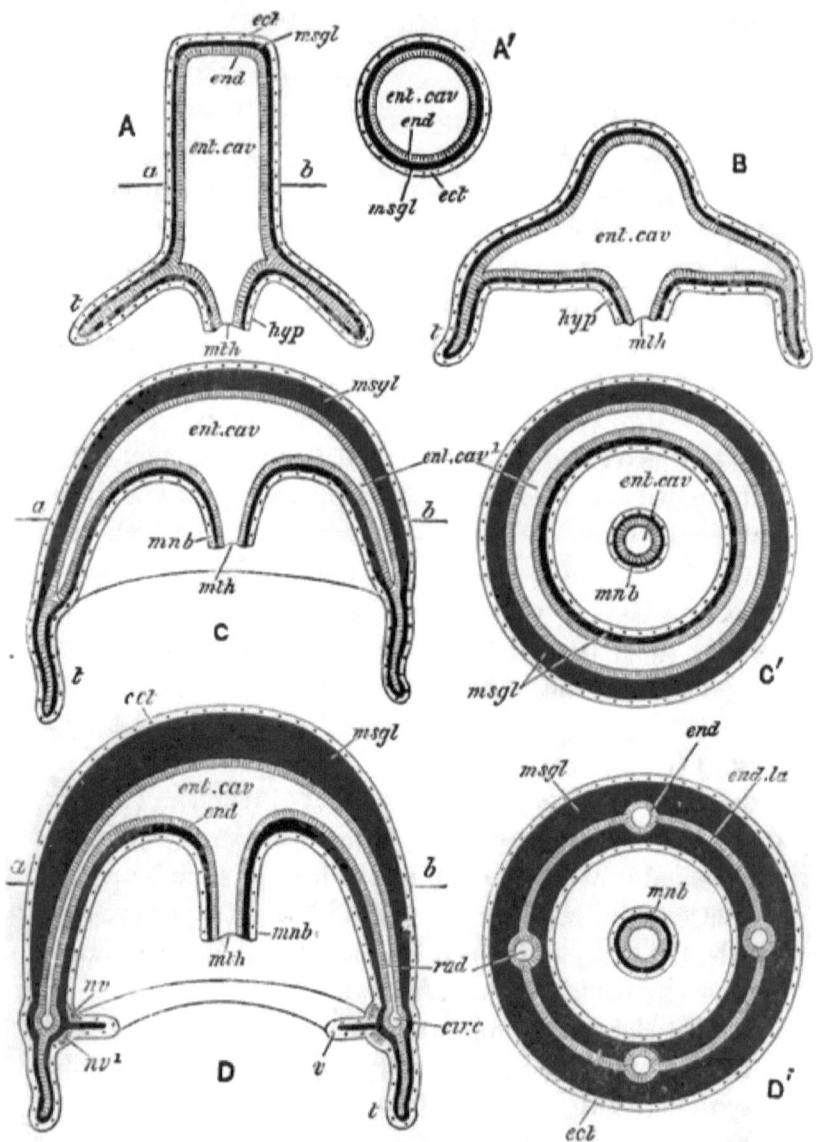

FIG. 55.—Diagrams illustrating the derivation of the medusa from the hydranth. In the whole series of figures the ectoderm (*ect*) is dotted, the endoderm (*end*) striated, and the mesogloea (*msgl*) black.

A, longitudinal section of a Hydra-like body, showing the tubular body with enteric cavity (*ent. cav*), hypostome (*hyp*), mouth (*mth*), and tentacles (*t*).

A′, transverse section of the same through the plane *a b*.

B, the tentacular region is extended into a hollow disc.

C, the tentacular region has been further extended and bent into a bell-like form, the enteric cavity being continued into the bell (*ent. cav′*): the hypostome now forms a manubrium (*mnb*).

C′, transverse section of the same through the plane *a b*, showing the continuous cavity (*ent. cav′*) in the bell.

D, fully formed medusa: the cavity in the bell is reduced to the radiating (*rad*) and circular (*cir. c*) canals, the velum (*v*) is formed, and a double nerve-ring (*nv, nv′*) is produced from the ectoderm.

D′, transverse section of the same through the plane *a b*, showing the four radiating canals (*rad*) united by the endoderm-lamella (*end. la*), produced by partial obliteration of the continuous cavity *ent. cav′* in C′

hypostome, and to undergo a great thickening of its mesoglœa. A form would be produced like c, *i.e.* a medusa-like body with bell and manubrium, but with a continuous cavity (C′, *ent. cav′*) in the thickness of the bell instead of four radial canals. Finally, suppose the inner and outer walls of this cavity to grow towards one another and meet, thus obliterating the cavity, except along four narrow radial areas (D, *rad*) and a circular area near the edge of the bell (D, *cir. c*). This would result in the substitution for the continuous cavity of four radial canals opening on the one hand into a circular canal and on the other into the cavity of the manubrium (*ent. cav*), and connected with one another by a membrane—the endoderm-lamella (*end. la*)—indicating the former extension of the cavity.

It follows from this that the inner and outer layers of the manubrium are respectively endoderm and ectoderm: that the gelatinous tissue of the bell is an immensely thickened mesoglœa: that the layer of cells covering both inner and outer surfaces of the bell is ectodermal: and that the layer of cells lining the system of canals, together with the endoderm-lamella, is endodermal.

Thus the medusa and the hydranth are similarly constructed or *homologous* structures, and the hydroid colony, like Zoothamnium (p. 136), is dimorphic, bearing zooids of two kinds.

The ectoderm cells of the hydranth bear muscle-processes like those of Hydra (p. 227, Fig. 51, c): in the medusæ similar processes are found on the inner concave side of the bell and in the velum. Sometimes, however, the place of these processes is taken by a layer of spindle-shaped fibres (Fig. 56, A), many times longer than broad, and provided each with a nucleus. Such *muscle-fibres* are obviously cells greatly extended in length, so that the ectoderm cell of Hydra with its continuous muscle-*process* is here represented by an ectoderm cell with an adjacent muscle-*cell*. We thus get a partial intermediate layer of cells between the ectoderm and endoderm, in addition to the gelatinous mesoglœa, and so, while a hydroid polyp is, like Hydra, *diploblastic* (p. 225), it shows a tendency towards the assumption of a three-layered or *triploblastic* condition. Both the muscle-processes and muscle-cells of the medusæ differ from those of the hydranths in exhibiting a delicate transverse striation (Fig. 56).

Sooner or later the medusæ separate from the hydroid colony and begin a free existence. Under these circumstances the rhythmical contraction—*i.e.* contraction taking place at regular intervals—of the muscles of the bell causes an alternate contraction and expansion of the whole organ, so that water is alternately pumped out of and drawn into it. The obvious result of this is that the medusa is propelled through the water by a series of jerks.

There is still another important matter in the structure of the medusa which has not been referred to. At the junction of the velum with the edge of the bell there lies, immediately beneath the ectoderm, a layer of peculiar branched cells (Fig. 56, B, *n. c*), containing large nuclei and produced into long fibre-like processes. These *nerve-cells* (see p. 230) are so disposed as to form a double ring round the margin of the bell, one ring (Fig. 55, D, *nv*) being immediately above, the other (*nv'*) immediately below the insertion of the velum. An irregular network of similar cells and fibres

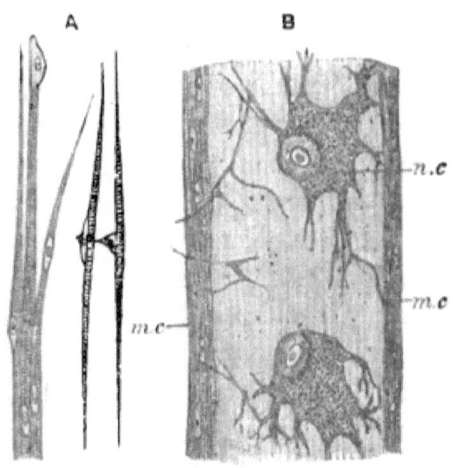

FIG. 56.—A, Muscle fibres from the inner face of the bell of the medusa of a hydroid polype (*Eucopella campanularia*), showing nucleus and transverse striation.

B, portion of the nerve-ring of the same, showing two large nerve-cells (*n. c*) and muscle-fibres (*m. c*) on either side. (After von Lendenfeld.)

occurs on the inner or concave face of the bell, between the ectoderm and the layer of muscle-fibres. The whole constitutes the *nervous system* of the medusa; the double nerve-ring is the *central*, the network the *peripheral* nervous system.

Some of the processes of the nerve-cells are connected with ordinary ectoderm-cells, which thus as it were connect the nervous system with the external world: others, in some instances at least, are probably directly connected with muscle-fibres.

We thus see that while the manubrium of a medusa has the same simple structure as a hydranth, or what comes to the same thing, as a Hydra, the bell has undergone a very remarkable differentiation of its tissues. Its ordinary ectoderm cells instead of being large and eminently contractile form little more than a thin cellular skin or *epithelium* over the gelatinous mesogloea: they have largely given up the function of contractility to the muscle processes or fibres, and have taken on the functions of a protective and sensitive layer.

Similarly the function of automatism, possessed by the whole body of Hydra, is made over to the group of specially modified ectodermal cells which constitute the central nervous system. If a Hydra is cut into any number of pieces, each of them is able to perform the ordinary movements of expansion and contraction, but if the nerve-ring of a medusa is removed by cutting away the edge of the bell, the rhythmical swimming movements stop dead: the bell is in fact permanently paralysed.

It is not, however, rendered incapable of movement, for a sharp pinch, *i.e.* an external stimulus, causes a single contraction, showing that the muscles still retain their irritability. But no movement takes place without such external stimulus, each stimulus giving rise infallibly to one single contraction: the power possessed by the entire animal of independently originating movement, *i.e.* of supplying its own stimuli, is lost with the central nervous system.

Another instance of morphological and physiological differentiation is furnished by the pigment spots or ocelli (Fig. 54, c, *oc*) situated at the bases of the tentacles. They consist of groups of ectoderm cells in which are deposited granules of deep red pigment. Their function is proved by the following experiment.

If a number of medusæ are placed in a glass vessel of

water in a dark room, and a beam of light from a lantern is allowed to pass through the water, the animals are all found to crowd into the beam, thus being obviously sensitive to and attracted by light. If however the ocelli are removed this is no longer the case : the medusæ do not make for the beam of light, and are incapable of distinguishing light from darkness. The ocelli are therefore organs of sight.

In Zoothamnium we saw that the two forms of zooid were respectively nutritive and reproductive in function, the reproductive zooids becoming detached and swimming off to found a new colony elsewhere (p. 135).

This is also the case with Bougainvillea : the hydranths are purely nutritive zooids, the medusæ, although capable of feeding, are specially distinguished as reproductive zooids. The gonads are found in the walls of the manubrium, between the ectoderm and endoderm, some medusæ reproducing ovaries, others spermaries only. Thus while Hydra is monœcious, both male and female gonads occurring in the same individual, Bougainvillea is *diœcious*, certain individuals producing only male, others only female products.

In some Hydroids it has been found that the sexual cells from which the ova and sperms are developed do not originate in the manubrium of a medusa, but apparently arise in the endoderm of the stem of the hydroid colony, afterwards migrating, while still small and immature, to their permanent situation where they undergo their final development.[1] In Bougainvillea, however, the reproductive products are said to originate in the manubrium.

---

[1] This migration of the sexual cells renders the question of their origin in many cases a very difficult one. In some Hydroids, at any rate, they arise in the ectoderm, but migrate into the endoderm at a very early stage.

The medusæ, when mature, become detached and swim away from the hydroid colony. The sperms of the males are shed into the water and carried to the ovaries of the females, where they fertilize the ova, converting them, as usual, into oosperms.

The changes by which the oosperm or unicellular embryo of a hydroid polype is converted into the adult are very remarkable.

The process is begun by the oosperm, still enclosed within the body of the parent (Fig. 57, A), undergoing binary fission, so that a two-celled embryo is formed (B). Each of the two cells again divides (C), and the process is repeated, the embryo consisting successively of 2, 4, 8, 16, 32, &c., cells, until a solid globular mass of small cells is produced (D, E) by the repeated division of the one large cell which forms the starting-point of the series. The embryo in this stage has been compared to a mulberry, and is called the *morula* or *polyplast*.

So far all the cells of the polyplast are alike—globular nucleated masses of protoplasm squeezed into a polyhedral form by mutual pressure. But before long the cells lying next the surface alter their form, becoming cylindrical, with their long axes disposed radially (F). In this way a superficial layer of cells, or *ectoderm*, is differentiated from an internal mass, or *endoderm*.

The embryo now assumes an elongated form (G) and begins to exhibit slow, worm-like movements, finally escaping from the parent and beginning a free existence (H). The ectoderm cells are now found to be ciliated, and before long a cavity appears in the previously solid mass of endoderm cells: this is the first appearance of the enteron or digestive cavity. In this stage the embryo is called a *planula:* it swims slowly through the water by means of its cilia, the

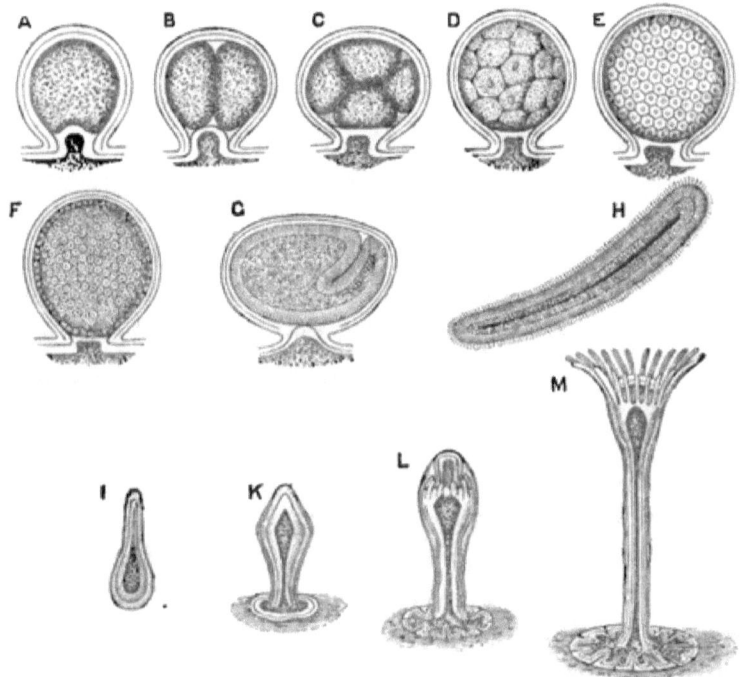

Fig. 57.—Stages in the development of two hydroid polypes, *Laomedea flexuosa* (A–H) and *Eudendrium ramosum* (I–M).

A, oosperm.

B, two-celled, and C, four-celled stage.

D, E, polyplast.

F, G, formation of planula by differentiation of ectoderm and endoderm.

In A–G the embryo is embedded in the maternal tissues.

H, free swimming planula, showing ciliated ectoderm, and endoderm enclosing a narrow enteric cavity.

I, planula, after loss of its cilia, about to affix itself.

K, the same after fixation.

L, Hydra-like stage, still enclosed in cuticle.

M, the same after rupture of the cuticle and liberation of the tentacles. (After Allman.)

broader end being directed forwards in progression. It then loses its cilia and settles down on a rock, shell, sea-weed, or other submarine object, assuming a vertical position with its broader end fixed to the support (I).

The attached or proximal end widens into a disc of attachment, a dilatation is formed a short distance from the free or distal end, and a thin cuticle is secreted from the whole surface of the ectoderm (K). From the dilated portion short buds arise in a circle: these are the rudiments of the tentacles: the narrow portion beyond their origin becomes the hypostome (L). Soon the cuticle covering the distal end is ruptured so as to set free the growing tentacles (M): an aperture, the mouth, is formed at the end of the hypostome, and the young hydroid has very much the appearance of a Hydra with a broad disc of attachment, and with a cuticle covering the greater part of the body.

Extensive budding next takes place, the result being the formation of the ordinary hydroid colony.

Thus from the oosperm or impregnated egg-cell of the medusa the hydroid colony arises, while the medusa is produced by budding from the hydroid colony. The analogy with Nitella (p. 219) will be at once obvious: in each case there is an alternation of generations, the asexual generations or agamobium (hydroid colony, pro-embryo of Nitella) giving rise by budding to the sexual generation or gamobium (medusa, Nitella-plant), which in its turn produces the agamobium by a sexual process, *i.e.* by the conjugation of ovum and sperm.

Two other Hydroids must be briefly referred to in concluding the present lesson.

Floating on the surface of the ocean in many parts of the world is found a beautiful transparent organism called *Diphyes*. It consists of a long, slender stem (Fig. 58, A, *a*), at one end of which are attached two structures called *swimming-bells* (*m, m*) in form something like the bowl of a German pipe, while all along the stem spring at intervals groups of structures (*e*), one of which is shown on an enlarged scale at B.

Each group contains, first, a tubular structure (B, *n*) with an expanded, trumpet-like mouth, through which food is taken: this is clearly a hydranth. From the base of the hydranth proceeds a single, long, branched tentacle or "grappling-line" (*i*), abundantly provided with nematocysts. Springing from the stem near the base of the hydranth is a body called a medusoid (*g*), very like a sort of imperfect medusa, and like it containing gonads. Lastly, enclosing all these structures, much as the white petaloid bract of the common Arum-lily encloses the flower-stalk, is a delicate folded membranous plate, to which the name *bract*, borrowed from botany, is applied. The whole organism is propelled through the water by the rhythmical contraction of the swimming-bells.

Microscopic examination shows that the stem consists, like that of Bougainvillea, of ectoderm, mesogloea, and endoderm, but without a cuticle. The hydranth has a similar structure to that of Bougainvillea, only differing in shape and in the absence of tentacles round the mouth: the medusoids are merely simplified medusæ: the swimming-bells are practically medusæ in which the manubrium is absent: and both the bracts and grappling-lines are shown by comparison with allied forms to be greatly modified medusa-like structures.

Diphyes is in fact a free-swimming hydroid colony which, instead of being dimorphic like Bougainvillea, is *polymorphic*. In addition to nutritive zooids or hydranths, it possesses locomotive zooids or swimming-bells, protective zooids or bracts, and tentacular zooids or grappling-lines. Morpho-

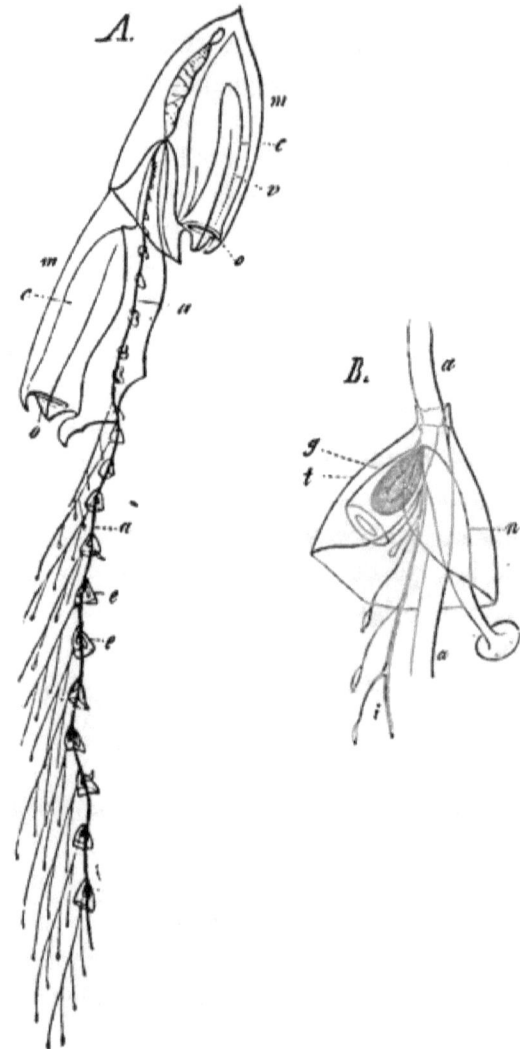

FIG. 58.—*Diphyes campanulata.*

A, the entire colony, natural size, showing stem (*a*) bearing groups of zooids (*e*) and two swimming bells (*m, m*), the apertures of which are marked *o*.

B, one of the groups of zooids marked *e* in A, showing common stem (*a*), hydranth (*n*), medusoid (*g*), bract (*t*), and branched tentacle or grappling line (*i*). (From Gegenbaur.)

logical and physiological differentiation are thus carried much further than in such a form as Bougainvillea.

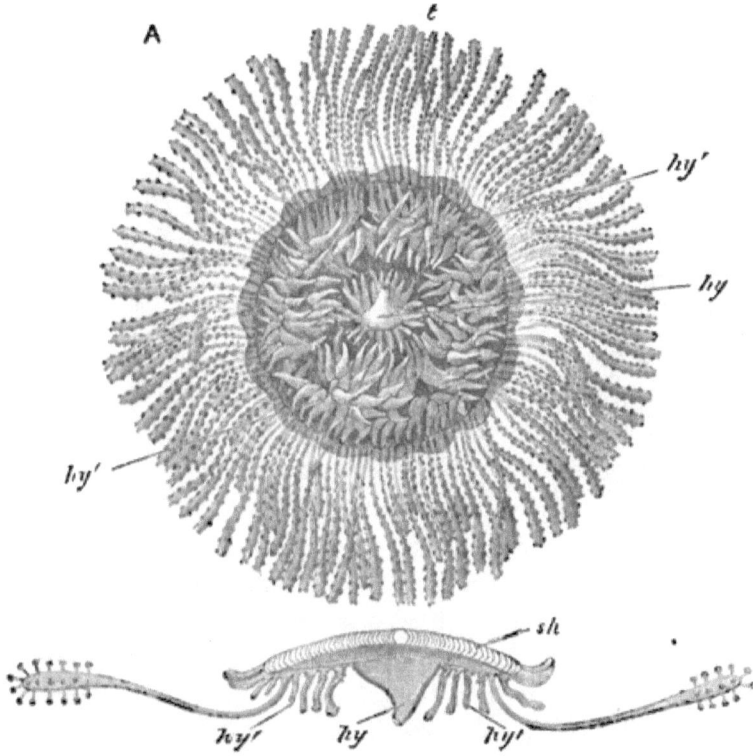

FIG. 59.—A, *Porpita pacifica* (nat. size), from beneath, showing disc-like stem surrounded by tentacles (*t*), a single functional hydranth (*hy*), and numerous mouthless hydranths (*hy'*).

B, vertical section of *P. mediterranea*, showing the relative positions of the functional (*hy*) and mouthless (*hy'*) hydranths, the tentacles, and the chambered shell (*sh*). (A after Duperrey; B from Huxley after Kölliker.)

Porpita is another free-swimming Hydroid, presenting at first sight no resemblance whatever to Diphyes. It has much the appearance of a flattened medusa (Fig. 59), consisting of a circular disc, slightly convex above and concave below, bearing round its edge a number of close-set tentacles, and on its under side a central tubular organ (hy) with a terminal mouth, like the manubrium of a medusa, surrounded by a great number of structures like hollow tentacles (hy'). The discoid body is supported by a sort of shell having the consistency of cartilage and divided into chambers which contain air (B, sh).

Accurate examination shows that the manubrium-like body (hy) on the under surface is a hydranth, that the short, hollow, tentacle-like bodies (hy') surrounding it are mouthless hydranths, and that the disc represents the common stem of Diphyes or Bougainvillea. So that Porpita is not what it appears at first sight, a single individual, like a Medusa or a Hydra, but a colony in which the constituent zooids have become so modified in accordance with an extreme division of physiological labour, that the entire colony has the character of a single physiological individual.

It was pointed out in the previous lesson (p. 233) that Hydra, while morphologically the equivalent of an indefinite number of unicellular organisms, was yet physiologically a single individual, its constituent cells being so differentiated and combined as to form one whole. A further stage in this same process of individuation is seen in Porpita, in which not cells but zooids, each the morphological equivalent of an entire Hydra, are combined and differentiated so as to form a colony which, from the physiological point of view, has the characters of a single individual.

LESSON XXIV

SPERMATOGENESIS AND OOGENESIS. THE MATURATION AND IMPREGNATION OF THE OVUM. THE CONNECTION BETWEEN UNICELLULAR AND DIPLOBLASTIC ANIMALS

IN the preceding lessons it has more than once been stated that sperms arise from ordinary undifferentiated cells in the spermary, and that ova are produced by the enlargement of similar cells in the ovary. Fertilization has also been described as the conjugation or fusion of ovum and sperm. We have now to consider in greater detail what is known as to the precise mode of development of sperms (*spermatogenesis*) and of ova (*oogenesis*), as well as the exact steps of the process by which an oosperm or unicellular embryo is formed by the union of the two sexual elements. The following description applies to animals: recent researches show that essentially similar processes take place in plants.

Both ovary and spermary are at first composed of cells of the ordinary kind, the *primitive sex-cells*, and it is only by the further development of these that the sex of the gonad is determined.

In the spermary the sex cells (Fig. 60, A) undergo repeated fission, forming what are known as the sperm-mother-cells (B). These have been found in several instances to be distinguished by a peculiar condition of the nucleus. We saw (p. 65) that the number of chromosomes is constant in

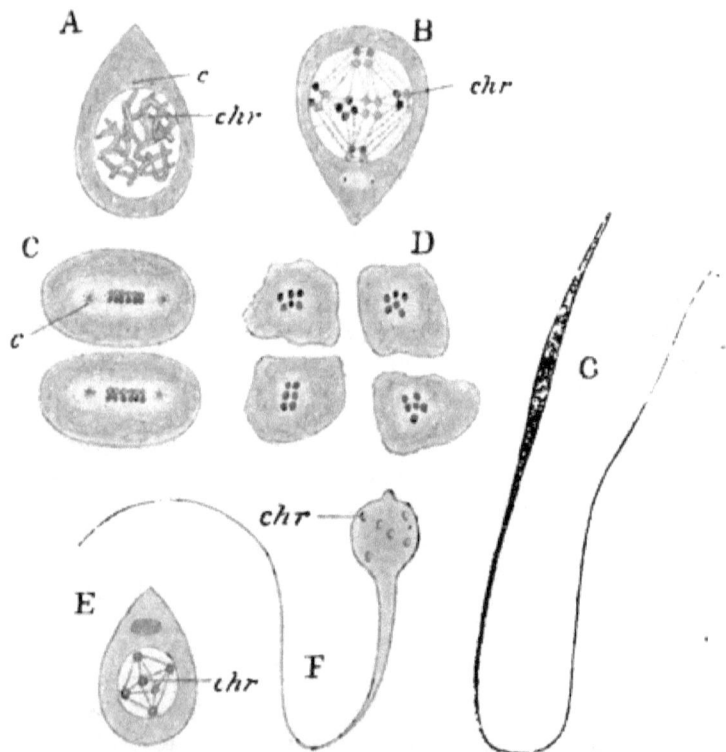

FIG. 60.—Spermatogenesis in the Mole-Cricket (*Gryllotalpa*).
A. Primitive sex-cell, just preparatory to division, showing twelve chromosomes (*chr*); *c*, the centrosome.
B. Sperm-mother-cell, formed by the division of A, and containing twenty-four chromosomes. The centrosome has divided into two.
C. The sperm-mother-cell has divided into two by a reducing division, each daughter cell containing twelve chromosomes.
D. Each daughter cell has divided again in the same manner, a group of four sperm-cells being produced, each with six chromosomes.
E. A single sperm-cell about to elongate to form a sperm.
F. Immature sperm; the six chromosomes are still visible in the head.
G. Fully formed sperm.
(After vom Rath.)

any given animal, though varying greatly in different species. In the formation of the sperm-mother-cells from the primitive sex-cells the number becomes doubled: in the case of the mole-cricket, for instance, shown in Fig. 60, while the ordinary cells of the body, including the primitive sex-cells, contain twelve chromosomes, the sperm-mother-cells contain twenty-four.

The sperm-mother-cell now divides (C), but instead of its chromosomes splitting in the ordinary way (p. 66, Fig. 10) half of their total number—in the present instance twelve—passes into each daughter cell: in this way two cells are produced having the normal number of chromosomes. The process of division is immediately repeated in the same peculiar way (D), the result being that each sperm-mother-cell gives rise to a group of four cells having half the normal number of chromosomes—in the present instance six. The four cells thus produced are the immature sperms (E): in the majority of cases the protoplasm of each undergoes a great elongation, being converted into a long vibratile thread, the *tail* of the sperm (F, G), while the nucleus becomes its more or less spindle-shaped *head*.

Thus the sperm or male gamete is a true cell, specially modified in most cases for active movement: its head, representing the nucleus, is directed forwards in progression, its long tail, formed from the protoplasm, backwards. The direction of movement is thus the precise opposite of that of a monad (p. 36) to which a sperm presents a certain resemblance. This actively motile tailed form is, however, by no means essential: in many animals the sperms are non-motile and in some they resemble ordinary cells.

The peculiar variety of karyokinesis described above, by which the number of chromosomes in the sperm-mother-cells is reduced by one-half, is known as a *reducing division*.

As already stated, the ova arise from primitive sex-cells, precisely resembing those which give rise to sperms. These divide and give rise to the *egg-mother-cells* in which, as in the sperm-mother-cells, the number of chromosomes is doubled. The egg-mother-cells do not immediately undergo division but remain passive and increase, often enormously, in size, by the absorption of nutriment from surrounding parts: in this way each egg-mother-cell becomes an *ovum*. Sometimes this nutriment is simply taken in by osmosis, in other cases the growing ovum actually ingests neighbouring cells after the manner of an Amœba. Thus in the developing egg the processes of constructive are vastly in excess of those of destructive metabolism.

We saw in the second lesson (p. 33) that the products of destructive metabolism might take the form either of waste products which are got rid of, or of plastic products which are stored up as an integral part of the organism. In the developing egg, in addition to increase in the bulk of the protoplasm itself, a formation of plastic products usually goes on to an immense extent. In plants the stored-up materials may take the form of starch, as in Nitella (p. 216), of oil, or of proteid substance: in animals it consists of rounded or angular grains of proteid material, known as *yolk-granules*. These being deposited, like plums in a pudding, in the protoplasm, have the effect of rendering the fully-formed egg opaque, so that its structure can often be made out only in sections. When the quantity of yolk is very great the ovum may attain a comparatively enormous size, as for instance in birds, in which, as already mentioned (p. 68), the "yolk" is simply an immense egg-cell.

When fully formed, the typical animal ovum (Fig. 61) consists of a more or less globular mass of protoplasm, generally exhibiting a reticular structure and enclosing a larger or smaller quantity of yolk-granules. Surrounding the cell-body is usually a cell-wall or cuticle, often of considerable thickness and known as the *vitelline membrane*. The nucleus is large and has the usual constituents (p. 63)—nuclear membrane, nuclear matrix, and chromatin. As a rule there is a very definite nucleolus, which is often known as the *germinal spot*, the entire nucleus being called the *germinal vesicle*.

Such a fully-formed ovum is, however, incapable of being fertilized or of developing into an embryo : before it is ripe for

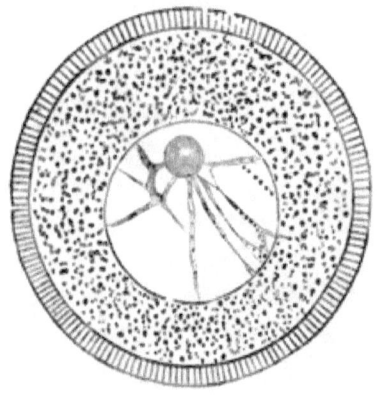

FIG. 61.—Ovum of a Sea-urchin (*Toxopneustes lividus*), showing the radially-striated cell-wall (vitelline membrane), the protoplasm containing yolk granules (vitellus), the large nucleus (germinal vesicle) with its network of chromatin, and a large nucleolus (germinal spot). (From Balfour after Hertwig.)

conjugation with a sperm or able to undergo the first stages of yolk division it has to go through a process known as the *maturation of the egg*.

Maturation consists essentially in a twice-repeated process of cell-division. The nucleus (Fig. 62, A, *nu*) loses its membrane, travels to the surface of the egg, and takes on the

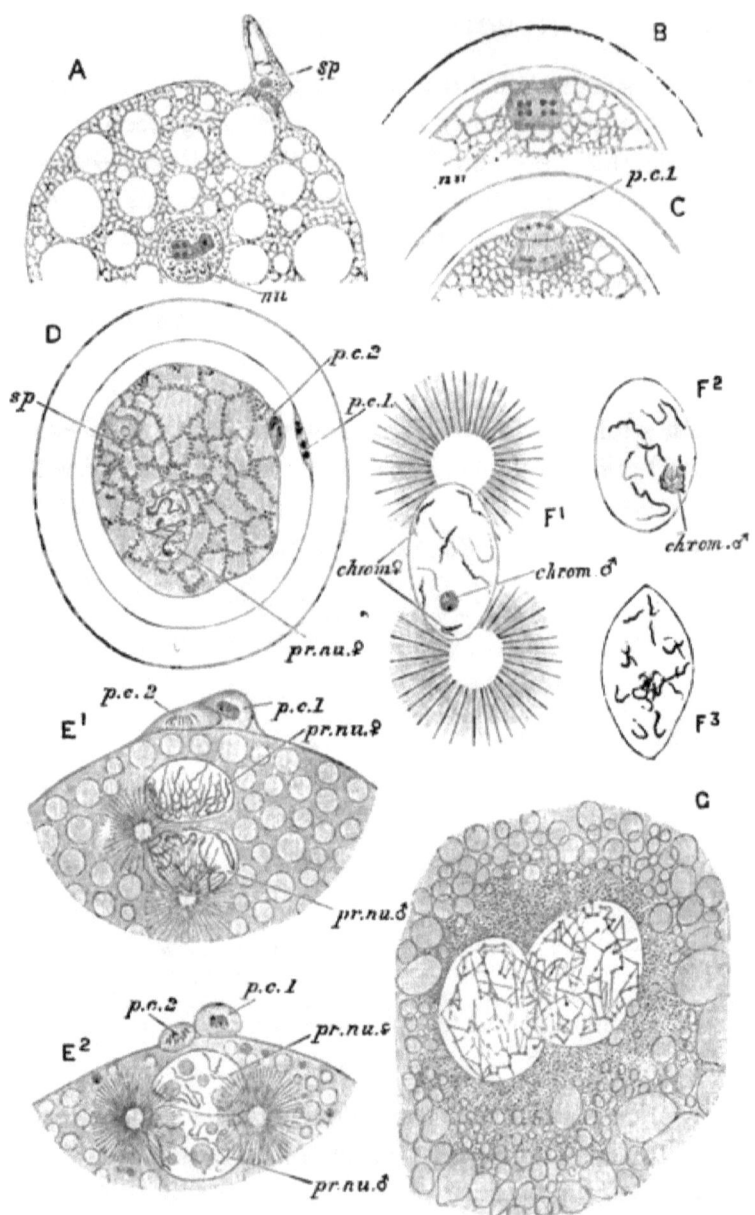

FIG. 62.—The Maturation and Impregnation of the Animal Ovum. A, portion of the ovum of a Round worm (*Ascaris megalocephala*), showing the sperm (*sp*) in the act of conjugation, and the unaltered

nucleus (*nu*) of the egg, Ascaris being an animal in which the conjugation of ovum and sperm takes place before the maturation of the former. In the nucleus, the nuclear membrane and matrix, and a band-like mass of chromatin are visible. The sperm of Ascaris is of peculiar form, and is non-motile.

B, the same at the commencement of maturation : the nucleus (*nu*) has travelled to the periphery of the egg and taken on the spindle form. In this and the two next figures the vitelline membrane is shown.

C, formation of the first polar cell (*p. c.* 1).

D, the entire egg after the completion of maturation, showing the two polar cells, the first (*p. c.* 1) adhering to the vitelline membrane, the second, (*p. c.* 2) to the surface of the protoplasm : the female pronucleus (*pr. nu.* ♀): and the sperm (*sp*), which has penetrated into the cell-protoplasm, but has not yet become converted into the male pronucleus.

E^1, E^2, two stages in the conjugation of the pronuclei in Molluscs (E^1, *Pterotrachea*, E^2, *Phyllirhoë*).

In E^1 the male (*pr. nu.* ♂) and female (*pr. nu.* ♀) pronuclei are separated : in E^2 they are applied by their flattened adjacent faces : in connection with each the cell-protoplasm has a radiating arrangement around one of the directive spheres ; the polar cells (*p.c.*1, *p.c.* 2) are shown.

F^1–F^3, three stages in the development of the nucleus of the oosperm in a Sea-urchin (*Echinus microtuberculatus*) : in F^1 the nucleus contains nine chromatin-fibres (*chrom.* ♀) derived from the female pronucleus, and a globular mass of the same (*chrom,* ♂) derived from the male pronucleus : the two directive spheres are now situated one at each end of the nucleus. In F^2 the male chromatin (*chrom.* ♂) has begun to unwind itself : in F there is no longer any distinction between male and female elements, the nucleus containing eighteen similar chromatin-threads.

G, central portion of the egg of a Hermit-Crab (*Eupagurus prideauxii*), showing the conjugation of the pronuclei. The male and female chromatin-networks appear to be fused along the plane of union. The pronuclei are surrounded by finely-granular protoplasm devoid of yolk-spheres.

(A–F after Boveri ; G after Weismann and Ischikawa.)

form of an ordinary nuclear spindle (B, *nu*, see p. 65). Next, the protoplasm grows out into a small projection or bud, into which one end of the spindle projects (c). The usual process of nuclear division then takes place (Fig. 10, p. 64), one of the daughter nuclei remaining in the bud, the other in the ovum itself. Nuclear division is followed as usual by division of the protoplasm, and the bud becomes separated as a small cell distinguished as the *first polar cell* (C—E *p.c.* 1).

It was mentioned in a previous lesson (p. 200) that in some cases development from an unfertilized female gamete took place, the process—which is not uncommon among insects and crustaceans—being distinguished as parthenogenesis. It has been proved in many instances and may be generally true that in such cases the egg begins to develop after the formation of the first polar cell. Thus in parthenogenetic ova it appears that maturation is completed by the separation of a single polar cell.

In the majority of animals, however, development takes place only after fertilization, and in such cases maturation is not complete until a *second polar cell* (D and E, *p.c.* 2) has been formed in the same manner as the first. The ovum has now lost a portion of its protoplasm together with three-fourths of its chromatin, half having passed into the first polar cell and half of what remained into the second: the remaining one-fourth of the chromatin takes on a rounded form and is distinguished as the *female pronucleus* (D, *pr. nu.* ♀).

The formation of both polar cells takes place by a reducing division, so that, while the immature ovum contains double the number of chromosomes found in the ordinary cells of the species, the mature ovum, like the sperm, contains only one-half the normal number.

In some animals the first polar body has been found to divide after separating from the egg. In such cases the egg-mother-cell or immature ovum gives rise to a group of four cells—the mature ovum and three polar-cells; just as the sperm-mother-cell gives rise to a group of four cells, all of which, however, become sperms.

Shortly after, or in some cases before maturation the

ovum is fertilized by the conjugation with it of a single sperm. As we have found repeatedly, sperms are produced in vastly greater numbers than ova, and it often happens that a single egg is seen quite surrounded with sperms, all apparently about to conjugate with it. It has however been found to be a general rule that only one of these actually conjugates: the others, like the drones in a hive, perish without fulfilling the one function they are fitted to perform.

The successful sperm (A, *sp*) takes up a position at right angles to the surface of the egg and gradually works its way through the vitelline membrane until its head lies within the egg protoplasm (D, *sp*). The tail is then cast off, and the head, penetrating deeper into the protoplasm, takes on the form of a rounded nucleus-like body, the *male pronucleus* (E^1, *pr. nu.* ♂).

The two pronuclei, each accompanied by its directive sphere and centrosome, approach one another (E^1, E^2) and finally unite to form the single nucleus (F^1—F^3) of what is now not the *ovum* but the *oosperm*—the impregnated egg or unicellular embryo. The fertilizing process is thus seen to consist of the union of two nuclear bodies, one contributed by the male gamete or sperm, the other by the female gamete or ovum. It follows from this that the essential nuclear matter or chromatin of the oosperm is derived in equal proportions from each of the two parents.

Moreover, as both male and female pronuclei contain only half the number of chromosomes found in the ordinary cells of the species, the union of the pronuclei results in the restoration of the normal number to the oosperm.

There is reason for thinking that the directive spheres of the sperm and ovum as well as their nuclei unite with one another: in this way the directive sphere of the oosperm is derived, like its nucleus, in equal proportions from the two parents.

Fertilization being thus effected, the process of segmentation or division of the oosperm takes place as described in the preceding lesson (p. 248).

In concluding the present lesson, we shall consider briefly a point which has probably already struck the reader. Among the plant-forms which have come under our notice there has been a very complete series of gradations from the simple cell, through the branched cell, linear aggregate, and superficial aggregate, to the solid aggregate, whilst among the animals already discussed there has so far been no attempt to fill up the very considerable gap between the unicellular Infusoria and Hydra, which is not only a solid aggregate, but has its cells arranged in two definite layers enclosing a digestive cavity.

When we say that no attempt has been made to fill up this gap, we mean as far as adult forms are concerned. If the reader will turn to the account, in the previous lesson, of the development of hydroid polypes (p. 248), he will see that the facts there described do as a matter of fact help us to see a possible connection between unicellular animals and multicellular two-layered forms with mouth and digestive cavity. The oosperm of the hydroid (Fig. 57, A) has the essential character of an Amœba, the polyplast (E) is practically a colony of Amœbæ, and the planula (H) a similar colony in which the zooids (cells) are dimorphic, being arranged in two layers with a central digestive cavity which finally communicates with the exterior by a mouth. In hydroids the mouth is not formed until after the appearance of the tentacles, but in a large proportion of the higher animals the polyplast stage is succeeded not by a mouthless planula but by a two-layered embryo with a mouth at one end, called a *gastrula* (Fig. 63). This is a very important stage, since it exhibits in the simplest possible way the essential characteristic of a diploblastic animal—a two-layered sac with mouth (*Blp*) and stomach (*U*), the outer layer of cells (*Ekt*) being protective and sensory, the inner (*Ent*) having a digestive function. The

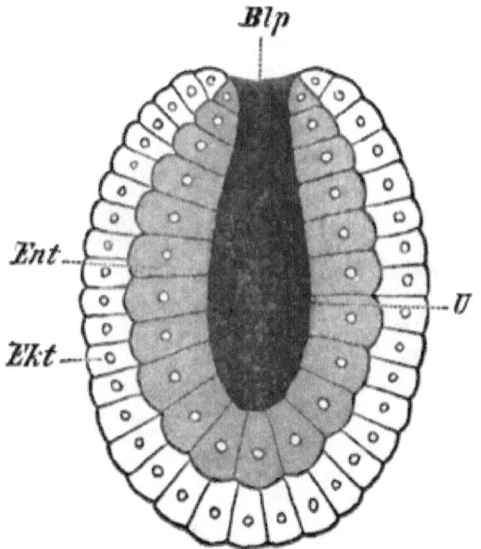

FIG. 63.—A typical animal gastrula in vertical section, showing ectoderm (*Ekt*), endoderm (*Ent*), enteron or digestive cavity (*U*), and mouth (*Blp*). (From Wiedersheim.)

planula of a hydroid may be looked upon as a gastrula in which the mouth has not yet appeared.

Another very important difference is the fact that in unicellular organisms reproduction is effected either asexually by the fission of the entire individual, or, in the case of sexual reproduction, by two entire individuals undergoing conjugation. In multicellular forms, on the other hand, single cells are set apart for sexual reproduction.

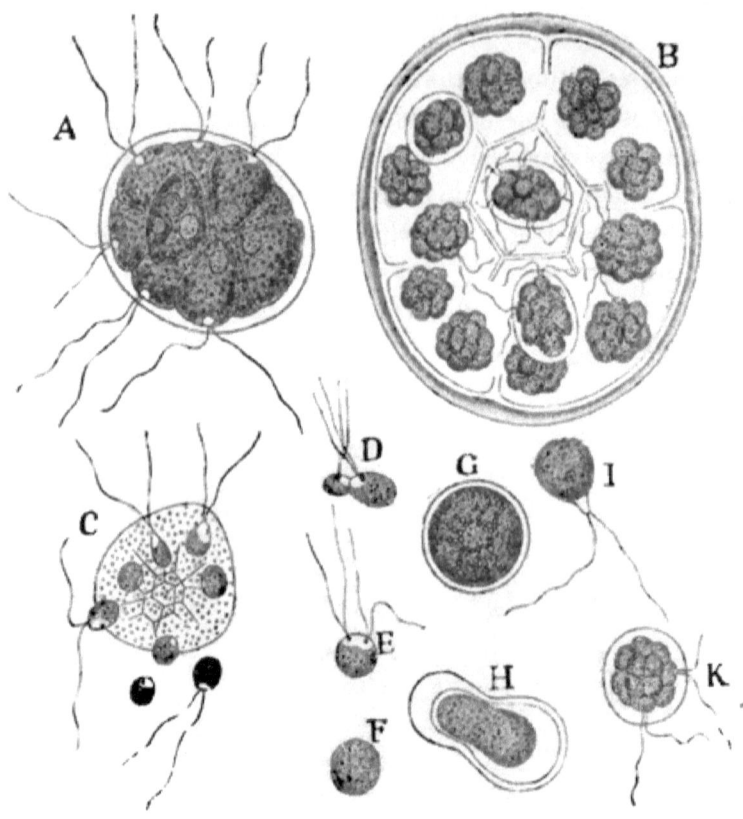

FIG. 64.—*Pandorina morum.*

A. The entire colony, consisting of sixteen flagellate zooids, enclosed in a gelatinous envelope.

B. Asexual reproduction; each zooid has divided into sixteen, forming as many daughter families, still enclosed within the original gelatinous envelope.

C. Sexual reproduction; zooids are being set free from the colony, forming gametes.

D. Conjugation of two gametes.

E. The same after complete fusion.

F. The immature zygote.

G. The fully-formed zygote.

H. Protoplasm of zygote escaping from cell-wall.

I. The same after acquisition of flagella.

K. The same undergoing division and forming a young colony. (From Goebel.)

There are several interesting organisms which help to bridge this gulf. Two of the more accessible and well-known forms will now be described.

Pandorina (Fig. 64, A) is a colony consisting of sixteen zooids closely packed in a gelatinous case of a globular form. Each zooid resembles in general characters a motile Hæmatococcus or Euglena, having an ovoid cell-body coloured green by chlorophyll, a red pigment spot, and two flagella, which protrude through the gelatinous wall of the colony, and by their action impart to it a rotatory movement.

In asexual reproduction each of the sixteen zooids divides and re-divides, forming at last a group of sixteen cells. In this way sixteen daughter colonies are produced within the gelatinous envelope of the original mother colony (B). By the solution of the envelope the daughter colonies are set free, and each begins an independent existence.

In sexual reproduction the zooids are set free singly from the colony (C). They swim about actively, approach one another in pairs, and conjugate (D), becoming completely fused together (E) to form a zygote (F). This increases in size and develops a thick cell wall (G). After a period of rest, the protoplasm escapes from the cell wall (H), puts out a pair of flagella (I), and swims about. Finally it settles down, divides and re-divides, and so gives rise to a new colony (K).

It is obvious that Pandorina resembles the polyplast stage of an embryo : moreover it is produced by the repeated fission of a flagellula, just as the polyplast is formed by the repeated fission of an oosperm.

The beautiful *Volvox* (Figs. 65 and 66), one of the favourite studies of microscopists, is a colony of Hæmatococcus-like zooids arranged in the form of a hollow sphere containing a

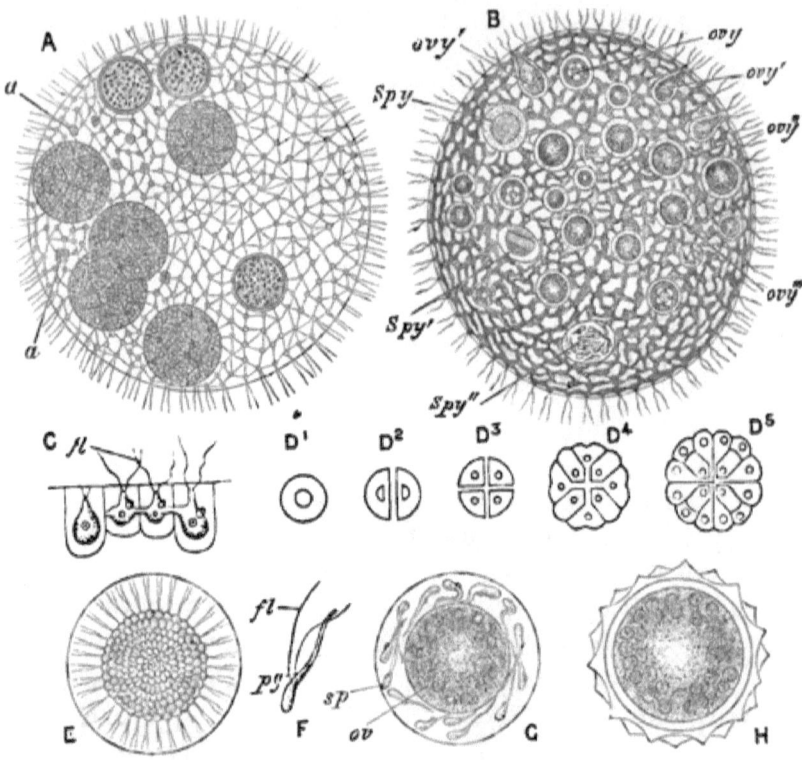

Fig. 65.—*Volvox globator.*

A, the entire colony, surface view, showing the biflagellate zooids and several daughter-colonies swimming freely in the interior; the latter are produced by the repeated fission of non-flagellate reproductive zooids (*a*).

B, the same during sexual maturity, showing spermaries from the surface (*spy*), in profile (*spy′*) and after complete formation of sperms (*spy″*): and ovaries from the surface (*ovy*, *ovy″*, *ovy‴*) and in profile (*ovy′*).

C, four zooids in optical section, showing cell-wall, nucleus, contractile vacuole, with adjacent pigment-spot, and flagella (*fl.*)

D^1-D^5, stages in the formation of a colony by the repeated binary fission of an asexual reproductive zooid.

E, a ripe spermary.

F, a single sperm, showing pigment-spot (*pg*) and flagella (*fl*).

G, an ovary containing a single ovum surrounded by several sperms.

H, oosperm enclosed in its spinose cell wall.

(A from Geddes and Thomson, after Kirchner; B–H after Cohn.)

transparent mucilage. Each cell (c) has a nucleus, a contractile vacuole, a large green chromatophore, a small red pigment-spot like that of Euglena (p. 47) and two flagella. The cells are surrounded by thick mucilaginous cell walls which do not give the reaction of cellulose, but are probably formed of an allied carbohydrate. By the combined movement of all the flagella a rotating movement is given to the entire colony.

Asexual reproduction takes place by certain of the zooids

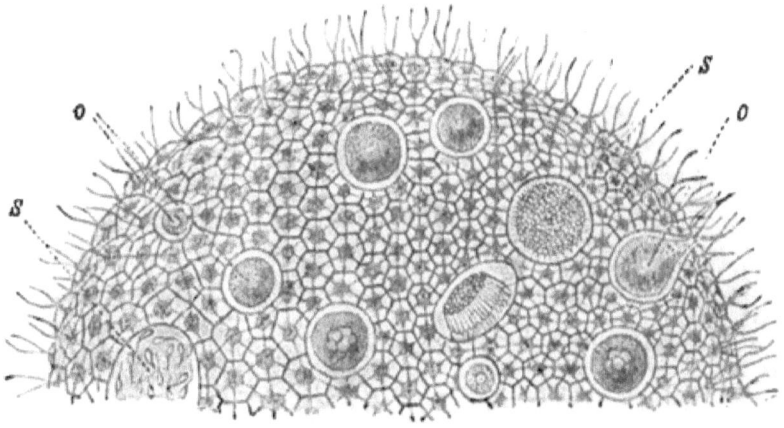

FIG. 66.

Part of a Volvox-colony showing the structure in greater detail than in Fig. 65 : s, spermaries ; o, ovaries. (After Lang.)

which are not ciliated, undergoing a process very like the segmentation of the hydroid egg (p. 248), dividing into 2, 4, 8, 16, &c. cells (A, a, and D^1—D^5), and so forming a daughter colony which becomes detached and swims freely in the interior of the parent colony (A), by the rupture of which it is finally liberated. In sexual reproduction certain cells enlarge and take on the characters of ovaries (B, ovy, ovy', ovy'', ovy''', and Fig. 66, o) the protoplasm of each forming a single ovum : the protoplasm of others divides repeatedly and forms aggregations of sperms (B, *spy*, *spy'*, *spy"*, and Fig. 66, *s*). By the conjugation of a sperm (F) with an ovum (G) an oosperm (H) is produced, and from this by continued division a new colony arises.

Volvox is clearly comparable to a hollow polyplast, and further resembles the higher or multicellular animals in that certain of its cells are differentiated to form true sexual products.

LESSON XXV

POLYGORDIUS

POLYGORDIUS is a minute worm, about 3 or 4 cm. in length, found in the European seas, where it lives in sand at a depth of a few fathoms. It has much the appearance of a tangle of pink thread with one end produced into two delicate processes (Fig. 67, A). These, which are the *tentacles*, mark the anterior end of the animal—the opposite extremity, which in some species also bears a pair of slender processes, is the posterior end. As the creature creeps along, one side is kept constantly upwards and is distinguished as the dorsal aspect ; the lower surface is called ventral.

The anterior end is narrower than the rest of the body, and is marked off behind by a groove (B and C); this division is called the *prostomium* (*Pr. st*) and bears the tentacles (*t*) already mentioned in front and above ; and on each side a small oval depression (*c. p*) lined with cilia. Immediately following the prostomium is a region clearly marked off in front, but ill-defined posteriorly, and known as the *peristomium* (*Per. st*); on its ventral surface is a transverse triangular aperture the mouth (*Mth*). The rest of the body is more or less distinctly marked by annular grooves (D and E, *gr*) into *body-segments* or *metameres*

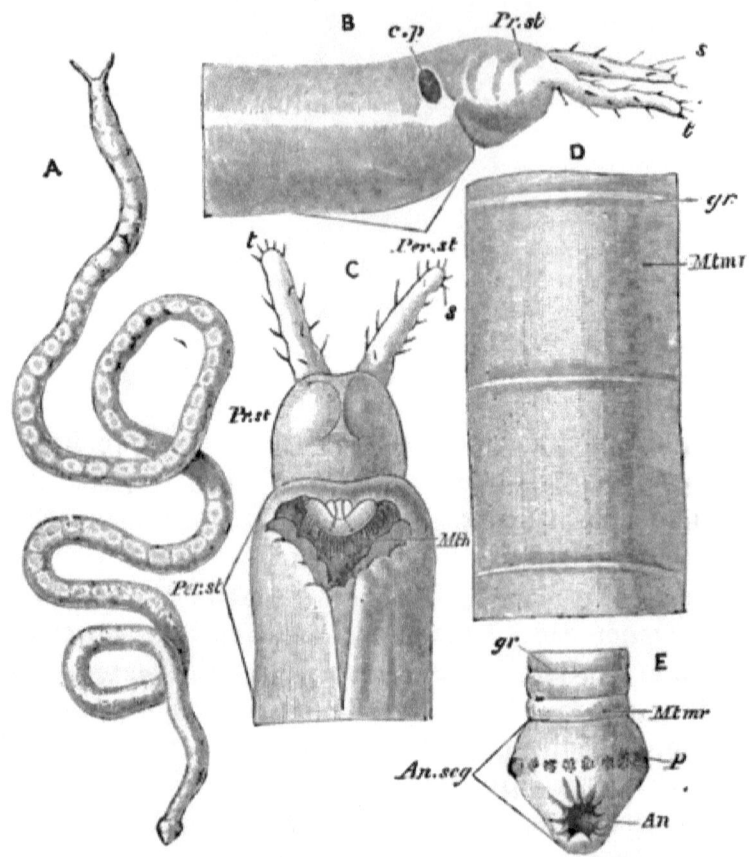

FIG. 67.—*Polygordius neapolitanus.*

A, the living animal, dorsal aspect, about five times natural size.

B, anterior end of the worm from the right side, more highly magnified, showing the prostomium (*Pr. st*), peristomium (*Per. st*), tentacles (*t*), with setæ (*s*) and ciliated pit (*c. p*).

C, ventral aspect of the same : letters as before except *Mth*, mouth.

D, portion of body showing metameres (*Mtmr*) separated by grooves (*gr*).

E, posterior extremity from the ventral aspect, showing the last three metameres (*Mtmr*) separated by distinct grooves (*gr*), the anal segment (*An. seg*) bearing the anus (*An*), and a circlet of papillæ (*p*). (After Fraipont.)

(*Mtmr*), the number of which varies considerably. Polygordius is thus the first instance we have met with of a transversely segmented animal. The last or *anal segment* (E, *An. seg*) differs from the others by its swollen form and by bearing a circlet of little prominences or papillæ (*p*); it is separated from the preceding segment by a deep groove, and bears at its posterior end a small circular aperture, the anus (*An*).

Polygordius may therefore be described as consisting of a number of more or less distinct *segments* which follow one another in longitudinal series; three of these, the *prostomium*, which lies altogether in front of the mouth, the *peristomium*, which contains the mouth, and the *anal segment*, which contains the anus, are constant; while between the peristomium and the anal segment are intercalated a variable number of *metameres* which resemble one another in all essential respects.

Polygordius feeds in much the same way as an earthworm: it takes in sand, together with the various nutrient matters contained in it, such as infusoria, diatoms, &c., by the mouth, and after retaining it for a longer or shorter time in the body, expels it by the anus. It is obvious, therefore, that there must be some kind of digestive cavity into which the food passes by the mouth, and from which effete matters are expelled through the anus. Sections (Fig. 68) show that this cavity is not a mere space excavated in the interior of the body, but a definite tube, the *enteric canal* (A, B), which passes in a straight line from mouth to anus, and is separated in its whole extent from the walls of the body (A, *B. IV.*) by a wide space, the *body cavity* or *cœlome* (*cœl*). So that the general structure of Polygordius might be imitated by taking a wide tube, stopping the ends of it with corks, boring a hole in each cork, and then inserting through

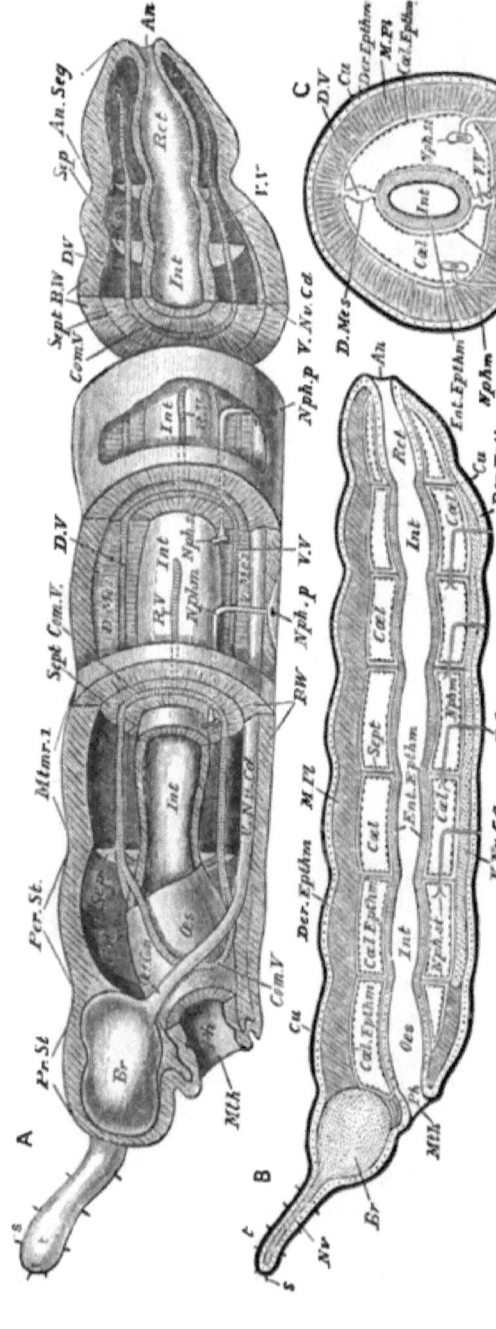

FIG. 68.[1]—A, semi-diagrammatic figure of the anatomy of Polygordius in the form of a dissection from the left side. The prostomium (*Pr. st*), peristomium (*Per. st*), and first three metameres (*Mtmr.* I, &c.) are shown to the left, the last two metameres and the anal segment (*An. seg*) to the right; the remaining metameres are supposed to be removed; the right tentacle (*t*) is shown with its setae (*s*).

The body-wall (*B. W.*) is shown mainly in longitudinal section, but in the anterior metameres part of it is seen from the surface and part in transverse section.

The mouth (*Mth*) leads into the enteric canal, which is somewhat dilated in each segment, and ends posteriorly in the anus (*An*); it is shown partly entire, partly in section: *Ph*, pharynx; *Oes*, oesophagus; *Int*, intestine; *Rct*, rectum.

[1] The diagrammatic figures in this and the following illustrations to Polygordius are founded upon Fraipont's figures, although not actually copied from them.

Between the enteric canal and the body-wall is the cœlome (*Cœl*), divided into right and left portions by the dorsal (*D. Mes*) and ventral (*V. Mes*) mesenteries, and into segmental compartments by the septa (*Sept*).

Lying in the mesenteries are the dorsal (*D. V*) and ventral (*V. V*) blood-vessels, connected by commissural vessels (*Com. V*) running in the septa; from the latter go off recurrent vessels (*R. V*)

Nephridia (*Nphm*) are shown in the second and third metameres, each consisting of a horizontal portion which perforates a septum and opens in the preceding segment by a nephrostome (*Nph. st*), and of a vertical portion which perforates the body-wall and opens externally by a nephridiopore (*Nph. p*).

The brain (*Br*) lies in the prostomium and is connected with the ventral nerve-cord (*V. Nv. Cd*) by a pair of œsophageal connectives (*Œs. Com*).

B, diagrammatic longitudinal section showing the cell-layers.

The cuticle is represented by a black line, the ectoderm is dotted the endoderm radially striated, the muscle-plates evenly shaded, the cœlomic epithelium represented by a beaded line, and the nervous system finely dotted.

The body-wall is composed of cuticle (*Cu*), deric epithelium (*Der. Epthm*), muscle-plates (*M. Pl*), and parietal layer of cœlomic epithelium (*Cœl. Epthm*).

The enteric canal is formed of enteric epithelium (*Ent. Epthm*) covered by the visceral layer of cœlomic epithelium (*Cœl. Epthm'*); in the neighbourhood of the mouth (*Mth*) and anus (*An*) the enteric epithelium is ectodermal, elsewhere it is endodermal; *Ph*, pharynx; *Oes*, oesophagus; *Int*, intestine; *Rct*, rectum.

The septa (*Sept*) are formed of muscle covered on both sides by cœlomic epithelium.

Four nephridia (*Nphm*) with nephrostome (*Nph. st*) and nephridiopore (*Nph. p*) are shown.

The brain (*Br*) and ventral nerve cord (*V. Nv. Cd*) are seen to be in contact with the ectoderm: from the brain a nerve (*nv*) passes to the tentacle.

C, diagrammatic transverse section showing the cell-layers as in B, viz: the cuticle (*Cu*), deric epithelium (*Der. Epthm*), muscle-plates (*M. Pl.*), and parietal layer of cœlomic epithelium (*Cœl. Epthm*), forming the body-wall; and the enteric epithelium (*Ent. Epthm*) and visceral layer of cœlomic epithelium (*Cœl. Epthm'*), forming the enteric canal.

The dorsal (*D. Mes*) and ventral (*V. Mes*) mesenteries are seen to be formed of a double layer of cœlomic epithelium, and to enclose respectively the dorsal (*D. V*) and ventral (*V. V*) blood-vessels.

A nephridium (*Nphm*) is shown on each side with nephrostome (*Nph. st*) and nephridiopore *Nph. p*).

The connection of the ventral nerve-cord with the ectoderm (deric epithelium) is well shown.

Fig. 71, A (p. 294), should be compared with this figure, as it is an accurate representation of the parts here shown diagrammatically.

the holes a narrow tube of the same length as the wide one. The outer tube would represent the body-wall, the inner the enteric canal, and the cylindrical space between the two the cœlome. The inner tube would communicate with the exterior by each of its ends, representing respectively mouth and anus; the space between the two tubes, on the other hand, would have no communication with the outside.

Polygordius is the first example we have studied of a *cœlomate* animal: one in which there is a definite body-cavity separating from one another the body-wall and the enteric canal, and in which therefore a transverse section of the body has the general character of two concentric circles (Fig 68, c).

It will be remembered that a transverse section of Hydra has the character of two concentric circles, formed respectively of ectoderm and endoderm (Fig. 55, A', p. 242), the two layers being, however, in contact or only separated by the thin mesoglœa. At first sight then, it seems as if we might compare Polygordius to a Hydra in which the ectoderm and endoderm instead of being in contact were separated by a wide interval; we should then compare the body-wall of Polygordius with the ectoderm of Hydra and its enteric canal with the endoderm. But this comparison would only express part of the truth.

A thin transverse section shows the body-wall of Polygordius to consist of four distinct layers. Outside is a thin transparent cuticle (Fig. 68, c, and Fig. 71, A, *cu*) showing no structure beyond a delicate striation. Next comes a layer of epithelial cells (*Der. Epthm*), their long axes at right angles to the surface of the body, and the boundaries between them very indistinct, so as to give the whole layer the character of a sheet of protoplasm with regularly disposed nuclei: this is the *deric epithelium* or *epidermis*. Within it is a rather thick layer of *muscle-plates* (*M. Pl.*), having the form of long flat spindles (Fig. 70, p. 287, *M. Pl.*) exhibiting a delicate longitudinal striation and covered on their free services with a fine network of protoplasm containing scattered nuclei. Each plate is arranged longitudinally, extending through several segments, and with its short axis perpendicular to the surface of the body (Fig. 71, *M. Pl.*). It is by the contraction of the muscle-plates that the movements of the body, which resemble those of an earthworm, are produced. Finally, within the muscular layer and lining the cœlome is a very thin layer of cells, the *cœlomic epithelium* (*Cæl. Epthm*).

A transverse section of the enteric canal shows only two layers. The inner consists of elongated cells (*Ent. Epthm*) fringed on their inner or free surfaces with cilia: these constitute the *enteric epithelium*. Outside these is a very thin layer of flattened cells (*Cæl. Epthm'*) bounding the cœlome, and hence called, like the innermost layer of the body-wall, cœlomic epithelium. We have, therefore, to distinguish two layers of cœlomic epithelium, an outer or *parietal layer* (*Cæl. Epthm.*) which lines the body-wall, and an inner or *visceral layer* (*Cæl. Epthm'*) which invests the enteric canal.

We are now in a better position to compare the transverse section of Hydra and of Polygordius (Fig. 55, A', and Fig. 68, c). The deric epithelium of Polygordius being the outermost cell-layer is to be compared with the ectoderm of Hydra, and its cuticle with the layer of the same name which, though absent in Hydra, is present in the stem of hydroid polypes such as Bougainvillea (p. 239). The enteric epithelium of Polygordius, bounding as it does the digestive cavity, is clearly comparable with the endoderm of Hydra. So that we have the layer of muscle-plates and the two layers of cœlomic epithelium not represented in Hydra, in which their position is occupied merely by the mesoglœa.

But it will be remembered that in Medusæ there is sometimes found a layer of separate muscle-fibres between the ectoderm and the mesogloea, and it was pointed out (p. 244) that such fibres represented a rudimentary intermediate cell-layer or mesoderm. We may therefore consider the muscular layer and the cœlomic epithelium of Polygordius as mesoderm, and we may say that in this animal the mesoderm is divisible into an outer or *somatic layer*, consisting of the muscle-plates and the parietal layer of cœlomic epithelium, and an inner or *splanchnic layer*, consisting of the visceral layer of cœlomic epithelium.[1]

The somatic layer is in contact with the ectoderm or deric epithelium, and with it forms the body-wall; the splanchnic layer is in contact with the endoderm or enteric epithelium and with it forms the enteric canal. The cœlome separates the somatic and splanchnic layers from one another, and is lined throughout by cœlomic epithelium.

The relation between the diploblastic polype and the triploblastic worm may therefore be expressed in a tabular form as follows—

Hydroid.		*Polygordius.*
Cuticle		Cuticle.
Ectoderm		Deric epithelium or epidermis.
Mesoderm (rudimentary)	Somatic layer	Muscle-plates. Cœlomic epithelium (parietal layer).
	Splanchnic layer	Cœlomic epithelium (visceral layer).
Endoderm		Enteric epithelium.

[1] In the majority of the higher animals there is a layer of muscle between the enteric and cœlomic epithelia: in such cases the body-wall and enteric canal consist of the same layers but in reverse order, the cœlomic epithelium being internal in the one, external in the other.

Strictly speaking, this comparison does not hold good of the anterior and posterior ends of the worm : at both mouth and anus the deric passes insensibly into the enteric epithelium, and the study of development shows (p. 298) that the cells lining both the anterior and posterior ends of the canal are, as indicated in the diagram (Fig. 68, B), ectodermal. For this reason the terms deric and enteric epithelium are not mere synonyms of ectoderm and endoderm respectively.

It is important that the student should, before reading further, understand clearly the general composition of a triploblastic animal as typified by Polygordius, which may be summarised as follows : It consists of two tubes formed of epithelial cells, one within and parallel to the other, the two being continuous at either end of the body where the inner tube (enteric epithelium) is in free communication with the exterior; the outer tube (deric epithelium) is lined by a layer of muscle-plates within which is a thin layer of cœlomic epithelium, the three together forming the body-wall; the inner tube (enteric epithelium) is covered externally by a layer of cœlomic epithelium which forms with it the enteric canal; lastly, the body-wall and enteric canal are separated by a considerable space, the cœlome.

The enteric canal is not, as might be supposed from the foregoing description, connected with the body-wall only at the mouth and anus, but is supported in a peculiar and somewhat complicated way. In the first place there are thin vertical plates, the *dorsal* and *ventral mesenteries* (Fig. 68, A and C, *D. Mes, V. Mes*), which extend longitudinally from the dorsal and ventral surfaces of the canal to the body wall, dividing the cœlome into right and left halves. The structure of the mesenteries is seen in a transverse section (Fig. 68, C, and Fig. 71, A) which shows that at the middle dorsal line the parietal layer of cœlomic epithelium becomes deflected downwards, forming a two-layered membrane, the dorsal mesentery; the two layers of this on reaching the enteric canal diverge and pass one on either side of it, forming the visceral layer of cœlomic epithelium; uniting again below the canal, they are continued downwards as the ventral mesentery, and on reaching the body-wall diverge once more to join the parietal layer. Thus the mesenteries are simply formed of a double layer of cœlomic epithelium, continuous on the one hand with the parietal and on the other with the visceral layer of that membrane.

Besides the mesenteries, the canal is supported by transverse vertical partitions or *septa* (Fig. 68, A and B, *Sept*) which extend right across the body-cavity, each being perforated by the canal. The septa are regularly arranged and correspond with the external grooves by which the body is divided into metameres. Thus the transverse or metameric segmentation affects the cœlome as well as the body-wall. Each septum is composed of a sheet of muscle covered on both sides with cœlomic epithelium (B, *Sept*).

Where the septa come in contact with the enteric canal, the latter is more or less definitely constricted so as to present a beaded appearance (A and B); thus we have segmentation of the canal as well as of the body-wall and cœlome.

The digestive canal, moreover, is not a simple tube of even calibre throughout, but is divisible into four portions. The first or *pharynx* (*Ph*) is very short, and can be protruded during feeding; the second, called the gullet or *œsophagus* (*Oes*), is confined to the peristomium and is distinguished by its thick walls and comparatively great diameter; the third or *intestine* (*Int*) extends from the first metamere to the last—*i.e.*, from the segment immediately following the peristomium to that immediately preceding the anal segment; it is laterally compressed so as to have an elongated form in cross section (C, and Fig. 71, A): the fourth portion or *rectum* (*Rct*) is confined to the anal segment; it is somewhat dilated and is not laterally compressed. The epithelium of the intestine is, as indicated in the diagram (B), endodermal; that of the remaining division of the canal is ectodermal. The large majority of the cells in all parts of the canal are ciliated.

The cells of the enteric canal and especially those of the gullet are very granular, and like the endoderm cells of the hypostome of Hydra (p. 231) are to be considered as gland cells. They doubtless secrete a digestive juice which, mixing with the various substances taken in by the mouth, dissolves the proteids and other digestible parts, so as to allow of their absorption. There is no evidence of intracellular digestion such as occurs in Hydra (p. 232), and it is very probable that the process is purely extra-cellular or enteric, the food being dissolved and rendered diffusible entirely in the cavity of the canal. By the movements of the canal—caused partly by the general movements of the body and partly by the contraction of the muscles of the septa—aided by the action of the cilia, the contents are gradually forced backwards and the sand and other indigestible matters are expelled at the anus.

The cœlome is filled with a colourless transparent *cœlomic fluid* in which are suspended minute, irregular, colourless bodies, as well as oval bodies containing yellow granules. From the analogy of the higher animals one would expect these to be leucocytes (p. 56), but their cellular nature has not been proved.

The function of the cœlomic fluid is probably to distribute the digested food in the enteric canal to all parts of the body. In Hydra, where the lining wall of the digestive cavity is in direct contact with the simple wall of the body, the products of digestion can pass at once by diffusion from endoderm to ectoderm, but in the present case a means of communication is wanted between the enteric epithelium and the comparatively complex and distant body-wall. The peptones and other products of digestion diffuse through the enteric epithelium into the cœlomic fluid, and by the continual movement of the latter—due to the contractions of the body-wall—are distributed to all parts. Thus the external epithelium and the muscles, as well as the nervous system and reproductive organs, not yet described, are wholly dependent upon the enteric epithelium for their supply of nutriment.

We have now to deal with structures which we find for the first time in Polygordius, namely blood-vessels. Lying in the thickness of the dorsal mesentery is a delicate tube (Fig. 68, A and C, *D.V*) passing along almost the whole length of the body: this is the *dorsal vessel*. A similar *ventral vessel* (V.V) is contained in the ventral mesentery,[1] and the two are placed in communication with one another in every segment by a pair of *commissural vessels* (A, *Com.v*) which spring right and left from the dorsal trunk, pass downwards in or close behind the corresponding septum, following the contour of body-wall, and finally open into the ventral vessel. Each commissural vessel, at about the middle of its length, gives off a *recurrent vessel* (R.V.) which passes backwards and

[1] The statement that the dorsal and ventral vessels lie in the thickness of the mesenteries requires qualification. As a matter of fact, these vessels are simply spaces formed by the divergence of the two layers of epithelium composing the mesentery (Fig. 68, C, and Fig. 71, A): only their anterior ends have proper walls.

ends blindly. The anterior parts of the commissural vessels lie in the peristomium and have an oblique direction, one on each side of the gullet. The whole of these vessels form a single, closed vascular system, there being no communication between them and any of the remaining cavities of the body.

The vascular system contains a fluid, the *blood*, which varies in colour in the different species of Polygordius, being either colourless, red, green, or yellow. In one species corpuscles (? leucocytes) have been found in it.

The function of the blood has not been actually proved in Polygordius, but is well known in other worms. In the common earthworm, for instance, the blood is red, the colour being due to the same pigment, *hæmoglobin*, which occurs in our own blood and in that of other vertebrate animals.

Hæmoglobin is a nitrogenous compound, containing, in addition to carbon, hydrogen, nitrogen, oxygen, and sulphur, a minute quantity of iron. It can be obtained pure in the form of crystals which are soluble in water. Its most striking and physiologically its most important property is its power of entering into a loose chemical combination with oxygen. If a solution of hæmoglobin is brought into contact with oxygen it acquires a bright scarlet colour, and the solution is then found to have a characteristic spectrum distinguished by two absorption-bands, one in the yellow, another in the green. Loss of oxygen changes the colour from scarlet to purple, and the spectrum then presents a single broad absorption-band intermediate in position between the two of the oxygenated solution.

This property is of use in the following way. All parts of the organism are constantly undergoing destructive metabolism and giving off carbon dioxide : this gas is absorbed by the blood, and at the same time the hæmoglobin gives up its oxygen to the tissues. On the other hand, whenever the blood is brought sufficiently near the external air—or water in the case of an aquatic animal—the opposite process takes place, oxygen being absorbed and carbon dioxide given off. Hæmoglobin is therefore to be looked upon as a respiratory or oxygen-carrying pigment; its function is to provide the various parts of the body with a constant supply of oxygen, while the carbon dioxide formed by their oxidation is given up to the blood. The particular part of the body in which the carbon dioxide accumulated in the blood is exchanged for the oxygen of the surrounding medium is called a respiratory organ; in Polygordius, as in the earthworm and many others of the lower animals, there is no specialised respiratory organ—lung or gill—but the necessary exchange of gases is performed by the entire surface of the body.

In discussing in a previous lesson the differences between plants and animals, we found (p. 178) that in the unicellular organisms previously studied, the presence of an excretory organ in the form of a contractile vacuole was a characteristic feature of such undoubted animals as the ciliate infusoria, but was absent in such undoubted plants as Vaucheria and Mucor. But the reader will have noticed that Hydra and its allies have no specialised excretory organ, waste products being apparently discharged from any part of the surface. In Polygordius we meet once more with an animal in which excretory organs are present, although, in correspondence with the complexity of the animal itself, they are very different from the simple contractile vacuoles of Paramœcium or Vorticella.

The excretory organs of Polygordius consist of little tubes called *nephridia*, of which each metamere possesses a pair, one on either side (Fig. 68, A, B, and C, *Nphm*). Each nephridium (Fig. 69) is an extremely delicate tube consisting of two divisions bent at right angles. The outer division is placed vertically, lies in the thickness of the body-wall, and opens externally by a minute aperture, the *nephridiopore* (Figs. 68 and 69, *Nph. p*). The inner division is horizontal and lies in the cœlomic epithelium; passing forward it pierces the septum which bounds the segment in front (Fig. 68, A and B), and then dilates into a funnel-shaped extremity or *nephrostome* (*Nph. st*), which places its cavity in free communication with the cœlome. The whole interior of the tube as well as the inner face of the nephrostome is lined with cilia which work outwards.

FIG. 69.—A nephridium of Polygordius, showing the cilia lining the tube, the ciliated funnel or nephrostome (*Nph. st*), and the external aperture or nephridiopore (*Nph. p*). (After Fraipont.)

A nephridium may therefore be defined as a ciliated tube, lying in the thickness of the body-wall and opening at one end into the cœlome and at the other on the exterior of the body.

In the higher worms, such as the earthworm, the nephridia are lined in part by gland-cells, and are abundantly supplied with blood-vessels. Water and nitrogenous waste from all parts of the body pass by diffusion into the blood and are conveyed to the nephridia, the gland-cells of which withdraw the waste products and pass them into the cavities of the tubes, whence they are finally discharged into the surrounding medium. In all probability some such process as this takes place in Polygordius.

In discussing the hydroid polypes we found that one of the most important points of difference between the locomotive medusa and the fixed hydranth was the presence in the former of a well-developed nervous system (p. 244) consisting of an arrangement of peculiarly modified cells, to which the function of automatism was assigned. It is natural to expect in such an active and otherwise highly-organized animal as Polygordius a nervous system of a considerably higher degree of complexity than that of a medusa.

The central nervous system consists of two parts, the *brain* and the *ventral nerve-cord*. The brain (Fig. 68, A and B, *Br.*) is a rounded mass occupying the whole interior of the prostomium and divided by a transverse groove into two lobes, the anterior of which is again marked by a longitudinal groove. The ventral nerve-cord (*V. Nv. Cd.*) is a longitudinal band extending along the whole middle ventral line of the body from the peristomium to the anal segment. The posterior lobe of the brain is connected with the anterior end of the ventral nerve-cord by a pair of nervous bands, the *œsophageal connectives* (*Œs. Con.*) which pass respectively right and left of the gullet.

It is to be noted that one division of the central nervous system—the brain—lies altogether above and in front of the enteric canal, the other division—the ventral nerve-cord—altogether beneath it, and that, in virtue of the union of the two divisions by the œsophageal connectives, the enteric canal perforates the nervous system.

It is also important to notice that the nervous system is throughout in direct contact with the epidermis or ectoderm, the ventral cord appearing in sections (Fig. 68, C, and Fig. 71, A) as a mere thickening of the latter.

Both brain and cord are composed of delicate nerve-fibres

(Fig. 70, *Nv. F.*) interspersed with nerve-cells (*Nv. C*). In the cord the fibres are arranged longitudinally, and the nerve-cells are ventral in position, forming a layer in imme-

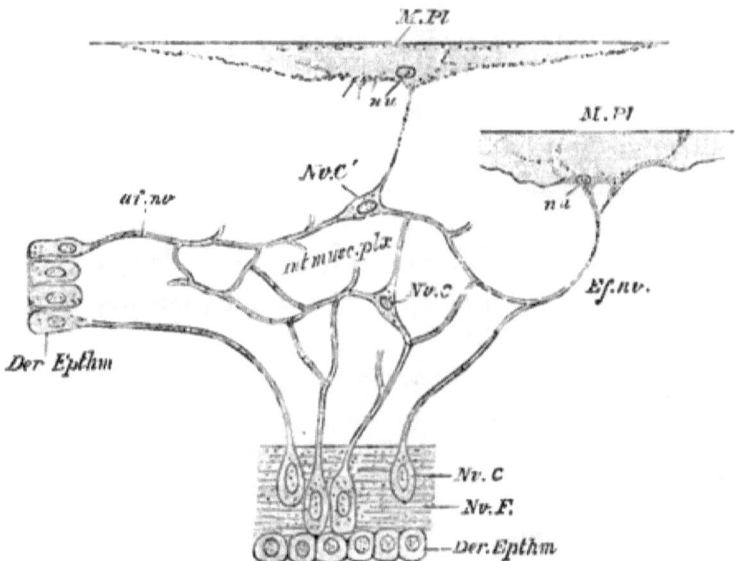

Fig. 70.—Diagram illustrating the relations of the nervous system of Polygordius.

The deric epithelium (*Der. Epthm*) is either indirect contact with the central nervous system (lower part of figure), or is connected by afferent nerves (*af. nv.*) with the inter-muscular plexus (*int. musc. plex.*): the latter is connected to the muscle-plates (*M. Pl*) by efferent nerves (*Ef. nv*).

The central nervous system consists of nerve-fibres (*Nv. F*) and nerve-cells (*Nv. C*): other nerve-cells (*Nv. C'*) occur at intervals in the inter-muscular plexus.

The muscle-plates (*M. Pl*), one of which is entire, while only the middle part of the other is shown, are invested by a delicate protoplasmic network, containing nuclei (*nu*), to which the efferent nerves can be traced. (The details copied from Fraipont.)

diate contact with the deric epithelium. In the posterior lobe of the brain the nerve-cells are superficial and the central part of the organ is formed of a finely punctate substance in which neither cells nor fibres can be made out.

Ramifying through the entire muscular layer of the body-wall is a network of delicate nerve-fibres (*int. musc. plx.*) with nerve-cells (*Nv. C*) at intervals, the *inter-muscular plexus*. Some of the branches of this plexus are traceable to nerve-cells in the central nervous system, others (*af. nv.*) to epidermic cells, others (*Ef. nv.*) to the delicate protoplasmic layer covering the muscle-plates. The superficial cells of both brain and cord are also, as has been said, in direct connection with the overlying epidermis, and from the anterior end of the brain a bundle of nerve-fibres (Fig. 68, B, *t., Nv.*) is given off on each side to the corresponding tentacle, constituting the *nerve* of that organ, to the epidermic cells of which its fibres are distributed.

We see then that, apart from the direct connection of nerve-cells with the epidermis, the central nervous system is connected, through the intermediation of nerve-fibres (*a*) with the sensitive cells of the deric epithelium and (*b*) with the contractile muscle-plates. And we can thus distinguish two sets of nerve-fibres, (*a*) *sensory* or *afferent* (*af. nv.*) which connect the central nervous system with the epidermis, and (*b*) *motor* or *efferent* (*Ef. nv.*) which connect it with the muscles.

Comparing the nervous system of Polygordius with that of a medusa (p. 244) there are two chief points to be noticed. Firstly, the concentration of the central nervous system in the higher type, and the special concentration at the anterior end of the body to form a brain. Secondly, the important fact that the inter-muscular plexus is not, like the peripheral nervous system of a medusa which it resembles, situated immediately beneath the epidermis (ectoderm) but lies in the muscular layer, or, in other words, has sunk into the mesoderm.

It is obvious that direct experiments on the nervous system

would be a very difficult matter in so small an animal as Polygordius. But numerous experiments on a large number of other animals, both higher and lower, allow us to infer with considerable confidence the functions of the various parts in this particular case.

If a muscle be laid bare or removed from the body in a living animal it may be made to contract by the application of various stimuli, such as a smart tap (mechanical stimulus), a drop of acid or alkali (chemical stimulus), a hot wire (thermal stimulus), or an electric current (electric stimulus). If the motor nerve of the muscle is left intact the application to it of any of these stimuli produces the same effect as its direct application to the muscle, the stimulus being conducted along the eminently irritable but non-contractile nerve.

Further, if the motor nerve is left in connection with the central nervous system, *i.e.*, with one or more nerve-cells, direct stimulation of these is followed by a contraction, and not only so, but stimulation of a sensory nerve connected with such cells produces a similar result. And finally, stimulation of an ectoderm cell connected, either directly or through the intermediation of a sensory nerve, with the nerve-cells, is also followed by muscular contraction. An action of this kind, in which a stimulus applied to the free sensitive surface of the body is transmitted along a sensory nerve to a nerve-cell or group of such cells and is then, as it were, reflected along a motor nerve to a muscle, is called a *reflex action*; the essence of the arrangement is the interposition of nerve-cells between sensory or afferent nerves connected with sensory cells, and motor or efferent nerves connected with muscles.

The diagram (Fig. 70) serves to illustrate this matter. The muscle-plate (*M. Pl.*) may be made to contract by a stimulus applied (*a*) to itself directly, (*b*) to the motor fibre

(*Ef. nv*), (*c*) to the nerve-cells (*Nv. C*) in the central nervous system, or to those (*Nv. C'*) in the inter-muscular plexus, (*d*) to the sensory fibre (*af. nv.*), or (*e*) to the epidermic cells (*Der. Epthm.*).

In all probability the whole central nervous system of Polygordius is capable of automatic action. It is a well-known fact that if the body of an earthworm is cut into several pieces each performs independent movements; in other words, the whole body is not, as in the higher animals, paralysed by removal of the brain. There can, however, be little doubt that complete co-ordination, *i.e.*, the regulation of the various movements to a common end, is lost when the brain is removed.

The nervous system is thus an all-important means of communication between the various parts of the organism and between the organism and the external world. The outer or sensory surface is by its means brought into connection with the entire muscular system with such perfection that the slightest touch applied to one end of the body may be followed by the almost instantaneous contraction of muscles at the other.

In some species of Polygordius the prostomium bears a pair of eye-specks, but in the majority of species the adult animal is eyeless, and, save for the ciliated pits (Fig. 67, B, *c.p*), the function of which is not known, the only definite organs of sense are the tentacles, which have a tactile function, their abundant nerve-supply indicating that their delicacy as organs of touch far surpasses that of the general surface of the body. They are beset with short, fine processes of the cuticle called *setæ* (Figs. 67 and 68, *s*), which probably, like the whiskers of a cat, serve as conductors of external stimuli to the sensitive epidermic cells.

There are two matters of general importance in connection with the structure of Polygordius to which the student's attention must be drawn in concluding the present lesson.

Notice in the first place how in this type, far more than in any of those previously considered, we have certain definite parts of the body set apart as *organs* for the performance of particular functions. There is a mouth for the reception of food, an enteric canal for its digestion, and an anus for the extrusion of fæces : a cœlomic fluid for the transport of the products of digestion to the more distant parts of the body : a system of blood-vessels for the transport of oxygen to and of carbon dioxide from all parts : an epidermis as organ of touch and of respiration : nephridia for getting rid of water and nitrogenous waste : and a definite nervous system for regulating the movements of the various parts and forming a means of communication between the organism and the external world. It is clear that differentiation of structure and division of physiological labour play a far more obvious and important part than in any of the organisms hitherto studied.

Notice in the second place the vastly greater complexity of microscopic structure than in any of our former types. The adult organism can no longer be resolved into more or less obvious cells. In the deric, enteric, and cœlomic epithelia we meet with nothing new, but the muscle-plates are not cells, the nephridia show no cell-structure, neither do the nerve-fibres nor the punctate substance of the brain. The body is thus divisible into *tissues* or fabrics each clearly distinguishable from the rest. We have epithelial tissue, cuticular tissue, muscular tissue, and nervous tissue : and the blood and cœlomic fluid are to be looked upon as liquid tissues. One result of this is that, to a far greater extent that in the foregoing types, we can study the morphology of Polygordius under two distinct heads: *anatomy*, dealing with the general structure of the parts, and *histology*, dealing with their minute or microscopic structure.

One point of importance must be specially referred to in connection with certain of the tissues. It has been pointed out (p. 276) that the epidermis has rather the character of a sheet of protoplasm with regularly-arranged nuclei than of a layer of cells, and that the muscle-plates are covered with a layer of protoplasm with which the ultimate nerve-fibres are continuous (p. 277). Thus certain of the tissues of Polygordius exhibit *continuity of the protoplasm*, a phenomenon which appears to be of wide occurrence both in animals and in plants.

LESSON XXVI

Polygordius (*Continued*)

Asexual reproduction is unknown in Polygordius, and the organs of sexual reproduction are very simple. The animal is dioecious, gonads of one sex only being found in each individual.

In the species which has been most thoroughly investigated (*P. neapolitanus*) the reproductive products are formed in each metamere from the fourth to the last. Crossing these segments obliquely are narrow bands of muscle (Fig. 71, A, *O.M*) and certain of the cells of cœlomic epithelium covering these bands multiply by fission and form little heaps of cells (*Spy*), each of which is to be looked upon as a gonad. There is thus a pair of gonads to each segment with the exception of the prostomium, the peristomium, the first three metameres, and the anal segment, the reproductive organs exhibiting the same simple metameric arrangement as the digestive, excretory, and circulatory organs. It will be noticed that the primitive sex-cells, arising as they do from cœlomic epithelium, are mesodermal structures, not ectodermal as in hydroids (pp. 234 and 247).

In the male the primitive sex-cells divide and sub-divide, the ultimate products being converted into sperms (Fig. 71,

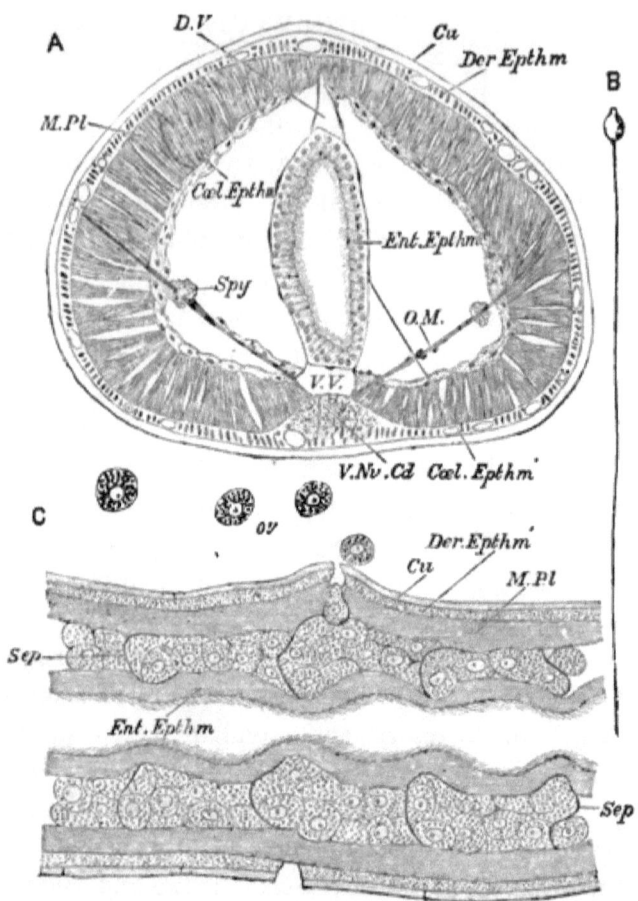

FIG. 71.—*Polygordius neapolitanus.*

A, transverse section of a male specimen to show the position of the immature gonads (*spy*) and the precise form and arrangement of the various layers represented diagrammatically in Fig. 68, c.

The body-wall consists of cuticle (*Cu*), deric epithelium (*Der. Epthm*), muscle-plates (*M. Pl*), and parietal layer of cœlomic epithelium (*Cœl. Epthm*). The ventral nerve cord (*V. Nv. Cd*) is shown to be continuous with the deric epithelium.

The enteric canal consists of ciliated enteric epithelium (*Ent. Epthm*) covered by the visceral layer of cœlomic epithelium (*Cœl. Epthm'*): connecting it with the body-wall are the dorsal and ventral mesenteries formed of a double layer of cœlomic epithelium, and containing respectively the dorsal (*D. V*) and ventral (*V. V*) blood-vessels.

Passing obliquely across the cœlome are the oblique muscles (*O. M*) covered with cœlomic epithelium : by differentiation of groups of cells of the latter the spermaries (*Spy*) are formed.

B, a single sperm, showing expanded head and delicate tail.

C, horizontal section of a sexually mature female.

The body-wall (*Cu, Der. Epthm, M. Pl*) has undergone partial histological degeneration, and is ruptured in two places to allow of the escape of the ova (*ov*) which still fill the cœlomic spaces enclosed between the body-wall, the enteric canal (*Ent. Epthm*), and the septa (*Sep*). (After Fraipont.)

B : see p. 255) : in the female they enlarge immensely, and take on the character of ova (c, *ov*). Multiplication of the sexual products takes place to such an extent that the whole cœlome becomes crammed full of either sperms or ova (c).

In the female the growth of the eggs takes place at the expense of all other parts of the body, which undergo more or less complete atrophy : the epidermis, for instance, becomes liquefied and the muscles lose their contractility. Finally rupture of the body-wall takes place in each segment (c), and through the slits thus formed the eggs escape. So that Polygordius, like an annual plant, produces only a single brood : death is the inevitable result of sexual maturity. Whether or not the same dehiscence of the body-wall takes place in the male is not certain : it has been stated that the sperms make their escape through the nephridia.

Thus while there are no specialized *gonaducts*, or tubes for carrying off the sexual products, it is possible that the nephridia may, in addition to their ordinary function, serve the purpose of male gonaducts or *spermiducts*. Female gonaducts or *oviducts* are however entirely absent.

The ova and sperms being shed into the surrounding water, impregnation takes place, and the resulting oosperm undergoes segmentation or division (see p. 248), a polyplast being formed. By the arrangement of its cells into two layers and the formation of an enteron or digestive cavity the polyplast becomes a gastrula (see p. 265) which by further development is converted into a curious free-swimming creature shown in Fig. 72, A, and called a *trochosphere*.

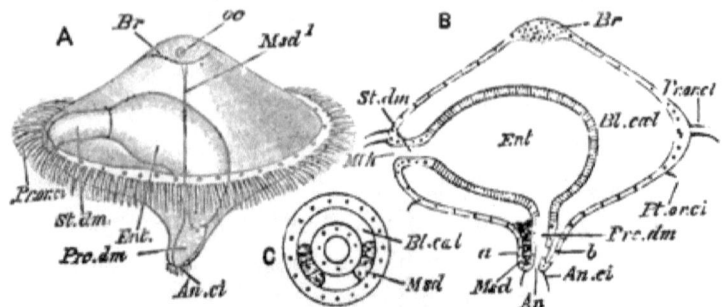

FIG. 72.—A, larva of *Polygordius neapolitanus* in the trochosphere stage ; from a living specimen.

B, diagrammatic vertical section of the same : the ectoderm is dotted, the endoderm radially striated, the mesoderm evenly shaded, and the nervous system finely dotted.

C, transverse section through the plane ab in B.

The body-wall consists of a single layer of ectoderm cells, which, at the apex of the prostomium (upper hemisphere) are modified to form the brain (*Br*) and a pair of ocelli (*oc*).

The enteric canal consists of three parts : the stomodæum (*St. dm*), opening externally by the mouth (*Mth*), and lined by ectoderm ; the enteron (*Ent*) lined by endoderm ; and the proctodæum (*Prc. dm*), opening by the anus (*An*) and lined by ectoderm.

Between the body-wall and the enteric canal is the larval body-cavity or blastocœle (*Bl. cœl*).

The mesoderm is confined to two narrow bands of cells (B and C, *Msd*) in the blastocœle, one on either side of the proctodæum ; slender mesodermal bands (*Msd¹*) are also seen in the prostomium in A.

The cilia consists of a præ-oral circlet (*Pr. or. ci*) above the mouth, a post-oral circlet (*Pt. or. ci*) below the mouth, and an anal circlet (*An. ci*) around the anus.

(A after Fraipont.)

The trochosphere, or newly-hatched larva of Polygordius (Fig. 72, A) is about $\frac{1}{4}$ mm. in diameter, and has something the form of a top, consisting of a dome-like upper portion, the *prostomium*, produced into a projecting horizontal rim ;

of an intermediate portion or *peristomium*, having the form of an inverted hemisphere; and of a lower somewhat conical *anal region*. Around the projecting rim is a double circlet of large cilia (*Pr. or. ci*) by means of which the larva is propelled through the water.

Beneath the edge of the ciliated rim is a rounded aperture, the mouth (*Mth*); this leads by a short, nearly straight gullet (*St. dm*), into a spacious stomach (*Ent*), from the lower side of which proceeds a short slightly curved intestine (*Prc. dm*), opening at the extremity of the conical inferior region by an anus (*An*). Between the body-wall and the enteric canal is a space filled with fluid (*Bl. cœl*), but, as we shall see, this does not correspond with the body-cavity of the adult. The body-wall and the enteric canal consist each of a single layer of epithelial cells, all the tissues included in the adult under the head of mesoderm (p. 278) being absent or so poorly developed that they may be neglected for the present.

Leaving aside all details, it will be seen that the trochosphere of Polygordius is comparable in the general features of its organization to a medusa (compare Fig. 55, p. 242), consisting as it does of an outer layer of cells forming the external covering of the body and of an inner layer lining the digestive cavity. There are, however, two important differences: the space between the two layers is occupied by the mesoglœa in the medusa, while in the worm it is a cavity filled with fluid; and the digestive cavity of the trochosphere has two openings instead of one.

But in order to compare more accurately the medusa with the trochosphere, it is necessary to fill up, by the help of other types, an important gap in our knowledge of the development of Polygordius—the passage from the gastrula to the trochosphere. From what we know of the development of other worms, the process, in its general features, is probably as follows:—

The ectoderm and endoderm of the gastrula (Fig. 73, A) are not in close contact with one another as in Fig. 63 (p. 265), but are separated by a space filled with fluid—the *blastocœle* or larval body-cavity. The mouth of the gastrula closes (*B*), the enteron (*Ent*), being thus converted into a shut sac. At about the same time the ectoderm is tucked

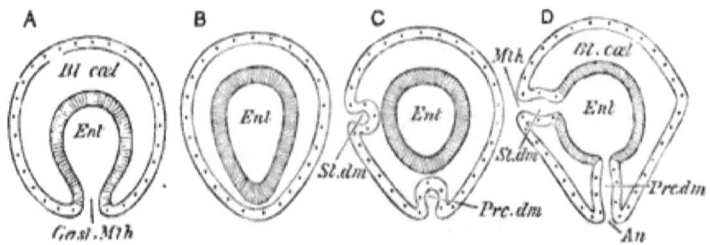

Fig. 73.—Diagram illustrating the origin of the trochosphere from the gastrula. The ectoderm is dotted, the endoderm striated.

A, gastrula, with enteron (*Ent*) and gastrula-mouth (*Gast. Mth*), and with the ectoderm and endoderm separated by the larval body-cavity or blastocœle (*Bl. cœl*).

B, the gastrula-mouth has closed, the enteron (*Ent*) becoming a shut sac.

C, two ectodermal pouches, the stomodæum (*St. dm*) and proctodæum (*Prc. dm*) have appeared.

D, the stomodæum (*St. dm*) and proctodæum (*Prc. dm*) have opened into the enteron (*Ent*), forming a complete enteric canal with mouth (*Mth*) and anus (*An*).

in or invaginated at two places (*C*), and the two little pouches (*St. dm, Prc. dm*) thus formed grow inwards until they meet with the closed enteron and finally open into it (*D*), so that a complete enteric canal is formed—formed, we must not fail to notice, of three distinct parts: (1) an anterior ectodermal pouch, opening externally by the mouth, and distinguished as the *stomodæum*; (2) the enteron, lined with endoderm; and (3) a posterior ectodermal pouch, opening externally by the anus, and called the *proctodæum*.

In the trochosphere (Fig. 72) the gullet is derived from the stomodæum, the stomach from the enteron, and the intestine from the proctodæum; so that only the stomach of the worm-larva corresponds with the digestive cavity of a medusa: the gullet and intestine are structures not represented in the latter form.

Two or three other points in the anatomy of the trochosphere must now be referred to.

At the apex of the dome-shaped prostomium the ectoderm is greatly thickened, forming a rounded patch of cells (Figs. 72 and 74, *Br*), the rudiment of the brain. On the surface of the same region and in close relation with the brain is a pair of small patches of black pigment, the eye-spots or ocelli (*Oc*).

On either side of the intestine, between its epithelium and the external ectoderm, is a row of cells forming a band which partly blocks up the blastocœle (B and C, *Msd*). These two bands are the rudiments of the whole of the mesodermal tissues of the adult—muscle, cœlomic epithelium, &c.—and hence called mesodermal bands.

Finally on either side of the lower or posterior end of the stomach is a delicate tube (Fig. 74, A, *Nph*) opening by a small aperture on to the exterior, and by a wide funnel-shaped extremity into the blastocœle: it has all the relations of a nephridium, and is distinguished as the *head-kidney*.

As the larva of Polygordius is so strikingly different from the adult, it is obvious that development must, in this, as in several cases which have come under our notice, be accompanied by a metamorphosis.

The first obvious change is the elongation of the conical anal region of the trochosphere into a tail-like portion which may be called the *trunk* (Fig. 74, A). The stomach (enteron), which was formerly confined to the pro- and peristomium, has now grown for a considerable distance into

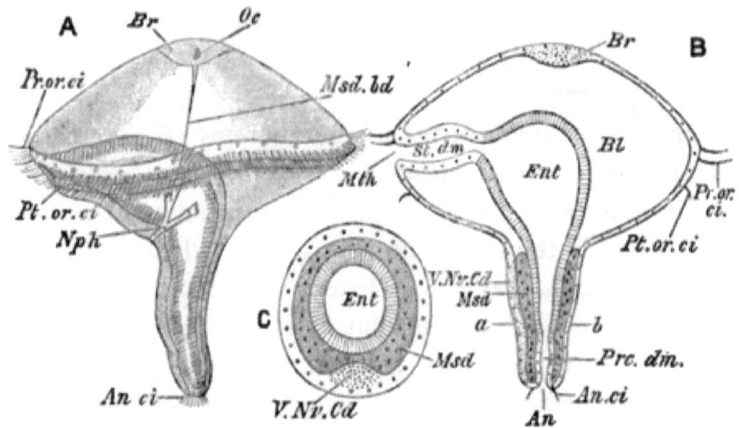

Fig. 74.—A, living specimen of an advanced trochosphere-larva of Polygordius neapolitanus, showing the elongation of the anal region to form the trunk.

B, diagrammatic vertical section of the same: the ectoderm is coarsely, the nervous system finely, dotted, the endoderm radially striated, and the mesoderm evenly shaded.

C, transverse section through the plane ab in B.

The pre-oral (*Pr. or. ci*), post-oral (*Pt. or. ci.*), and anal (*An. ci*) cilia, brain (*Br*), ocelli (*Oc*), blastocœle (*Bl.*), mouth (*Mth*), stomodæum (*St. dm*), proctodæum (*Prc. dm*), and anus (*An*) as in Fig. 72, A: the enteron (*Ent*) has extended some distance into the trunk.

In A, slender mesodermal bands (*Msd. bd*) in the prostomium, and the branched head-kidney (*Nph*) are shown.

In B and C the mesoderm (*Msd*) is seen to have obliterated the blastocœle in the trunk-region: the ectoderm has undergone a thickening, forming the ventral nerve-cord (*V. Nv. Cd*).

(A after Fraipont.)

the trunk (B, *ent*), so that the proctodæum (*Prc. dm*) occupies only the portion in proximity to the anus.

Important internal changes have also taken place. The deric epithelium or external ectoderm is for the most part composed, as in the preceding stage, of a single layer of cells; but on that aspect of the trunk which lies on the same side as the mouth—*i.e.*, to the left in Fig. 74, A and B—this layer has undergone a notable thickening, being now composed of several layers of cells. This ectodermal thickening is the rudiment of the ventral nerve-cord (*V. Nv. Cd*), and the side of the trunk on which it appears is now definitely marked out as the ventral aspect of the future worm, the opposite aspect—that to the right in the figures—being dorsal. At a later stage two ectodermal cords—the œsophageal connectives—are formed, connecting the anterior end of the ventral nerve-cord with the brain. Note that the two divisions of the central nervous system are originally quite distinct.

The mesodermal bands, which were small and quite separate in the preceding stage (Fig. 72, B and C, *Msd*), have now increased to such an extent as to surround completely the enteron and obliterate the blastocœle (Fig. 74, B and D, *Msd*). At this stage therefore there is no body-cavity in the trunk, but the space between the deric and enteric epithelia is occupied by a solid mass of mesoderm. In a word, the larva is at present, as far as the trunk is concerned, triploblastic but *acœlomate*.

Development continues, and the larva assumes the form shown in Fig. 75, A. The trunk has undergone a great increase in length and at the same time has become divided, by a series of annular grooves, into segments or metameres, like those of the adult worm but more distinct (compare Fig. 67, D, p. 272). By following the growth of the larva from the preceding to the present stage, it is seen that these segments are formed from before backwards, *i.e.*, the segment next the peristomium is the oldest, and new ones are continually being added between the last formed and the extremity of the trunk, or what may now be called the anal segment. By this process the larva has assumed the appearance of a worm with an immense head and a very slender trunk.

The original larval stomach (enteron) has extended, with the formation of the metameres, so as to form the greater portion of the intestine: the proctodæum (*Prc. dm*) is confined to the anal segment.

Two other obvious changes are the appearance of a pair of small slender processes (A, *t*)—the rudiments of the tentacles—on the apex of the prostomium, and of a circlet of cilia (*Pr. an. ci*) round the posterior end of the trunk.

The internal changes undergone during the assumption of the present form are very striking. In every fully formed metamere the mesoderm—solid, it will be remembered, in the previous stage—has become divided into two layers, a somatic layer (B and C, *Msd* (*som*)) in contact with the ectoderm and a splanchnic layer (*Msd* (*spl*)) in contact with the endoderm. The space between the two layers (*Cœl*) is the permanent body-cavity or cœlome, which is thus quite a different thing from the larval body-cavity or blastocœle, being formed, not as a space between ectoderm and endoderm, but by the splitting of an originally solid mesoderm.

The division of the mesoderm does not however extend quite to the middle dorsal and middle ventral lines: in both these situations a layer of undivided mesoderm is left (C), and in this way the dorsal and ventral mesenteries are formed. Spaces in these, apparently the remains of the blastocœle, form the dorsal and ventral blood-vessels. Moreover the splitting process takes place independently in each segment and a transverse vertical layer of undivided mesoderm (B, *Sep*) is left separating each segment from the

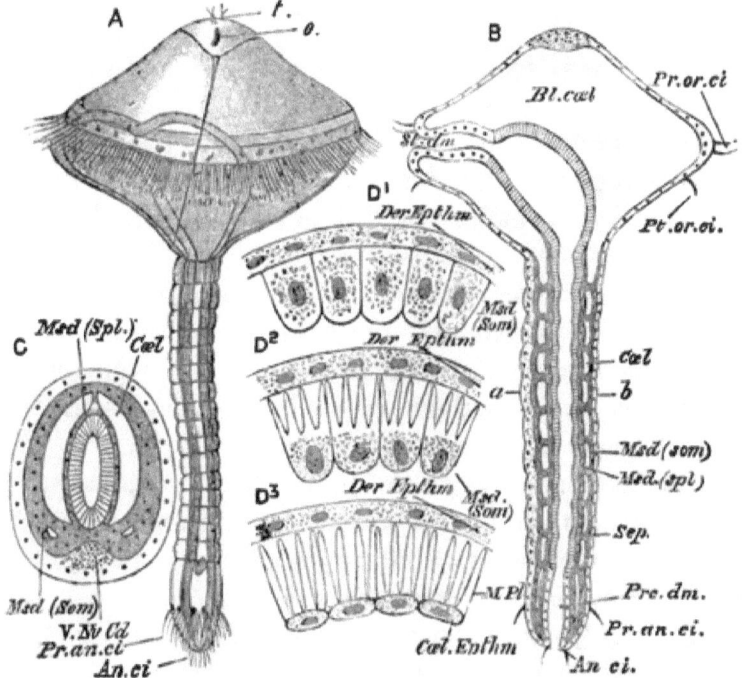

FIG. 75.—A, larva of Polygordius neapolitanus in a condition intermediate between the trochosphere and the adult worm, the trunk-region being elongated and divided into metameres.

B, diagrammatic vertical section of the same: the ectoderm is coarsely, the nervous system finely, dotted, the endoderm radially striated, and the mesoderm evenly shaded.

C, transverse section along the plane ab in B.

The pre-oral (*Pr. or. ci*), post-oral (*Pt. or. ci*), and anal (*An. ci*) cilia, the blastocœle (*Bl. cœl*), stomodæum (*St. dm*), and proctodæum (*Prc. dm*) are as in Fig. 72, A and B: the enteron now extends throughout the segmented region of the trunk.

A pair of tentacles (*t*) has appeared on the prostomium near the ocelli (*o*), and a pre-anal circlet of cilia (*Pr. an. ci*) is developed.

The mesoderm has divided into somatic (*Msd (som)*) and splanchnic (*Msd (spl)*) layers with the cœlome (*Cœl*) between: the septa (*Sep*) are formed by undivided plates of mesoderm separating the segments of the cœlome from one another.

D^1–D^3, three stages in the development of the somatic mesoderm. In D^1 it (*Msd (Som)*) consists of a single layer of cells in contact with the deric epithelium (*Der. Epthm*): in D^2 the cells have begun to split up in a radial direction: in D^3 each has divided into a number of radially arranged sections of muscle-plates (*M. Pl*) and a single cell of cœlomic epithelium (*Cœl. Epthm*).

(A after Fraipont.)

adjacent ones before and behind: in this way the septa arise.

The nephridia appear to have a double origin, the superficial portion of each being formed from ectoderm, the deep portion, including the nephrostome, from the somatic layer of mesoderm.

In the ventral nerve-cord the cells lying nearest the outer surface have enlarged and formed nerve-cells, while those on the dorsal aspect of the cord have elongated longitudinally and become converted into nerve-fibres. This process has already begun in the preceding stage.

But the most striking histological changes are those which gradually take place in the somatic layer of mesoderm. At first this layer consists of ordinary nucleated cells (D^1, *Msd* (*Som*)), but before long each cell splits up in a radial direction (D^2) from without inwards—*i.e.*, from the ectoderm (*Der. Epthm*) towards the cœlome—finally taking on the form of a book with four or more slightly separated leaves directed outwards or towards the surface of the body, and with its back—the undivided portion of the cell—bounding the cœlome. The cells being arranged in longitudinal series, we have a number of such books placed end to end in a row with the corresponding leaves in contact—page one of the first book being followed by page one of the second, third, fourth, &c., page two by page two, and so on through one or more segments of the trunk. Next, what we have compared with the leaves of the books—the divided portions of the cells—become separated from the backs—the undivided portions (D^3)—and each leaf (*M. Pl*) fuses with the corresponding leaves of a certain number of books in the same longitudinal series. The final result is that the undivided portions of the cells (backs of the books, *Cœl. Epthm*) become the parietal layer of cœlomic epithelium, the longitudinal bands formed by the union of the leaves (*M. Pl*) becoming the muscle-plates, which are thus *cell-fusions*, each being formed by the union of portions of a series of longitudinally arranged cells.

At the same time the cells of the splanchnic layer of mesoderm thin out and become the visceral layer of cœlomic epithelium.

We see then that by the time the larva has reached the stage shown in Fig. 75, it is no longer a mere aggregate of simple cells arranged in certain layers. The cells themselves have undergone differentiation, some becoming modified into nerve-fibres, others by division and subsequent fusion with their neighbours forming muscle-plates, while others, such as the epithelial cells, remain almost unaltered.

Thus, in the course of the development of Polygordius, cell-multiplication and cell-differentiation go hand in hand, the result being the formation of those complex tissues the presence of which forms so striking a difference between the worm and the simpler types previously studied.

It is important to notice that this comparatively complex animal is in one stage of its existence—the oosperm—as simple as an Amœba; in another—the polyplast—it is comparable to a Pandorina or a Volvox; in a third—the gastrula—it corresponds in general features with a Hydra; while in a fourth—the trochosphere—it resembles in many respects a Medusa. As in other cases we have met with, the comparatively highly-organized form passes through stages in the course of its individual development similar in general characters to those which, on the theory of evolution, its ancestors may be considered to have passed through in their gradual ascent from a lower to a higher stage of organization.

The rest of the development of Polygordius may be summarized very briefly. The trunk grows so much faster than the head (pro-*plus* peri-stomium)—that the latter undergoes a relative diminution in size, finally becoming of equal diameter with the trunk, as in the adult. The ciliated rings are lost, the tentacles grow to their full size, the eye-spots atrophy, and thus the adult form is assumed.

LESSON XXVII[1]

THE GENERAL CHARACTERS OF THE HIGHER ANIMALS

THE student who has once thoroughly grasped the facts of structure of such typical unicellular animals as Amœba and the Infusoria, of such typical diploblastic animals as Hydra and Bougainvillea, and of such a typical triploblastic animal as Polygordius, ought to have no difficulty in understanding the general features of the organization of any other members of the animal kingdom. When once the notions of a cell, a cell-layer, a tissue, body-wall, enteron, stomodæum, proctodæum, cœlome, somatic and splanchnic mesoderm, are fairly understood, all other points of structure become hardly more than matters of detail.

If we turn to any text-book of Zoology we shall find that the animal kingdom is divisible into seven primary subdivisions, called sub-kingdoms, types, or phyla. These are as follows:—

>   *Protozoa.*   *Cœlenterata.*
>   *Vermes.*   *Echinodermata.*
>   *Arthropoda.*   *Mollusca.*
>   *Vertebrata*

[1] Readers who have not studied zoology, or at least examined a series of selected animal types, should omit this lesson and go on to the next.

With a few exceptions, the discussion of which would be out of place here, the vast number of animals known to us may be arranged in one or other of these groups.

The *Protozoa* are the unicellular animals : they have been represented in previous lessons by Amœba and Protamœba, Hæmatococcus, Heteromita, Euglena, the Mycetozoa, Paramœcium, Stylonychia, Oxytricha, Opalina, Vorticella, Zoothamnium, the Foraminifera, and the Radiolaria. According to many authors, Pandorina and Volvox are also included in this group. The reader will therefore have no difficulty in grasping the general features of this phylum.

The *Cœlenterata* are the diploblastic animals, and have also been well represented in the foregoing pages, namely, by Hydra, Bougainvillea, Diphyes, and Porpita. The sea-anemones, corals, and sponges also belong to this phylum.

The *Vermes*, or Worms, are a very heterogeneous assemblage. They are all triploblastic, but while some are cœlomate, others have no body-cavity; some, again, are segmented, others not. Still, if the structure of Polygordius is thoroughly understood, there will be little difficulty in understanding that of a fluke, a tape-worm, a round-worm, an earthworm, or one of the ordinary marine worms.

Of the remaining four sub-kingdoms we have, so far, studied no example, but a brief description of a single typical form of each will show how they all conform to the general plan of organization of Polygordius, being all triploblastic and cœlomate.

Under the *Echinodermata* are included the various kinds of starfishes—sand-stars, brittle-stars, and feather-stars, as well as sea-urchins, sea-cucumbers, &c. A starfish will serve as an example of the group.

The phylum *Arthropoda* includes crayfishes, lobsters, crabs, shrimps, prawns, wood-lice, and water-fleas ; scorpions, spiders, and mites; centipedes and millipedes; and all kinds of insects, such as cockroaches, beetles, flies, ants, bees, butterflies, and moths. A crayfish forms a very fair type of the group.

In the phylum *Mollusca* are included the ordinary bivalves, such as mussels and oysters; snails, slugs, and other univalves or one-shelled forms; sea-butterflies; and cuttlefish, squids, and Octopi. An account of a fresh-water mussel will serve to give a general notion of the character of this group.

Finally, under the head of *Vertebrata* are included all the backboned animals: the lampreys and hags; true fishes, such as the shark, skate, sturgeon, cod, perch, trout, &c.; amphibians, such as frogs, toads, newts, and salamanders; true reptiles, such as lizards, crocodiles, snakes, and tortoises; birds; and mammals, or creatures with a hairy skin which suckle their young, such as the ordinary hairy quadrupeds, whales and porpoises, apes, and man. The essential structure of a vertebrate animal will be understood from a brief description of a dog-fish.

THE STARFISH.[1]

A common starfish consists of a central disc-like portion, from which radiate five arms or rays. It crawls over the rocks with its ventral surface downwards, its dorsal surface upwards. It can move in any direction, so that, in the ordinary sense of the words, anterior and posterior extremities cannot be distinguished. Radial symmetry such as this, *i.e.*, the division of the body into similar parts radiating from a common centre, is characteristic of the Echinodermata generally.

[1] For a detailed description of a Starfish, see Rolleston and Hatchett Jackson, *Forms of Animal Life* (Oxford, 1888), pp. 190 and 311.

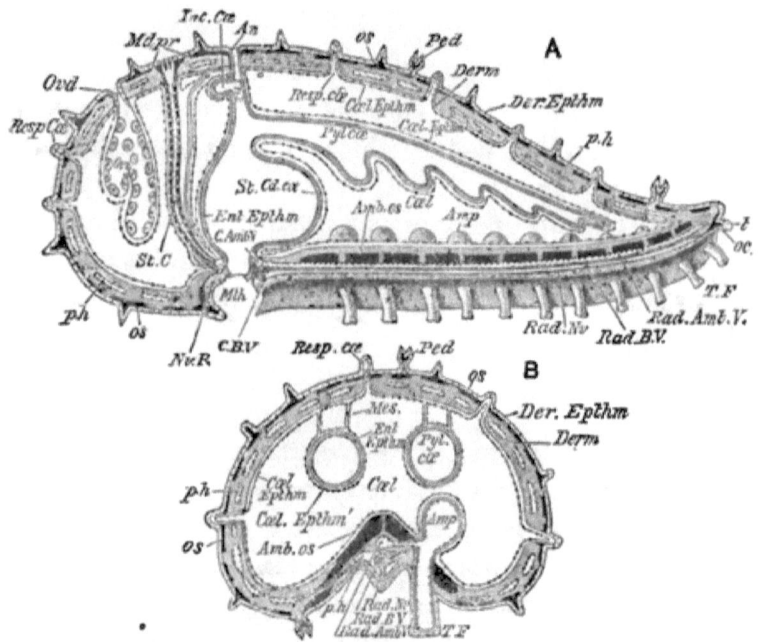

Fig. 76.—Diagrammatic sections of a Starfish.

A, vertical section passing on the right through a radius on the left through an inter-radius. The off side of the ambulacral groove with the tube feet (*T. F*) and ampullæ (*Amp*) are shown in perspective.

B, transverse section through an arm.

The ectoderm is coarsely dotted, the nervous system finely dotted, the endoderm radially striated, the mesoderm evenly shaded, the ossicles of the skeleton black, and the cœlomic epithelium represented by a beaded line.

The body-wall consists of deric epithelium (*Der. Epthm*), dermis (*Derm*), and the parietal layer of cœlomic epithelium (*Cœl. Epthm*).

To the body-wall are attached pedicellariæ (*Ped*), and the end of the arm bears a tentacle (*t*) with an ocellus (*oc*) at its base.

The skeleton consists of ossicles (*os*) imbedded in the derm: large ambulacral ossicles (*Amb. os*) bound the ambulacral grooves on the ventral surfaces of the arms.

The mouth (*Mth*) leads by a short gullet into a stomach (*St*), which gives off a cardiac cæcum (*Cd. cæ*) and a pair of pyloric cæca (*Pyl. cæ*) to each arm, and passes into an intestine (*Int*) which gives off intestinal cæca (*Int. cæ*) to the inter-radii, and ends in the anus (*An*). The pyloric cæca are connected to the dorsal body-wall by mesenteries (*Mes.* in B). The wall of the enteric canal consists of enteric epithelium covered by the visceral layer of cœlomic epithelium (*Cœl. Epthm'*).

From the cœlome are given off respiratory cæca (*Resp. cæ*), which project through the body-wall: the latter contains peri-hæmal spaces (*p. h*) derived from the cœlome.

The circular blood-vessel (*C. B. V*) surrounds the gullet and gives off radial vessels (*Rad. B. V*) to the arms and an inter-radial plexus connected with a pentagonal ring round the intestine.

The circular ambulacral vessel (*C. Amb. V*) gives off radial vessels (*Rad. Amb. V*) to the arms connected with the ampullæ (*Amp*) and tube-feet (*T. F*) : it is also connected with the stone-canal (*St. C*), which opens externally by the madreporite (*Mdpr*).

The nerve-ring (*Nv. R*) gives off radial nerves (*Rad. Nv*) to the arms.

The ovary (*Ovy*) is inter-radial, and opens by a dorsal oviduct (*Ovd*).

In the centre of the disc on the ventral surface is the large mouth (Fig. 76, A, *Mth*), and from it radiate five grooves, one along the ventral surface of each arm (A and B). In the living animal numerous delicate semi-transparent cylinders, the *tube-feet* (*T. F*), are protruded from these grooves; they are very extensible and each ends in a sucker. It is by moving these structures in various directions, protruding some and withdrawing others, that the starfish is able to move along either a horizontal or a vertical surface, and even to turn itself over when placed with the ventral side upwards.

Near the middle of the disc, on the dorsal surface, is the very minute anus (A, *An*); it is situated on a line drawn from the centre of the disc to the re-entering angle between two of the rays, and is therefore said to be inter-radial in position. Near the anus, and also inter-radially situated, is a circular calcareous plate, the *madreporite* (*Mdpr*), perforated by numerous microscopic apertures. Innumerable other calcareous plates, or *ossicles* (*os*), are embedded in the body-wall, and constitute a skeleton, to which the firm and resistant character of the starfish is due.

Sections show that there is a well-marked cœlome, separating the body-wall from the enteric canal and containing the gonads, blood-vessels, &c. The body-wall consists externally of a very thin cuticle, then of a layer of deric epithelium or epidermis (*Der. Epthm*), then of a thick fibrous layer (*Derm*)—the *dermis* or deep layer of the skin, then of a thin and interrupted layer of muscle, and finally, of a layer of cœlomic epithelium (*Cœl. Epthm*) bounding the body cavity.

The dermis is formed of *connective tissue*, a substance not met with in Polygordius, formed by the elongation of mesoderm cells into wavy fibres. The ossicles of the skeleton (*os*) are formed by deposits of calcium carbonate in the dermis; the skeleton is therefore a dermal exoskeleton. The large ambulacral ossicles (*Amb. os*), however, which bound the ambulacral grooves, lie internal to the vessels (*Rad. B. V., Rad. Amb. V.*) and have an endoskeletal character.

The enteric canal passes vertically from mouth (A, *Mth*) to anus (*An*), and is divisible into gullet, stomach (*St*), and intestine (*Int*). The stomach gives off five wide pouches (*Cd. cæ*), one extending into the base of each arm, and above these five other pouches (*Pyl. cæ*), each of which divides into two (B, *Pyl. cæ*) and extends to the extremity of the corresponding arm. The intestine gives off smaller pouches (*Int. cæ*) which are inter-radial in position. Thus the enteric canal, like the body as a whole, exhibits radial symmetry. The canal is lined by enteric epithelium, mostly endodermal, and is covered externally by cœlomic epithelium (*Cœl. Epthm'*).

Respiration is affected by blind, finger-like offshoots of the cœlome, the *respiratory cæca* (*Resp. cæ*), which pass between the ossicles of the skeleton and project on the surface of the body, thus bringing the cœlomic fluid into close relation with the surrounding water.

The blood-system consists of a circular vessel (A, *C. B. V*) round the gullet, connected with a pentagonal vessel round the intestine by an elongated network or plexus of vessels. From the circular vessel five radiating trunks (*Rad. B. V*) pass to the arms.

Parallel with and above the circular blood-vessel is a similar but larger structure, the *ambulacral ring* (*C. Amb. V*) which also sends off five radiating vessels (*Rad. Amb. V*) to the arms. These give off a branchlet to each tube-foot (B, *T.F.*), the branchlet having a sac or *ampulla* (*Amp*) at its base. From the ambulacral ring a tube with calcareous walls, the stone-canal (*St. C*) passes upwards and ends in the madreporite (*Mdpr*), by the apertures in which the fluid filling the whole of the *ambulacral system* of vessels is placed in communication with the surrounding water.

The function of the ambulacral system is mainly locomotive. By the contraction of the ampullæ fluid is forced into the tube feet, and by the action of the muscles of the tube-feet it is sent back into the ampullæ, and in this way the tube-feet are protruded and retracted at the will of the animal. The system, which is peculiar to the Echinodermata, is lined with epithelium, continuous, in the larva, with the cœlomic epithelium. It has been compared to a gigantic and greatly modified nephridium.

The nervous system is very simple. It consists of a pentagonal ring (A, *Nv. R*) round the mouth giving off five radial nerves (A and B, *Rad. Nv*) which pass along the ambulacral grooves, below the blood-vessels, to the extremities of the arms, where each is connected with an eye-spot. Both nerve-ring and radial nerves are mere thickenings of the deric epithelium.

The gonads (A, *Ovy*) are branched organs, five in number, which lie inter-radially near the bases of the arms, and open by gonaducts (*Ovd*) on the dorsal surface of the disc. The sexes are lodged in distinct individuals.

Both eggs and sperms are shed into the water, and after impregnation the oosperm becomes a gastrula, which is converted into a peculiar free-swimming larva; this undergoes metamorphosis and is converted into the adult form.

THE CRAYFISH.[1]

In a crayfish or lobster the body is bilaterally symmetrical and is distinctly segmented, consisting of a prostomium and of nineteen metameres. The anterior twelve metameres are united with one another and with the prostomium to form an unjointed portion of the body, the *cephalothorax* (Fig. 77, A, *C. Th.*): the seven posterior segments are free and constitute the abdomen (*Abd. Seg.* 1, *Abd. Seg.* 7). It is very generally characteristic of Arthropods to have the metameres limited and constant in number, and for more or fewer of them to undergo concrescence.

Another distinctive arthropod character illustrated by the Crayfish is the possession of lateral *appendages* of the body. These are given off from the ventral region, two pairs being borne by the prostomium and one by each of the metameres, except the last. Moreover the appendages themselves are segmented, being divided into freely articulated limb-segments or *podomeres*.

In the Crayfish there is a marked differentiation of the appendages. Those of the prostomium are a pair of eye-stalks, and one of small feelers or antennules which perform

[1] For detailed descriptions of the Crayfish see Huxley, *The Crayfish* (London, 1880): Huxley and Martin, *Elementary Biology*, new ed. (London, 1888), p. 173: Rolleston and Jackson, *Forms of Animal Life* (Oxford, 1888), pp. 162 and 307: Marshall and Hurst, *Practical Zoology*, 3rd. ed. (London, 1892), p. 130: and Parker, *The Skeleton of the New Zealand Crayfishes* (Wellington, N.Z., 1889).

an olfactory function and also contain the organ of hearing.[1] The metameres of the cephalothorax bear one pair of tactile appendages or antennæ, six pairs acting as jaws (mandibles, first and second maxillæ, and first, second, and third maxillipedes), and five pairs of legs, the first of which are—in the fresh-water crayfishes and in lobsters—much larger than the rest. The abdomen bears small fin-like swimmerets on its first five metameres, the sixth bearing larger appendages which, together with the seventh segment or telson, constitute the tail-fin.

Sections show the body-wall to consist of a layer of deric epithelium (*Der. Epthm*) secreting a thick cuticle (*Cu*), a layer of connective tissue forming the Dermis (*Derm*), and a very thick layer of large and complicated muscles (*M*), which fill up a great part of the interior of the body.

The cuticle (*Cu*) is of great thickness, and except at the joints between the various segments of the body and limbs, is impregnated with lime salts so as to form a hard, jointed armour. It thus constitutes a skeleton which, unlike that of the starfish (p. 312), is a cuticular exoskeleton, forming a continuous investment over the whole body but discontinuously calcified.

The mouth (*Mth*) is on the ventral surface of the head, in the segments of the mandibles or first pair of jaws. It has therefore, as compared with the mouth of Polygordius, undergone a backward shifting, the appendages of the first metamere (antennæ) being altogether in front of it. The enteric canal consists of a short gullet (*Gul*), a large stomach (*St*), and a straight intestine divisible into a short anterior division or small intestine (*S. Int*) and a long posterior division or large intestine (*L. Int*) : the latter

---

[1] The antennules are frequently considered as belonging to the first metamere, the number of segments being then reckoned as twenty.

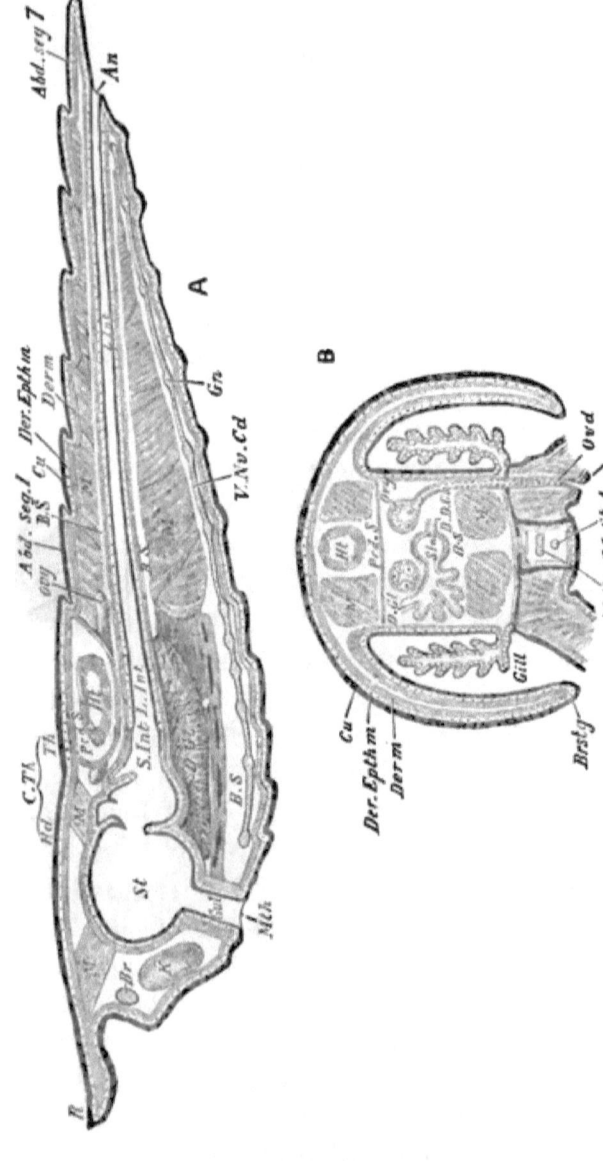

FIG. 77.—Diagrammatic sections of a Fresh-water Crayfish.

A, longitudinal section: the right digestive gland (*D. Gl*) and ovary (*Ovy*) are shown in perspective. B, transverse section through the thorax: the digestive gland (*D. Gl*) is shown only on the left, the oviduct (*Ovd*) only on the right side.

The cuticle is black, the ectoderm coarsely dotted, the nervous system finely dotted, the endoderm radially striated, and the mesoderm evenly shaded.

The body is divided into a head (*Hd*) and thorax (*Th*), together constituting the cephalothorax (*C. Th*), and seven free abdominal segments (*Abd. seg.* 1, *Abd. seg.* 7): the head is produced in front into a rostrum.

The body-wall consists of cuticle (*Cu*), partly calcified to form the exoskeleton, deric epithelium (*Der. Epthm*), dermis (*Derm*), and a very thick layer of muscle (*M*) which in the abdomen is distinctly segmented.

The mouth (*Mth*) leads by a short gullet (*Gul*) into a large stomach (*St*), from which a short small intestine (*S. Int*) leads into a large intestine (*L. Int*), ending in the anus (*An*). Opening into the small intestine are the digestive glands (*D. Gl*). The epithelium of the small intestine and digestive glands is endodermal, that of the rest of the canal is ectodermal and secretes a cuticle: the outer layer throughout is mesodermal (connective tissue and muscle).

The cavity (*B. S*) between the enteric canal and the body-muscles is a blood-sinus.

The heart (*Ht*) is enclosed in the pericardial sinus (*Per. S*): the chief ventral blood-vessel or sternal artery (*St. A*) is shown in B.

The gills (B, *Gill*) are enclosed in a cavity formed by a fold of the thoracic body-wall called the branchiostegite (*Brstg*): they are formed of the same layers as the body-wall, of which they are offshoots.

The kidneys (A, *K'*) are situated in the head.

The brain (*Br*) lies in the prostomium: the ventral nerve-cord (*V. Nv. Cd*) consists of a chain of ganglia (*Gn*) united by connectives.

The ovary (*ovy*) is a hollow organ opening by an oviduct (B, *ovd*) on the base of one of the legs (*Leg*).

opens by an anus (*An*) on the ventral surface of the last segment. The study of development shows that the only part of the canal derived from the enteron of the embryo is the small intestine: the gullet and stomach arise from the stomodæum, the large intestine from the proctodæum. Thus the only portion of the enteric epithelium which is endodermal is that of the small intestine: the epithelium of gullet, stomach, and large intestine is ectodermal, and like the deric epithelium secretes a cuticle. The outer layer of the whole enteric canal consists of connective tissue and muscle: there is no cœlomic epithelium.

On each side of the small intestine is a large organ, the *digestive gland* (*D. Gl*): it consists of numberless glove-finger-like processes or *cæca* which open by a short tube or *duct* into the small intestine (B, *D. Gl*). Both cæca and duct are lined with epithelium derived from the endoderm, and the whole digestive gland is to be looked upon as a branched lateral outgrowth of the enteron. The secretion of digestive juice is performed exclusively by the epithelium of the digestive glands.

Between the enteric canal and the body-wall are a series of spaces (*B..S*) containing blood and having the general relations of a cœlome, but very probably only representing a number of enlarged blood-spaces or *sinuses*.

Respiration is performed by special organs, the *gills* (B, *Gill*, see p. 317), developed in the thoracic region as outgrowths of the body-wall and containing the same layers (cuticle, epithelium, and connective tissue) as the latter. They have a brush-like form and are protected by a fold of the body-wall (*Brstg*).

The blood-system is constructed on the same general lines as those of Polygordius, but is greatly modified. A portion of the dorsal vessel is enlarged to form a muscular dilatation, the *heart* (*Ht*), and the rest of the vessels, now called arteries (B, *St. A*), instead of forming by themselves a closed system, ramify extensively over the body, their ultimate branches opening into larger cavities or *sinuses* between the muscles. One of these cavities—the pericardial sinus *Pcd. S*)—surrounds the heart. The heart, arteries, and sinuses together form a closed system through which the blood is propelled in a definite direction by the contractions of the heart.

Renal excretion is performed by a pair of glandular bodies, the *kidneys* (A, *K*), situated in the front part of the head and enclosed in spacious sacs which open by ducts on the bases of the antennæ. They consist of convoluted tubes lined by epithelium, and are probably to be looked upon as greatly modified nephridia.

The Crayfish is diœcious. The ovaries (*ovy*) are a pair of hollow organs, united in the middle line in some genera, situated in the thorax, and opening by oviducts (B, *ovd*) on the bases of the third pair of legs. The spermaries (testes) are also frequently united in the middle line and open by spermiducts (vasa deferentia) on the bases of the fifth pair of legs. There is some reason for thinking that the gonaducts represent modified nephridia, and the cavities of the hollow gonads a greatly reduced cœlome from the epithelium of which the sex-cells are produced.

The nervous system is formed on quite the same plan as that of Polygordius, consisting of a dorsal brain (*Br*) united by œsophageal connectives to a ventral nerve-cord (*V. Nv. Cd*). In the cord, however, the nerve-cells, instead of being evenly distributed, are aggregated into little enlargements or *ganglia* (*Gn*), of which there is primatively a pair to each metamere, the number being reduced in the adult by concrescence. The portions of the ventral nerve-cord between the ganglia consist of nerve-fibres only, and are called connectives. In the embryo the nervous system is, as in Polygordius, in direct connection with the epidermis, but in the adult it has sunk inwards so as to be entirely surrounded by mesoderm.

A striking feature in the histology of the Crayfish, and one in which it agrees with the vast majority of Arthropoda, is the entire absence of cilia. Another peculiarity—also shared by the greater part of the phylum—is that the sperms are non-motile.

The laid eggs become attached to the swimmerets of the mother, and in this situation undergo their development. In the fresh-water crayfish the young is hatched in a condition closely resembling the adult, but in the lobster and the sea-crayfish there is a metamorphosis.

The Fresh-Water Mussel.[1]

The body is bilaterally symmetrical, and is greatly compressed from side to side. Its dorsal margin is produced into paired flaps, the *mantle-lobes* (Fig. 78, A and B, *Mant*), which pass downwards one on either side of the body. Closely applied to the outer surface of the mantle-lobes, and formed as a cuticular secretion of their deric epithelium, are the two *valves* of the bivalved, strongly calcified *shell* (B., *Sh*). The ventral region of the body is produced into a laterally compressed muscular structure, the *foot* (A and B, *Foot*), by the contraction of which the animal can move slowly through the sand or mud in which it lives partly buried.

The possession of a mantle formed as a prolongation of the dorsal region, of a calcareous shell secreted by the mantle, and of a muscular foot formed as an unpaired prolongation of the ventral region, are the most characteristic features of the Mollusca generally.

Posteriorly the edges of the mantle-lobes are greatly thickened, and are connected with one another in such a way as to form two apertures, a large ventral *inhalent* (*Inh. Ap*), and a small dorsal exhalent aperture (*Exh. Ap*). By means of the cilia of the gills (see below) a current of water is produced which enters at the inhalent aperture, carrying abundant oxygen and the minute organisms used as food, and makes its escape at the exhalent aperture, taking with it the various products of excretion and fæcal matter.

The mouth (*Mth*) is anterior and ventral, lying just in front of the foot: it is bounded on either side by a pair of

---

[1] For detailed descriptions of the fresh-water Mussel see Rolleston and Jackson, *Forms of Animal Life*, pp. 124 and 285: Huxley and Martin, *Elementary Biology*, p. 305: and Marshall and Hurst, *Practical Zoology*, p. 80.

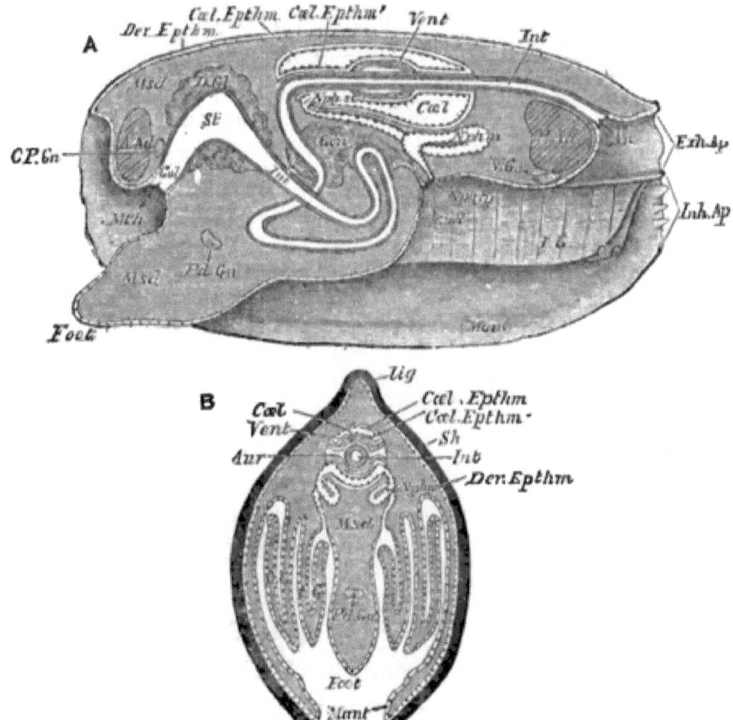

Fig. 78.—Diagrammatic sections of the Fresh-water Mussel.

A, longitudinal section : the right mantle-lobe (*Mant*) and gills (*I. G, O. G*) are shown in perspective.

B, transverse section.

The cuticular shell (*Sh*), shown only in B, is black, the ectoderm dotted, the nervous system finely dotted, the endoderm radially striated, the mesoderm evenly shaded, and the cœlomic epithelium represented by a beaded line.

The dorsal region is produced into the right and left mantle-lobes (*Mant*), attached to which are the valves of the shell (*Sh*) joined dorsally by an elastic ligament (*lig*).

The mantle-lobes are partly united so as to form the inhalent (*Inh. Ap*) and exhalent (*Exh. Ap*) apertures at the posterior end.

The body is produced ventrally into the foot (*Foot*), on each side of which are the gills, an inner (*I. G*) and an outer (*O. G*), each formed of an inner and an outer lamella.

The body is covered externally by deric epithelium (*Der. Epthm*), within which is mesoderm (*Msd*) largely differentiated into muscles, of which the anterior (*A. Ad*) and posterior (*P. Ad*) adductors are indicated in A.

The mouth (*Mth*) leads by the short gullet (*Gul*) into the stomach (*St*), from which proceeds the coiled intestine (*Int*), ending in the anus

Y

(*An*): the enteric epithelium is mostly endodermal. The digestive gland (*D. Gl*) surrounds the stomach. The cœlome (*Cœl*) is reduced to a small dorsal chamber enclosing part of the intestine and the heart; the parietal (*Cœl. Epthm*) and visceral (*Cœl. Epthm²*) layers of cœlomic epithelium are shown.

The heart consists of a median ventricle (*Vent*), enclosing part of the intestine, and of paired auricles (*Aur*).

The paired nephridia (*Nphm*) open by apertures into the cœlome (*Nph. st*) and on the exterior (*Nph. p*).

The gonads (*Gon*) are imbedded in the solid mesoderm, and open on the exterior by gonaducts (*Gnd*).

The nervous system consists of a pair of cerebro-pleural ganglia (*C. P. Gn*) above the gullet, a pair of pedal ganglia (*Pd. Gn*) in the foot, and a pair of visceral ganglia (*V. Gn*) below the posterior adductor muscle.

triangular bodies, the labial palpi, and leads by a short gullet (*Gul*) into a stomach (*St*) from which proceeds a long, coiled intestine (*Int*): this makes several turns in the ventral region of the trunk, then passes to the dorsal region, and finally backwards in the median plane to open by an anus (*An*) at the posterior end of the body, just within the exhalent aperture. The enteric canal is formed almost exclusively from the enteron, the stomodæum and proctodæum being both insignificant: hence the enteric epithelium is almost wholly endodermal. There is a large digestive gland (*D. Gl*) surrounding the stomach and opening into it by several ducts.

The cœlome (*Cœl*) is a small cavity in the dorsal region containing a portion of the intestine: the rest of the enteric canal is embedded in solid mesoderm.

The mesoderm, as usual, is largely differentiated into muscle. There are numerous muscles connected with the foot, and two very large ones (*A. Ad, P. Ad*) pass transversely from valve to valve of the shell, one immediately above the gullet, the other immediately below the anal end of the intestine; these latter are called *adductors*, and serve to close the shell.

On either side of the body, between the trunk and the mantle, are two gills (*I. G, O. G*), each having the form of a double plate (B) nearly as long as the body. They serve, in conjunction with the mantle, as respiratory organs, but their main function is to produce the current of water referred to above by means of the cilia with which they are covered.

There is an extensive system of blood-vessels. The heart lies in the cœlome, and consists of three chambers, a median *ventricle* (*Vent*), which surrounds the intestine, and paired *auricles* (*Aur*).

Excretion is performed by a single pair of nephridia (*Nphm*) which open at one end (*Nph. st*) into the cœlome and at the other (*Nph. p*) on to the exterior.

The nervous system consists of three pairs of ganglia, the two ganglia of each pair being united by transverse commissures. The *cerebro-pleural* ganglia (*C. P. Gn*) lie above the gullet, and represent, in a general way, the brain of Polygordius and the crayfish; they are united by longitudinal connectives with the *pedal* ganglia (*P. Gn*), which lie in the foot and may be taken as representing the ventral nerve-cord of worms and arthropods, and with the *visceral* ganglia (*V. Gn*) which are placed beneath the posterior adductor muscle.

The gonads (*Gon*) are large irregular organs, very similar in appearance in the two sexes, situated among the coils of the intestine and opening by a duct (*Gnd*) on either side of the trunk, close to the nephridiopore. The impregnated eggs are passed into the cavity of the outer gill of the female, where they undergo the early stages of their development. The larva of the fresh-water mussel is a peculiar bivalved form, very unlike the adult, and called a *glochidium* ; but in the more typical molluscs the embryo leaves the egg as a trochosphere, closely resembling that of Polygordius.

The Dog-Fish.[1]

A dog-fish is bilaterally symmetrical, the nearly cylindrical body (Fig. 79, A) terminating in front in a blunt snout and behind passing insensibly into an upturned tail. Externally there is no appearance of segmentation.

The mouth (Mth) is on the ventral surface of the head or anterior region of the body; it is transversely elongated, and is supported by jaws which are respectively anterior (upper) and posterior (lower). They thus differ fundamentally from the jaws of arthropods, which are modified appendages and are therefore disposed right and left.

A short distance behind the mouth are five vertical slits (B, *Ext. br. ap*) arranged in a longitudinal series, the *external branchial-apertures* or gill-clefts. The vent, or cloacal aperture (An) is situated on the ventral surface a considerable distance from the end of the tail. That part of the body lying in front of the last gill-cleft is counted as the head, all behind the vent as the tail, the intermediate portion as the trunk.

Appendages are present, but in a very different form from those of the crayfish. They consist of flat processes of the body-wall called *fins*. Two of them ($D.F^1, D.F^2$) are situated in the middle line of the back (*dorsal fins*): one ($V.F$) in the middle ventral line behind the cloacal aperture (*ventral fin*), and one ($C.F$) is attached to the up-turned end of the tail (*caudal fin*): all these being unpaired structures or median fins. Then there is a pair of *pectoral fins* situated

[1] For a detailed description of a dog-fish see Marshall and Hurst, *Practical Zoology* (London, 1892), p. 206. For descriptions of other fishes, equally suitable in some respects as types of Vertebrata, see Rolleston and Jackson, *Forms of Animal Life* (Oxford, 1888), pp. 83 and 273: and Parker, *Zootomy* (London, 1884), pp. 1, 27, 86.

one on each side just behind the last gill-cleft, and a pair of *pelvic fins* placed one on either side of the vent : these are the lateral or paired fins. It is characteristic of Vertebrata that the number of lateral appendages never exceeds two pairs.

The skin or external layer of the body-wall consists of an outer epidermis (*Der. Epthm*) composed of several layers of cells, and of an inner connective tissue layer or dermis (*Derm*). In the latter are found innumerable bony scales (*Derm. Sp*) constituting a dermal exoskeleton. The muscular layer of the body-wall (*M*) is of great thickness, especially in the dorsal region, and is distinctly segmented, indicating that the body of the dog-fish, like that of Polygordius and the crayfish, is divisible into metameres, although there is no indication of them externally.

The large cœlome (*Cœl*) is confined to the trunk : it is characteristic of vertebrates that both head and tail are acœlomate in the adult. The cœlomic epithelium (*Cœl. Epthm, Cœl. Epthm'*) is underlaid by a distinct layer of connective tissue, the two together forming the *peritoneum*.

Another important vertebrate character is that the dorsal region of the body-wall contains a median longitudinal canal (*C. Sp. Cav.*) extending from shortly behind the snout to near the end of the tail. This is the *cerebro-spinal cavity* and contains the central nervous system.

Still another characteristic feature is the presence, in addition to the dermal exoskeleton, of an *endoskeleton*, or system of internal supporting structures. Between the cerebro-spinal cavity above and the cœlome below is a longitudinal series of biconcave discs or *vertebral centra* (*V. Cent*) : they are formed of a peculiar tissue called *cartilage* or gristle, and are strongly impregnated with lime-salts : in the young condition their place is occupied by a

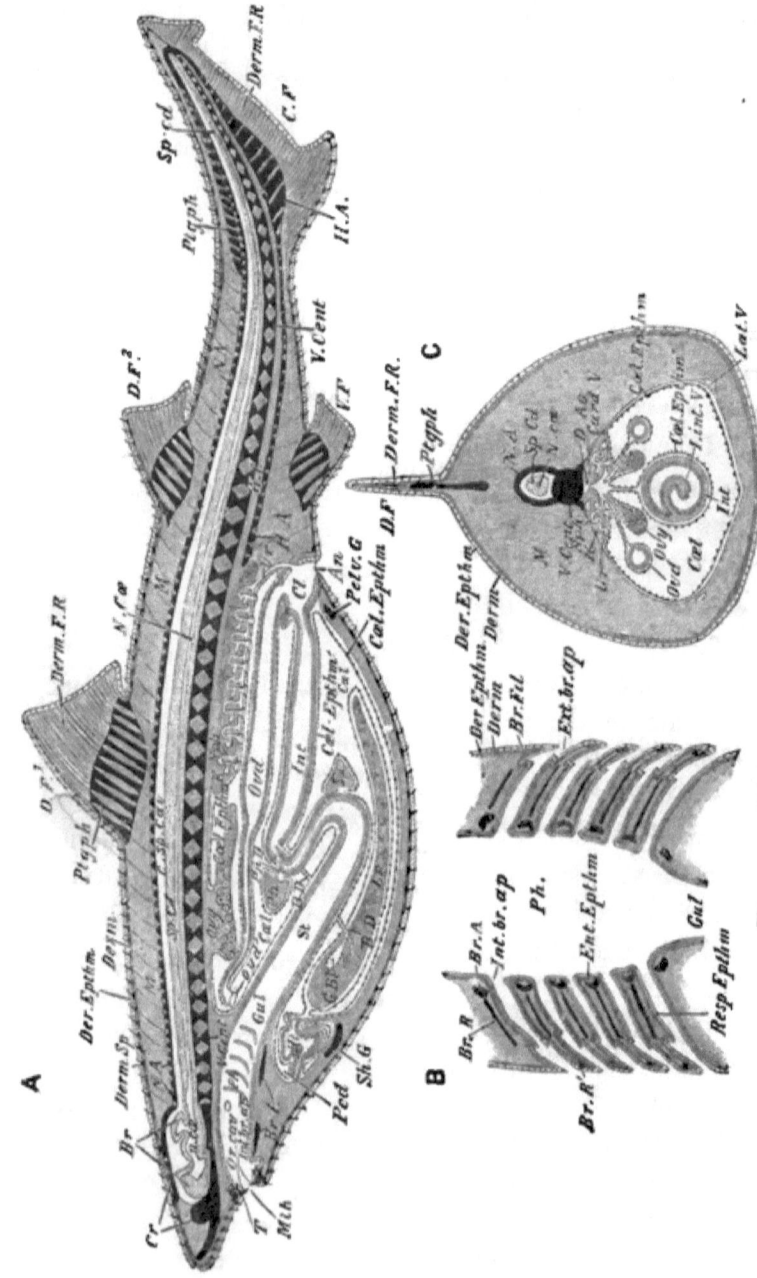

FIG. 79.—Diagrammatic sections of a Dog-fish.

A, longitudinal vertical section.
B, horizontal section through the pharynx and gills.
C, transverse section through the trunk.

The ectoderm is dotted, the nervous system finely dotted, the endoderm radially striated, the mesoderm evenly shaded, the cœlomic epithelium represented by a beaded line, and all skeletal structures black.

The body gives origin to the dorsal ($D. F^1, D. F^2$), ventral ($V. F$), and caudal ($C. F$) fins; the paired fins are not shown.

The body-wall consists of deric epithelium (*Der. Epthm*), dermis *Derm*), and muscle (*M*): the latter is metamerically segmented and is very thick, especially dorsally, where it forms half the total vertical height (C).

The exoskeleton consists of calcified dermal spines (*Derm. Sp*) in the dermis, and of dermal fin-rays (*Derm. F. R*) in the fins.

The endoskeleton consists of a row of vertebral centra (*V. Cent*) below the spinal cord (*Sp. Cd*), giving rise to neural arches (*N. A*), which enclose the cord, and in the caudal regions to hæmal arches (*H. A.*): a cranium (*Cr*) enclosing the brain (*Br*): upper and lower jaws: branchial arches (*Br. A*) and rays (*Br. R, Br. R'*), shown only in B, supporting the gills: shoulder (*Sh. G*) and pelvic (*Pelv. G*) girdles: and pterygiosphores (*Ptgph*) supporting the fins.

The mouth (*Mth*) leads into the oral cavity (*Or. cav*), from which the pharynx (*Ph*) and gullet (*Gul*) lead to the stomach (*St*): this is connected with a short intestine (*Int*) opening into a cloaca (*Cl*) which communicates with the exterior by the vent (*An*). The oral cavity and cloaca are the only parts of the canal lined by ectoderm.

Connected with the enteric canal are the liver (*Lr*) with the gall-bladder (*G. Bl*) and bile-duct (*B. D*), the pancreas (*Pn*), and the spleen (*Spl*). The mouth is bounded above and below by teeth (*T*).

The respiratory organs consist of pouches (shown in B) communicating with the pharynx by internal (*Int. br. ap*) and with the exterior by external (*Ext. br. ap*) branchial apertures, and lined by mucous membrane raised into branchial filaments (*Br. Fil*).

The heart (*Ht*) is ventral and anterior, and is situated in a special compartment of the cœlome (*Pcd*). Six of the most important blood-vessels, the dorsal vessel (dorsal aorta, *D. Ao*), the cardinal veins (*Card. V*), the lateral vessels (lateral veins, *Lat. V*), and the ventral vessel (intra-intestinal vein, *I. int. V*) are shown in C.

The whole cœlome is lined by epithelium, showing parietal (*Cœl. Epthm*) and visceral (*Cœl. Epthm'*) layers.

The ovaries (*Ovy*) are connected with the dorsal body-wall: the oviducts (*Ovd*) open anteriorly into the cœlome (*ovd'*) and posteriorly into the cloaca.

The kidneys (*K*) are made up of nephridia (*Nph*) and open by ureters (*Ur*) into the cloaca.

The nervous system is lodged in the cerebro-spinal cavity (*C. Sp. Cav*) hollowed out in the dorsal body-wall: it consists of brain (*Br*) and spinal cord (*Sp. Cd*), and contains a continuous cavity, the neurocœle *n. cœ*).

gelatinous rod, the *notochord*. The centra, which alternate with the muscle-segments, are connected with a series of cartilaginous arches (*N.A*), which extend over the cerebro-spinal cavity and with the centra constitute the *vertebral column*. In the tail there is also a ventral series of arches (*H.A.*) enclosing a space (*H.C*) which indicates a backward extension of the cœlome in the embryo.

Anteriorly the vertebral column is continued into a cartilaginous box, the *cranium* (*Cr*) which encloses the brain and the organs of smell and hearing. The jaws, referred to above, are cartilaginous rods which bound the mouth above and below. The gills are supported by a complicated system of cartilages (*Br. A, Br. R., Br. R'*), and both median and paired fins by parallel rods of the same material (*Ptgph*). All these cartilages are strengthened by a more or less extensive superficial deposit of bony matter.

The mouth (*Mth*) leads into a large *oral cavity* (*Or. cav*) which passes insensibly into a wide throat or *pharynx* (*Ph*): from this a short gullet (*Gul*) leads into a large U-shaped stomach (*St*), whence is continued a short wide intestine (*Int*) opening on to the exterior through the intermediation of a small chamber, the *cloaca* (*Cl*). From the gullet backwards the enteric canal is contained in the cœlome. The greater part of the enteric epithelium is endodermal : only the oral cavity arises from the stomodæum and the cloaca from the proctodæum.

In the skin covering the jaws dermal ossicles of unusual size are developed and constitute the teeth (*T*). The chief digestive glands are two in number, an immense *liver* (*Lr*) occupying the whole anterior and ventral region of the cœlome, and a small *pancreas* (*Pn*), attached to the anterior end of the intestine. The ducts of both glands open into the intestine, and their secreting cells are, as in former cases, endodermal. Gland-cells are also found in the walls of the stomach and intestine.

The respiratory organs or gills (B) consist of five pairs of pouches opening on the one hand into the pharynx (*Ph*) and on the other to the exterior by the branchial clefts already noticed: they have their walls raised into ridges, the branchial filaments (*Br. Fil*), which are covered with epithelium (*Resp. Epthm*) and are abundantly supplied with blood-vessels. The gills are developed as offshoots of the pharynx, and the respiratory epithelium is therefore endodermal, not ectodermal as in the crayfish and mussel.

The heart (*Ht*) lies below the pharynx in a separate anterior compartment of the cœlome, the pericardial cavity. It is composed of four chambers arranged in a single longitudinal series (sinus venosus, auricle, ventricle, and conus arteriosus), and is to be looked upon as a muscular dilatation of a ventral blood-vessel. The blood is propelled by the heart from the conus arteriosus into a paired series of hoop-like vessels (aortic arches) resembling the transverse commissures of Polygordius (Fig. 69, A, p. 282), which take it through the gills and pour it, in a purified condition, into the dorsal vessel (dorsal aorta, *D. Ao*) whence it is taken to all parts of the body to be finally returned by thin-walled vessels, called *veins*, to the sinus venosus. The ventral position of the heart and the fact that the blood is sent directly from the heart to the respiratory organs are characteristic vertebrate features: so also is the circumstance that the blood from the stomach, intestine, &c., is taken by a specially modified portion of the ventral vessel (portal vein) through the liver on its way to the heart. The blood is red, containing, in addition to leucocytes, oval corpuscles coloured by hæmoglobin (see p. 58).

The excretory organs are a pair of kidneys (*K*) situated at the posterior end of the dorsal region of the cœlome, and opening by ducts, the *ureters* (*Ur*), into the cloaca. Development shows that they consist of an aggregation of nephridia (*Nph*), the nephrostomes of which open in the young and sometimes throughout life, into the cœlome, while the nephridiopores discharge not directly on the exterior, but into a common tube.

The gonads (ovaries, *Ovy*, or spermaries) are situated in the anterior part of the cœlome, attached by peritoneum to its dorsal wall. The sex-cells are differentiated from cœlomic epithelium. The gonaducts of both sexes (*Ovd*) are developed from the nephridial system of the embryo.

As already stated, the central nervous system is contained in a cavity (*C. Sp. Cav*) of the dorsal body-wall, and is therefore far removed from the ectoderm from which it originates. It consists of a long cylindrical rod, the *spinal cord* (*Sp. Cd*) which is continued in front into a complicated brain (*Br*). It has the further peculiarity of being hollow, a more or less cylindrical cavity, the *neurocœle* (*n. cœ*) extending through its whole length.

The possession of a hollow nervous system lying altogether dorsal to the enteric canal and cœlome, of either a notochord or a chain of vertebral centra below the nervous system, and of pharyngeal pouches communicating with the exterior, are the three most characteristic features of the vertebrate phylum.

The organs of sense are highly developed, and consist of paired olfactory sacs, eyes, and auditory sacs situated in the head, together with an extensive system of integumentary organs. Their sensory cells are in every case ectodermal.

The eggs are very large, and are impregnated within the

body of the female. In the common Dog-fish (*Scyllium*) they are laid shortly after impregnation, each enclosed in a horny egg-shell: in the Piked Dog-fish (*Acanthias*) and the Smooth Hound (*Mustelus*) they are retained in the oviduct until the adult form is assumed.

LESSON XXVIII

MOSSES.

In the four previous lessons we have traced the advance in organization of animals from the simple diploblastic Hydra to the complicated triploblastic forms which constitute the five higher phyla of the animal kingdom. We have now to follow in the same way the advance in structure of plants. The last member of the vegetable kingdom with which we were concerned was Nitella (p. 206), a solid aggregate, exhibiting a certain differentiation of form and structure, but yet composed of what were clearly recognizable as cells, there being, as in Hydra, none of that formation of well-marked tissues which is so noticeable a feature in Polygordius as in other animals above the Cœlenterata.

Taking Nitella as a starting point, we shall see that among plants, as among animals, there is an increasing differentiation in structure and in function as we ascend the series. The first steps in the process are well illustrated by a consideration of that very abundant and beautiful group of plants, the Mosses. In spite of the variations in detail met with in different genera of the group, the essential features of their organization are so constant that the following description will be found to apply to any of the common forms.

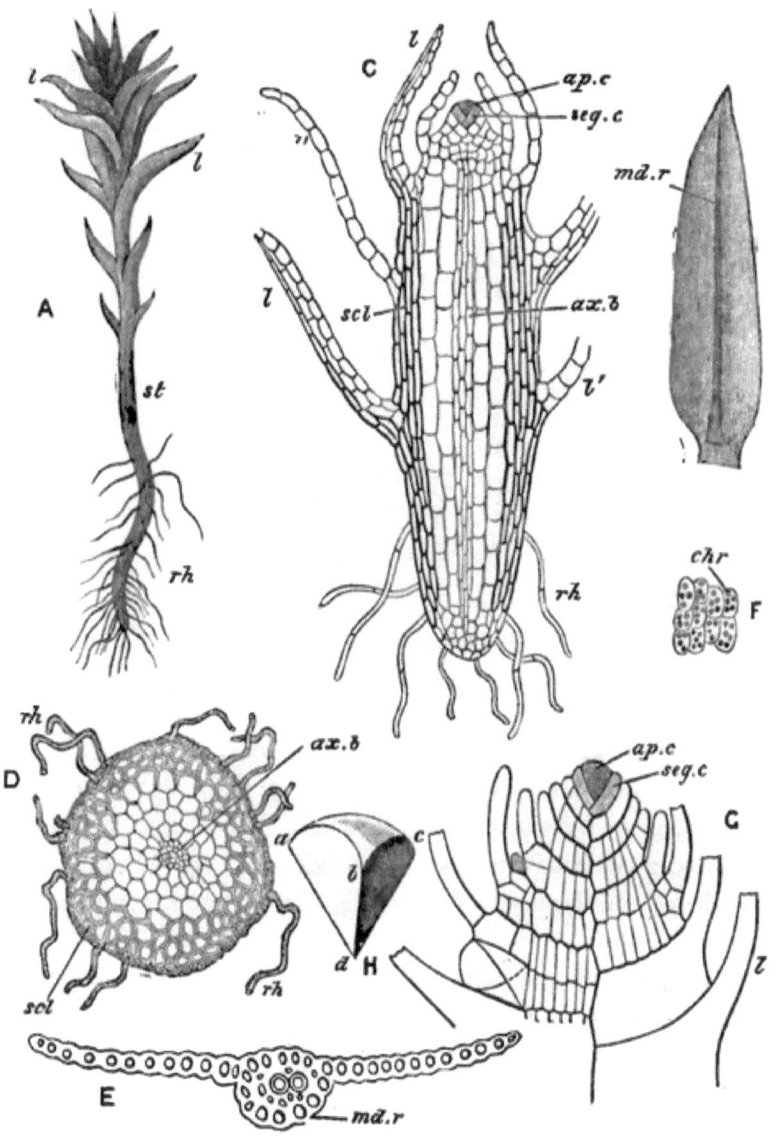

Fig. 80.—The Anatomy and Histology of Mosses.
A, Entire plant of *Funaria hygrometrica*, showing stem (*st*), leaves (*l*), and rhizoids (*rh*). (× 6.)
B, leaf of the same, showing midrib (*md. r*) and lateral portions. (× 25.)

C, semi-diagrammatic vertical section of a moss, showing the arrangement of the tissues. The stem is formed externally of sclerenchyma (*scl*), and contains an axial bundle (*ax. b*): in some of the leaves (*l*) the section passes through the midrib, in others (*l¹*) through the lateral portion: the stem ends distally in an apical cell (*ap. c*), from which segmental cells (*seg. c*) are separated.

D, transverse section of the stem of *Bryum roseum*, showing sclerenchyma (*scl*), axial bundle (*ax. b*), and rhizoids (*rh*). (× 60.)

E, transverse section of a leaf of *Funaria*, showing the midrib (*md. r*) formed of several layers of cells, and the lateral portions one cell thick. (× 150.)

F, small portion of the lateral region of the same, showing the form of the cells and the chromatophores (*chr*). (× 150.)

G, distal end of the stem of *Fontinalis antipyretica* in vertical section, showing the apical cell (*ap. c*) giving rise to segmental cells (*seg. c*), which by subsequent division form the segments of the stem with the leaves: the thick lines show the boundaries of the segments.

H, diagram of the apical cell of a moss in the form of a tetrahedron with rounded base *abc* and three flat sides *abd, bcd, acd*.

(D after Sachs; G after Leitgeb.)

The plant consists of a short slender stem (Fig. 80, A, *st*), from which are given off structures of two kinds, rhizoids or root-hairs (*rh*), which pass downwards into the soil, and leaves (*l*), which are closely set on the stem and its branches. As in Nitella (p. 208) the portion of the stem from which a leaf arises is called a node, and the part intervening between any two nodes an internode, while the name segment is applied to a node with the internode next below it. At the upper or distal end of the stem the leaves are crowded, forming a terminal bud.

Owing to the opacity of the stem, its structure can only be made out by the examination of thin sections (c and D). It is a solid aggregate of close-set cells which are not all alike, but exhibit a certain amount of differentiation. In the outer two or three rows the cells (*scl*) are elongated in the direction of the length of the stem, so as to have a spindle-shape, and their walls are greatly thickened and of a reddish colour. They thus form a protective and supporting tissue, to which the name *sclerenchyma* is applied. Running longitudinally through the centre of the stem is a mass of tissue (*ax. b*) distinguished by its small, thin-walled cells, and constituting the *axial bundle*.

The leaves (B) are shaped like a spear-head, pointed distally, and attached proximally by a broad base to the stem. The axial portion (B and E, *md. r.*, C. *l*) consists of several layers of somewhat elongated cells and is called the *midrib*: the lateral portions (E and F : C, *l'*) are formed of a single layer of short cells. Thus the leaf has, for the most part, the character of a superficial aggregate. The cells contain oval chromatophores (F, *chr*).

The rhizoids (C and D, *rh*) are linear aggregates, being formed of elongated cells, devoid of chlorophyll, arranged end to end.

In the terminal bud the leaves, as in Nitella (pp. 208 and 210), arch over the growing point of the stem, which in this case also is formed of a single apical cell (C and G, *ap. c*). But in correspondence with the increased complexity of the plant, the apical cell is not a hemisphere from which new segments are cut off parallel to its flat base, but has the form (H) of an inverted, three-sided pyramid or tetrahedron, the rounded base of which (*abc*) forms the apex of the stem while segments (*seg. c*) are cut off from each of its three triangular sides in succession.

The best way to understand the apical growth of a moss is to cut a tetrahedron with rounded base out of a carrot or turnip : this represents the apical cell (H) : then cut off a slice parallel to the side *abd*, a second parallel to *bcd*, and a third parallel to *acd*: these represent three successively formed segments. Now imagine that after every division the tetrahedron grows to its original size, and a very fair notion will be obtained of the way in which the successive segments of the moss-stem are formed by the fission in three planes of the apical cell. Each segment (c and g, *seg. c*) immediately after its separation divides and subdivides, producing a mass of cells from which a projection grows out forming a leaf, and in this way the stem increases in length and the leaves in number.

Asexual reproduction takes place in various ways: all of them are, however, varieties of budding, and the buds always arise in the form of a linear aggregate of cells called a *protonema*: from this the moss-plant develops in the same way as from the protonema arising from a spore (p. 339).

The gonads are developed at the extremity of the main stem or one of its branches, and are enclosed in a tuft of leaves often of a reddish colour—the terminal bud of the fertile shoot or so-called "flower" of the moss.

The spermary (Fig. 81, A^1, A^2) is an elongated club-shaped body consisting of a solid mass of cells, the outermost of which form the wall of the organ, while the inner (A^3) become converted into sperms. The latter (A^4) are spirally coiled and provided with two cilia: they are liberated by the rupture of the wall of the spermary at its distal end (A^2).

The ovaries [1] (see Preface, p. x, and p. 381) (B^1, B^2) may or may not occur on the same plant as the spermaries, some mosses being monœcious, others diœcious. Like the spermaries, they consist at first of a solid mass of cells which assumes the form of a flask, having a rounded basal portion or *venter* (v) and a long *neck* (n). The outer layer of cells in the neck and the two outer layers in the venter form the wall of the ovary, the internal cells are arranged in a single axial row at first similar to those of the wall. As the ovary develops, the proximal or lowermost cell of the axial row

---

[1] The ovary of mosses, ferns, &c., is usually called an *archegonium*: the spermary, as in the lower plants, an antheridium.

takes on the character of an ovum (B^2, *ov*); the others, called *canal cells* (*cn. c*) are converted into mucilage, which by its expansion forces open the mouth of the flask and thus makes a clear passage from the exterior to the ovum (B^3).

Through the passage thus formed a sperm makes it way and conjugates with the ovum, producing as usual an oosperm or unicellular embryo.

The development of the embryo is at first remarkably like what we have found to take place in Hydroids (p. 248). The oosperm divided into two cells by a wall at right angles to the long axis of the ovary: each of these cells divides again repeatedly, and there is produced a solid multicellular embryo or *polyplast* (c^1, *spgnm*).

Very early, however, the moss-polyplast exhibits a striking difference from the animal polyplast or morula: one of its cells—that nearest the neck of the ovary—takes on the character of an apical cell, and begins to form fresh segments like the apical cell of the stem. Thus the plant embryo differs almost from the first from the animal embryo. In the animal there is no apical cell: all the cells of the polyplast divide and take their share in the formation of the permanent tissues. In the plant one cell is at a very early period differentiated into an apical cell, and from it all cells thereafter produced are, directly or indirectly, derived.

The embryo continues to grow, forming a long rod-like body (c^2, *spgnm*) the base of which becomes sunk in the tissue of the moss-stem, while its distal end projects vertically upwards, covered by the distended venter (*v*) of the ovary. Gradually it elongates more and more and its distal end dilates: the embryo has now become a *sporogonium*, consisting of a slender stalk (c^4, *st*) bearing a vase-like capsule or urn (*u*) at its distal end. In the meantime the elongation of the stalk has caused the rupture of the enveloping

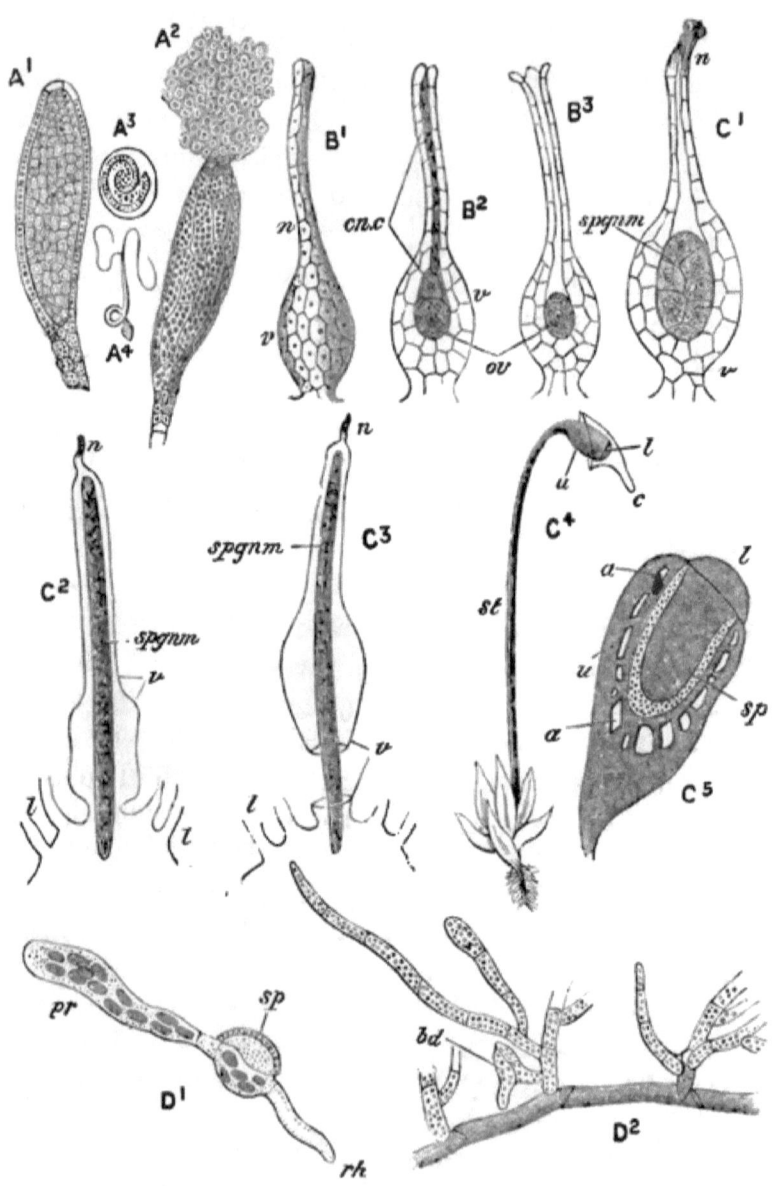

Fig. 81.—Reproduction and Development of Mosses.
A^1, A spermary of Funaria in optical section, showing the wall enclosing a central mass of sperm-cells: A^2, the same from the surface discharging its sperms. (× 300.)

A^3, a sperm-cell with enclosed sperm : A^4, a free-swimming sperm. (× 800.)

B^1, an ovary of Funaria, surface view, showing venter (v) and neck (n) : B^2, the same in optical section, showing ovum (ov) and canal cells ($cn. c$) : B^3, the same after disappearance of the canal cells : the neck is freely open, and the ovum (ov) exposed. (× 200.)

C^1, ovary with withered neck containing an embryo (*spgnm*) in the polyplast stage (× 200) : in C^2 the ovary, consisting of swollen venter (v) and shrivelled neck (n), encloses a young sporogonium (*spgnm*) ; the distal end of the stem is shown with bases of leaves (l) ; in C^3 the venter has ruptured, forming a proximal portion or sheath and a distal portion or calyptra which is carried up by the growth of the sporogonium. (× 10.)

C^4, a small plant of Funaria with ripe sporogonium consisting of seta (st), with urn (u) and lid (l) covered by the calyptra (c).

C^5, diagrammatic vertical section of urn (u), showing lid (l), air spaces (a), and spores (sp).

D^1, a germinating spore of Funaria, showing ruptured outer coat (sp) and young protonema (pr) with rhizoid (rh). (× 550.)

D^2, portion of protonema of the same, showing lateral bud (bd), from which the leafy plant arises. (× 90.)

(A and D after Sachs ; B, C^1, and C^3 altered from Sachs.)

venter of the ovary (C^3) : its proximal part remains as a sort of sheath round the base of the stalk, while its distal portion, with the shrivelled remains of the neck (n), is carried up by the elongation of the sporogonium and forms an extinguisher-like cap or *calyptra* (C^4, c) over the urn.

As development goes on, the distal end of the urn becomes separated in the form of a lid (C^4, C^5, l), and certain of the cells in its interior, called *spore-mother cells*, divide each into four daughter cells, which acquire a double cell-wall and constitute the *spores* (C^5, sp) of the moss.

When the spores are ripe the calyptra falls off or is blown away by the wind, the lid separates from the urn, and the spores are scattered.

In germination, the protoplasm of the spore covered by the inner layer of the cell-wall protrudes through a split in the outer layer (D^1, sp) and grows into a long filament, the *protonema* (pr.), divided by oblique septa into a row of cells. The protonema—which it will be observed is a simple linear aggregate—branches, and may form a closely-matted mass of filaments. Sooner or later small lateral buds (D^2, *bd*) appear at various places on the protonema: each of these takes on the form of a three-sided pyramidal apical cell, which then proceeds to divide in the characteristic way (p. 335), forming three rows of segments from which leaves spring. In this way each lateral bud of the protonema gives rise to a moss-plant.

Obviously we have here a somewhat complicated case of alternation of generations (see p. 220). The gamobium or sexual generation is represented by the moss-plant, which originates by budding and produces the sexual organs, while the agamobium consists of the sporogonium, developed from the oosperm and reproducing by means of spores. The protonema, arising from a spore and producing the leafy plant by budding, is merely a stage of the gamobium.

The nutrition of mosses is holophytic; but there is a striking differentiation of function correlated with terrestrial habits. In Nitella the entire organism is submerged in water and all the cells contain chlorophyll, so that decomposition of carbon dioxide and absorption of an aqueous solution of salts are performed by all parts alike, every cell being nourished independently of the rest. In the moss, on the other hand, the rootlets are removed from the influence of light and contain no chlorophyll: hence they cannot decompose carbon dioxide; but, being surrounded by moist soil, are in the most favourable position for absorbing water and mineral salts. The stem, again, is converted into an organ of support: the thickness of its external cells prevents absorption and it contains no chlorophyll. Hence the function of decomposing carbon dioxide is confined to the leaves.

We have thus as an important fact in the nutrition of an ordinary terrestrial plant that its carbon is taken in at one place, its water, nitrogen, sulphur, potassium, &c., at another. But as all parts of the plant require all these substances it is evident that there must be some means by which the root can obtain a supply of carbon, and the leaves a supply of elements other than carbon. In other words, we find for the first time in the ascending series of plants, just as we did in ascending from the simple Hydra to the complex Polygordius (p. 281) the need for some contrivance for the distribution of food-materials.

The way in which this distributing process is performed has been studied chiefly in the higher plants, but its essential features are probably the same for mosses.

Water is continually evaporating from the surface of the leaves, its place being as constantly supplied by water—with salts in solution—taken in by the rhizoids. This *transpiration*, or the giving off of water from the leaves, is one important factor in the process under consideration, since it ensures a constant upward current of water, or, more accurately, of an aqueous solution of mineral salts. The withering of a plucked moss-plant is of course due to the fact that when the roots are not embedded in moist soil or in water, transpiration is no longer balanced by absorption.[1] In the higher plants it has been found that the root-hairs have an absorbent action independent of transpiration, so that water may be absorbed in the absence of leaves.

By the transpiration current, then, the leaves are kept constantly supplied with a solution of mineral salts derived from the soil, and are thus nourished like any of the aquatic green plants considered in previous lessons : by the double

[1] Mosses, however, unlike most higher plants, can absorb water by their leaves.

decomposition of water and carbon dioxide a carbo-hydrate is formed: this, by further combination with the nitrogen of the absorbed ammonium salts or nitrates, forms simple nitrogenous compounds, and from these, probably through a long series of mesostates or intermediate products, protoplasm is finally manufactured.

In this way the food supply of the green cells of the leaves is accounted for, but we have still to consider that of the colourless cells of the stem and rhizoids, which, as we have seen, are supplied by the transpiration current with everything they require except carbon, and this, owing to their possessing no chlorophyll, they are unable to take in in the form of carbon dioxide.

As a matter of fact the chlorophyll-containing cells of the leaves have to provide not only their own food, but also that of their not-green fellows. In addition to making good the waste of their own protoplasm they produce large quantities of plastic products (see p. 33) such as grape sugar, and simple nitrogenous compounds like asparagin, and these pass by diffusion from cell to cell until they reach the uttermost parts of the plant, such as the centre of the stem and the extremities of the rhizoids. The colourless cells are in this way provided not only with the salts contained in the ascending transpiration current, but with carbo-hydrates and nitrogenous compounds. From these they derive their nutriment, living therefore like yeast-cells in Pasteur's solution, or like Bacteria in an organic infusion.

We see then that the colourless cells of the stem and rhizoids are dependent upon the green cells of the leaves for their supplies. Like other cells devoid of chlorophyll they are unable to make use of carbon dioxide as a source of carbon, but require ready-made carbo-hydrates, the manufacture of which is continually going on, during daylight, in the chlorophyll-containing cells of the leaves. This striking division of labour is the most important physiological difference between mosses and the more lowly organized green plants described in previous lessons.

LESSON XXIX

FERNS

WE saw in the previous lesson that in mosses there is a certain though small amount of histological differentiation, some cells being modified to form sclerenchyma, others to form axial bundles. We have now to consider a group of plants which may be considered to be, in this respect, on much the same morphological level as Polygordius, the adult organism being composed not of a mere aggregate of simple cells, but of various well-marked tissues.

A fern-plant has a strong stem which in some forms, such as the common Bracken (*Pteris aquilina*) is a horizontal underground structure, and is hence often incorrectly considered as a root: in others it creeps over the trunks of trees or over rocks: in others again, such as the tree-ferns, it is vertical, and may attain a height of three or four metres. From the stem are given off structures of two kinds—the leaves, which present an almost infinite variety of form in the various species, and the numerous slender roots. In some cases, such as the tree-ferns and the common Male Shield-fern (*Aspidium filix-mas*), the plant ends distally in a terminal bud, consisting, as in Nitella and mosses, of the growing end of the stem over-arched by leaves : in others such as Pteris, the stem ends in a blunt, knob-like extremity quite uncovered by leaves. On the proximal portion of the stem are usually found the withered remains of the leaves of previous seasons, or the scars left by their fall. The roots are given off from the whole surface of the stem, often covering it with a closely-matted mass of dark brown fibres.

When the stem is cut across transversely (Fig. 82, A) it is seen, even with the naked eye, to consist of three well marked tissues. The main mass of it is formed of a whitish substance, soft and rather sticky to the touch, and called *ground-parenchyma* (*par*): this is covered by an external layer of very hard tissue, dark brown or black in colour, the *hypodermis* (*hyp*): bands of a similar hard brown substance are variously distributed through the parenchyma, and constitute the *sclerenchyma* (*scl*): and interspersed with these are rounded or oval patches of a yellowish colour ($V.B$) harder than the parenchyma, but not so hard as the sclerenchyma, and called *vascular bundles*.

The general distribution of these tissues can be made out by making longitudinal sections of the stem in various planes or by cutting away the hypodermis, and then scraping the parenchyma from the vascular bundles and bands of sclerenchyma. The hypodermis is found to form a more or less complete hard sheath or shell to the stem, while the internal sclerenchyma and vascular bundles form longitudinal bands and rods imbedded in the parenchyma, and serve as a sort of supporting framework or skeleton.

The minute structure of the stem can be made out by the examination either of very thin longitudinal and transverse sections, or of a bit of stem which has been reduced to a pulp by boiling in nitric acid with the addition of a few crystals of potassium chlorate : by this process the various

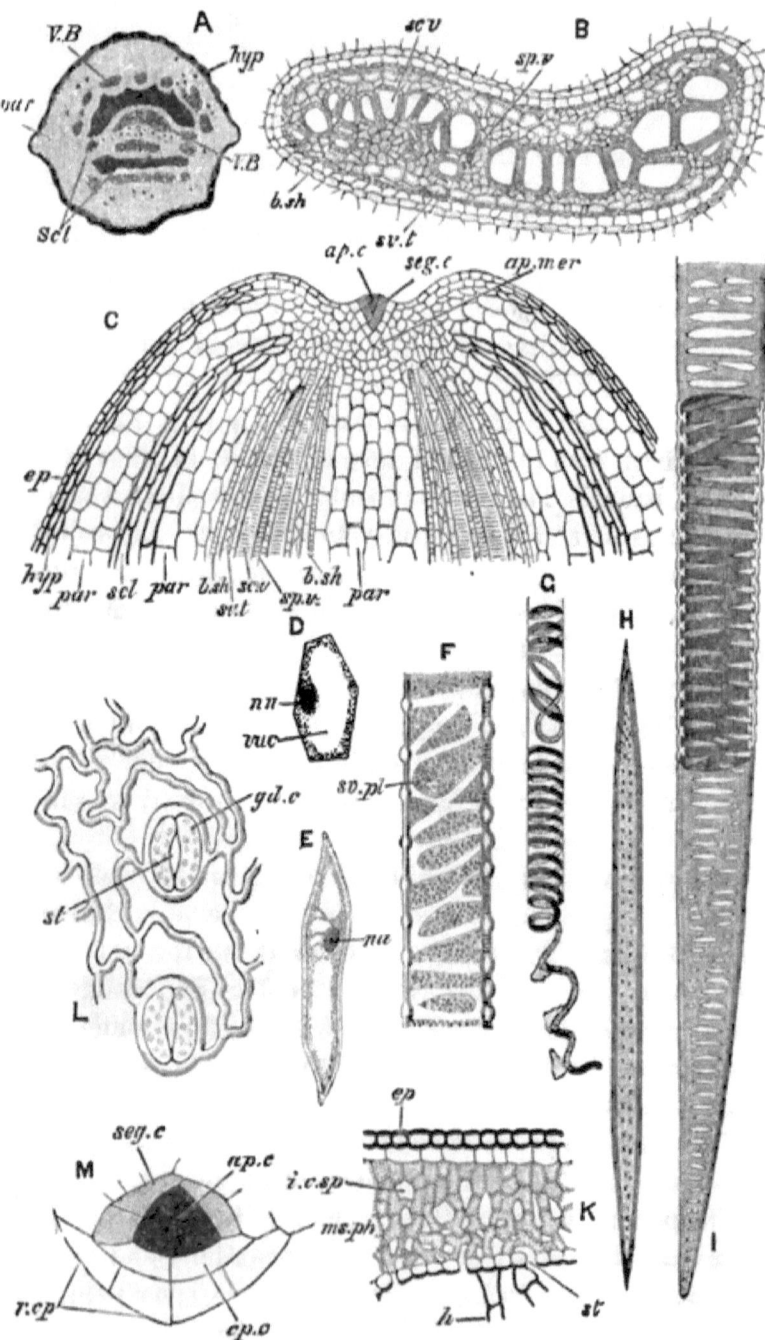

FIG. 82.—Anatomy and Histology of Ferns.

A, Transverse section of the stem of *Pteris aquilina*, showing hypodermis (*hyp*), ground parenchyma (*par*), sclerenchyma (*scl*), and vascular bundles (*V. B*). (× 2.)

B, transverse section of a vascular bundle, showing bundle-sheath (*b. sh*), sieve-tubes (*sv. t*), **scalariform vessels (*sc. v*)**, and spiral vessels (*sp. v*). (× 6.)

C, semi-diagrammatic vertical section of the growing point of the stem, showing apical cell (*ap. c*), segmental cells (*seg. c*), and apical meristem (*ap. mer*) passing into permanent tissue consisting of epidermis (*ep*), hypodermis (*hyp*), ground parenchyma (*par*), sclerenchyma (*scl*), and vascular bundles in which the sheath (*b. sh*), sieve-tubes (*sv. t*), scalariform vessels (*sc. v*), and spiral vessels (*sp. v*) are indicated.

D, a single parenchyma cell, showing nucleus (*nu*), and vacuole (*vac*).

E, cell of hypodermis.

F, portion of a sieve-tube, showing sieve-plates (*sv. pl*).

G, portion of a spiral vessel with the spiral fibre partly unrolled at the lower end.

H, fibre-like cell of sclerenchyma.

I, portion of a scalariform vessel, part of the wall being supposed to be removed.

K, vertical section of a leaf of Pteris, showing upper and lower epidermis (*ep*), mesophyll cells (*ms. ph*), with intercellular spaces (*i. c. sp*), a stoma (*st*) in the lower epidermis, and hairs (*h*).

L, surface view of epidermis of leaf of Aspidium, showing two stomata (*st*) with their guard-cells (*gd. c*).

M, vertical section of the end of a root, showing apical cell (*ap. c*), segmental cells (*seg. c*), and root-cap (*r. cp*) with its youngest cap-cells marked *cp. c*.

(A, B, and D–K after Howes; M from Sachs, slightly altered.)

tissue elements are separated from one another, and can be readily examined under a high power.

By combining these two methods of sectioning and dissociation, the parenchyma is found to consist of an aggregate of polyhedral cells (D) considerably longer than broad, their long axes being parallel with that of the stem itself. The cells are to be considered as right cylinders which have been converted into polyhedra by mutual pressure. They have the usual structure, and their protoplasm is frequently loaded with large starch-grains. They do not fit quite closely together, but spaces are left between them, especially at the angles, called *intercellular spaces*.

The cells of the hypodermis (E) are proportionally longer than those of the parenchyma, and are pointed at each end : they contain no starch. Their walls are greatly thickened, and are composed not of cellulose but of *lignin*, a carbohydrate allied in composition to cellulose, but containing a larger proportion of carbon Schulze's solution, which, as we have seen, stains cellulose blue, imparts a yellow colour to lignin.

Outside the hypodermis is a single layer of cells (C, *ep*) not distinguishable by the naked eye and forming the actual external layer of the stem : the cells have slightly thickened, yellowish-brown walls, and constitute the *epidermis*. From many of them are given off delicate filamentous processes consisting each of a single row of cells : these are called *hairs*.

In the sclerenchyma the cells (H) are greatly elongated, and pointed at both ends, so as to have the character rather of fibres than of cells. Their walls are immensely thickened and lignified, and present at intervals oblique markings due to narrow but deep clefts : these are produced by the deposition of lignin from the surface of the protoplasm (see p. 32) being interrupted here and there, instead of going on continuously as in the case of a cell-wall of uniform thickness.

The vascular bundles have in transverse section (B) the appearance of a very complicated network, with meshes of varying diameter. In longitudinal sections (C) and in dissociated specimens they are found to be partly composed of cells, but to contain besides structures which cannot be called cells at all.

In the centre of the bundle are a few narrow cylindrical tubes (B and C, *sp. v.*) characterized at once by a spiral marking, and hence called *spiral vessels*. Accurate examination shows that their walls (G) are for the most part thin, but are thickened by a spiral fibre, just as a paper tube might be strengthened by gumming a spiral strip of pasteboard to its inner surface. These vessels are of considerable length, and are open at both ends: moreover they contain no protoplasm, but are filled with either air or water: they have therefore none of the characteristics of cells. They are shown, by treatment with Schulze's solution, to be composed of lignin.

Surrounding the group of spiral vessels, and forming the large polygonal meshes so obvious in a transverse section, are wide tubes (B and C, *sc. v*) pointed at both ends and fitting against one another in longitudinal series by their oblique extremities. They have transverse markings like the rungs of a ladder, and are hence called *scalariform vessels*. The markings (1) are due to wide transverse pits in the otherwise thick lignified walls: in the oblique ends by which the vessels fit against one another the pits are frequently replaced by actual slits, so that a longitudinal series of such vessels forms a continuous tube containing, like the spiral vessels, air or water, but no protoplasm. In most ferns the terminal walls are not thus perforated, and the elements are then called *tracheides*.

The presence of these vessels—spiral and scalariform—is the most important histological character separating ferns and mosses. The latter group and all plants below them are composed exclusively of cells: ferns and all plants above them contain vessels in addition, and are hence called *vascular plants*.

The vessels, together with small parenchyma-cells interspersed among them, make up the central portion of the vascular bundle, called the *wood* or *xylem*. The peripheral portion is formed of several layers of cells composing the *bast* or *phloëm*, and surrounding the whole is a single layer of small cells, the *bundle-sheath* (*b. sh*).

The cells of the phloëm are for the most part parenchymatous, but amongst them are some to which special attention must be drawn. These (B and C, *sv. t*), are many times as long as they are broad, and have on their walls irregular patches or *sieve-plates* (F, *sv. pl.*) composed of groups of minute holes through which the protoplasm of the cell is continuous with that of an adjacent cell. The transverse or oblique partitions between the cells of a longitudinal series are also perforated, so that a row of such cells forms a *sieve-tube* in which the protoplasm is continuous from end to end. We have here, therefore, as striking an instance of protoplasmic continuity as in the deric epithelium and certain other tissues of Polygordius (see p. 276).

The distal or growing end of the stem terminates in a blunt *apical cone* or punctum vegetationis (C), surrounded by the leaves of the terminal bud in the case of vertical stems, or sunk in a depression and protected by close-set hairs in the underground stem of the bracken. A rough longitudinal section shows that, at a short distance from the apical cone, the various tissues of the stem—epidermis, parenchyma, sclerenchyma, and vascular bundles—merge insensibly into a whitish substance, resembling parenchyma to the naked eye, and called *apical meristem* (*ap. mer*).

Thin sections show that the summit of the apical cone is occupied by a wedge-shaped apical cell (*ap. c*) which in vertical stems is three-sided like that of mosses (Fig. 80, H, p. 335), while in the horizontal stem of Pteris it is two-sided. As in mosses, segmental cells (*seg. c*) are cut off from the three (or two) sides of the apical cell in succession, and by further division form the apical meristem (*ap. mer*), which consists of small, close-set cells without intercellular spaces. As the base of the apical cone is reached, the meristem is found to pass insensibly into the permanent tissues, the cells near the surface gradually merging into epidermis and hypodermis, those towards the central region into sclerenchyma and the various constituents of the vascular bundles, and those of the intermediate regions into parenchyma.

The examination of the growing end of the stem shows us how the process of apical growth is carried on in a complicated plant like the fern. The apical cell is continually undergoing fission, forming a succession of segmental cells; these divide and form the apical meristem, which is thus being constantly added to at the growing end by the formation and subsequent fission of new segmental cells: in this way the apex of the stem is continually growing upwards or forwards. But at the same time the meristem cells farthest from the apex begin to differentiate: some elongate but slightly, increasing greatly in size, and become parenchyma cells: others by elongation in the direction of length of the stem and by thickening and lignification of the cell-wall become sclerenchyma cells: others again elongate greatly, become arranged end to end in longitudinal rows, and, by the loss of their protoplasm and of the transverse partitions between the cells of each row, are converted into vessels—spiral or scalariform according to the character of their walls. Thus while the epidermis, parenchyma, and sclerenchyma are formed of cells, the spiral and scalariform vessels are *cell-fusions*, or more accurately cell-wall-fusions, being formed by the union in a longitudinal series of a greater or less number of cell-walls. It will be remembered that the muscle plates of Polygordius are proved by the study of development to be cell-fusions (p. 305).

We thus see that every cell in the stem of the fern was once a cell in the apical meristem, that every vessel has arisen by the concrescence of a number of such cells, and that the meristem cells themselves are all derived, by the ordinary process of binary fission, from the apical cell. In this way the concurrent processes of cell-division, cell-differentiation, and cell-fusion result in the production of the various and complex tissues of the fully-formed stem.

The leaves vary greatly in form in the numerous genera and species of ferns: they may consist of an unbranched *stalk* bearing a single expanded green *blade:* or the stalk may be more or less branched, its ramifications bearing the numerous subdivisions of the blade, or *pinnæ*.

The anatomy of the leaf, like that of the stem, can be readily made out by a rough dissection. The leaf-stalk and its branches have the same general structure as the stem, consisting of parenchyma coated externally with epidermis and strengthened internally by vascular bundles, which are continuous with those of the stem. But the blade, or in the case of a compound leaf, the pinna, has a different and quite peculiar structure. It is invested by a layer of epidermis which can be readily stripped off as an extremely thin, colourless membrane, exposing a soft, green substance, the leaf parenchyma or *mesophyll*. The leaf is marked externally by a network of delicate ridges, the *veins;* these are shown by dissection to be due to the presence of fine white threads which ramify through the mesophyll, and can be proved by tracing them into the leaf-stalk to spring from its vascular bundles, of which they are in effect the greatly branched distal ends.

Microscopic examination shows the epidermis of the leaf (K, *ep* and L) to consist of flattened, colourless cells of very irregular outline and fitting closely to one another like the parts of a child's puzzle. Amongst them are found at intervals pairs of sausage-shaped cells (*gd. c*) placed with their concavities towards one another so as to bound a narrow slit-like aperture (*st*). These apertures, which are the only intercellular spaces in the epidermis, are called *stomates:* the cells bounding them are the *guard-cells*, and are distinguished from the remaining epidermic cells by the possession of a few chromatophores.

The mesophyll, which as we have seen occupies the whole space between the upper and lower epidermis, is formed of thin-walled cells loaded with chromatophores (K, *ms.ph*) and therefore of a deep green colour. The cells in contact with the upper epidermis are cylindrical, and are arranged vertically in a single row: those towards the lower surface are very irregular both in form and arrangement. Large intercellular spaces (*i. c. sp*) occur between the mesophyll-cells and communicate with the outer air through the stomates.

The leaves arise as outgrowths of the distal or growing end of the stem, each originating from a single segmental cell of the apical cone.

The fern is the first plant we have yet considered which possesses true roots, the structures so-called differing fundamentally from the simple rhizoids of Nitella and the mosses. Instead of being mere linear aggregates of cells, they agree in general structure with the stem from which they spring, consisting of an outer layer of epidermis within which is parenchyma strengthened by bands of sclerenchyma and by a single vascular bundle in the centre. The epidermic cells give rise to unicellular prominences, the *root-hairs*.

The apex of the root, like that of the stem, is formed of a mass of meristem in which a single wedge-shaped apical cell (Fig. 82, M, *ap. c*) can be distinguished. But instead of the base of this cell forming the actual distal extremity, as in the stem (compare c), it is covered by several layers of cells which constitute the *root-cap* (*r.cp*). In fact the apical cell of the root divides not only by planes parallel to its three sides, but also by a plane parallel to its base, and in this way produces not only three series of segmental cells (*seg. c*) which afterwards subdivide to form the apical meristem, but also a series of cap-cells (*cp. c*) which form a protective sheath over the tender growing end of the root as it forces its way through the soil.

Roots are also peculiar in their development. Instead of being, like leaves, prominences of the superficial tissues of the stem, they arise from a layer of cells immediately external to the vascular bundles, and in growing force their way through the superficial portion of the stem, through a fissure from which they finally emerge. They are thus said to be *endogenous* in origin while leaves are *exogenous*.

The nutrition of ferns is carried on in much the same way as in mosses (see p. 340). Judging from the analogy of flowering plants it would seem that the ascending current of water from the roots passes mainly through the xylem of the vascular bundles, while the descending current of nitrogenous and other nutrient matters for the supply of the colourless cells of the stem and roots passes chiefly through the phloëm and especially through the sieve-tubes. The absorption of water is effected by the root-hairs.

In the autumn there are found on the under surfaces of the leaves brown patches called *sori*, differing greatly in form and arrangement in the various genera, and formed of innumerable, minute, seed-like, bodies, the *sporangia* (Fig. 83, A), just visible to the naked eye. Each sorus or group of sporangia is covered by a fold of the epidermis of the leaf, called the *indusium*.

A sporangium is attached to the leaf by a multicellular stalk (*st*), and consists of a sac resembling two watch-glasses placed with their concave surfaces towards one another and their edges united by a thick rim (*an*). The sides are formed by thin flattened cells with irregular outlines, the rim or *annulus* of peculiarly shaped cells which are thin and broad at one edge (to the left in A), but on the other (to the right) are thick, strongly lignified, and of a yellowish-brown colour. The whole internal cavity is filled with spores (B, *sp*) having the form of tetrahedra with rounded edges, and each consisting of protoplasm containing a nucleus, and surrounded by a double wall of cellulose. A spore is therefore, as in mosses, a single cell.

Each sporangium arises from a single epidermic cell of the leaf. This divides repeatedly so as to form a solid mass of cells, of which the outermost become the wall of the sporangium while the inner are the spore-mother-cells. The latter divide each into four spores, as in mosses (p. 339).

As the spores ripen, the wall of the sporangium dries, and as it does so the thickened part of the annulus straightens out, tearing the thin cells and producing a great rent through which the spores escape (B).

When the spores are sown on moist earth they germinate, by the protoplasm, covered by the inner coat, protruding through the ruptured outer coat (C, *sp*) in the form of a short filament. This divides transversely, forming two cells, the proximal of which sends off a short rhizoid (*rh*). The resemblance of this stage to the young protonema of a moss is sufficiently obvious (see Fig. 81, D^1., p. 338).

Further cell-division takes place, and before long the distal cells divide longitudinally, a leaf-like body being produced, which is called the *prothallus* (D). This is at first

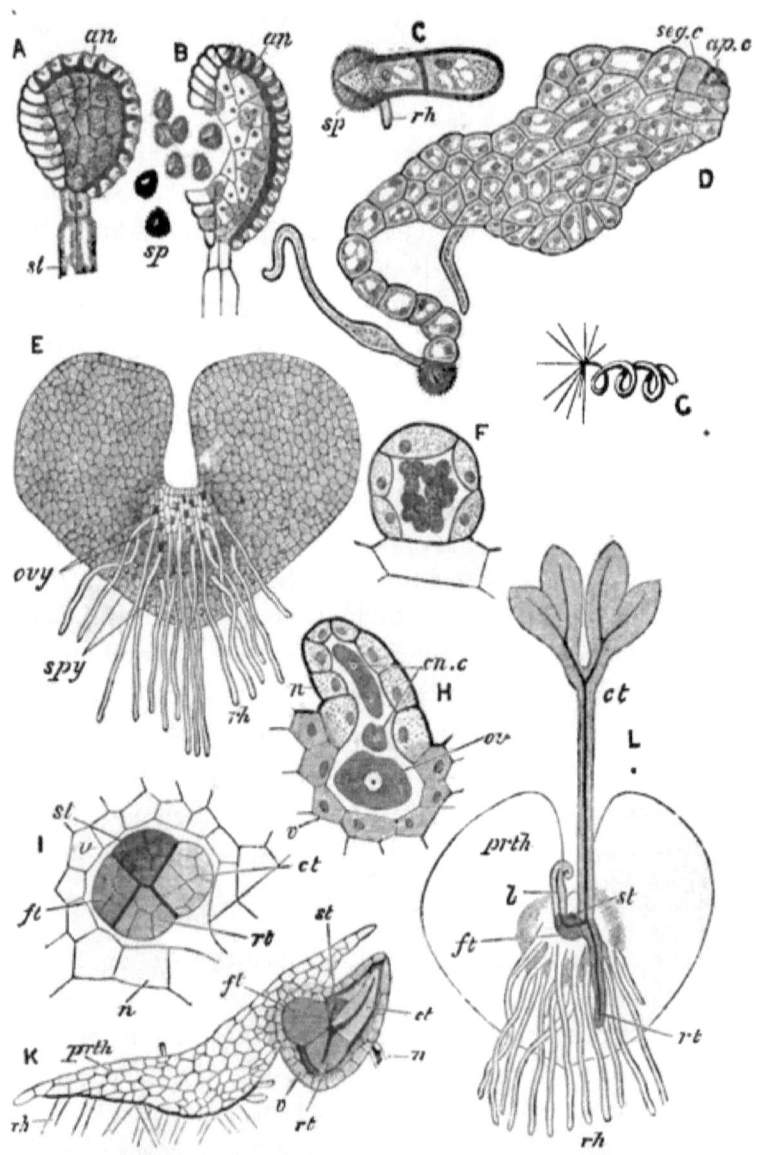

FIG. 83.—Reproduction and Development of Ferns.

A, Sporangium of *Pteris*, external view, showing stalk (*st*) and annulus (*an*).

B, the same, during dehiscence, the spores (*sp*) escaping.

C, a germinating spore, showing the ruptured outer coat (*sp*), and a rhizoid (*rh*) springing from the proximal cell of the rudimentary (two-celled) prothallus.

D, a young prothallus, showing spore, rhizoid (*rh*), apical cell (*ap. c*), and segmental cells (*seg. c*).

E, an advanced prothallus, from beneath, showing rhizoids (*rh*), ovaries (*ovy*), and spermaries (*spy*).

F, a mature spermary of *Pteris*, inverted (*i.e.* with its distal end directed upwards) so as to compare with Fig. 82, A.

G, a single sperm, showing coiled body and numerous cilia.

H, a mature ovary of *Aspidium*, inverted so as to compare with Fig. 82, B^2, showing venter (*v*), neck (*n*), ovum (*ov*), and canal cells (*cn. c*).

I, small portion of a prothallus of *Asplenium* in vertical section, showing the venter (*v*) and part of the neck (*n*) of a single ovary after fertilization. The venter contains an embryo just passing from the polyplast into the phyllula stage, and divided into four groups of cells, the rudiments respectively of the foot (*ft*), stem (*st*), root (*rt*), and cotyledon (*ct*).

K, vertical section of a prothallus (*prth*) of *Nephrolepis*, bearing rhizoids (*rh*), and a single ovary with greatly dilated venter (*v*) and withered neck (*n*). The venter contains an embryo in the phyllula stage, consisting of foot (*ft*), rudiments of stem (*st*), and root (*rt*), and cotyledon (*ct*) beginning to grow upwards.

L, prothallus (*prth*) with rhizoids (*rh*), bearing a young fern plant, consisting of foot (*ft*), rudiment of stem (*st*), first root (*rt*), cotyledon (*ct*), and first ordinary leaf (*l*). (After Howes.)

only one layer of cells thick, but it gradually increases in size, becoming more or less kidney-shaped (E), and as it does so its cells divide parallel to the surface, making it two and finally several cells in thickness. Thus the prothallus is at first a linear, then a superficial, and ultimately a solid aggregate. Root-hairs (*rh*) are produced in great number from its lower surface, and penetrating into the soil serve for the absorption of nutriment. At an early period a two-sided apical cell (D, *ap. c*) is differentiated, and gives off segmental cells (*seg. c*) in the usual way : an abundant formation of chromatophores also takes place at a very early period in the cells of the prothallus, which therefore resembles both in structure and in habit some very simple form of moss.

On the lower surface of the prothallus gonads (E, *spy*, *ovy*) are developed, resembling in their essential features those of mosses. The spermaries (*spy*) make their appearance first, being frequently found on very young prothalli. One of the lower cells forms a projection which becomes divided off by a septum : further division takes place, resulting in the differentiation (F) of an outer layer of cells forming the wall of the spermary, and of an internal mass of sperm-mother-cells in each of which a sperm is produced. The sperm (G) is a corkscrew-like body, probably formed from the nucleus of the cell, bearing at its narrow end a number of cilia which appear to originate from the protoplasm. To the thick end is often attached a globular body, also arising from the protoplasm of the mother-cell ; this is finally detached.

The ovaries (E and H, *ovy*) are not usually formed until the prothallus has attained a considerable size. Each arises, like a spermary, from a single cell cut off by a septum from one of the lower cells of the prothallus : the cell divides and forms a structure resembling in general characters the ovary of a moss (see Fig. 81, B, p. 338), except that the venter (H, *v*) is sunk in the prothallus, and is therefore a less distinct structure than in the lower type. As in mosses, also, an axial row of cells is early distinguished from those forming the wall of the ovary : the proximal of these becomes the ovum (*ov*), the others are the canal-cells (*cn. c*), which are converted into mucilage, and by their expansion force open the neck and make a clear passage for the sperm.

The sperms swarm round the aperture of the ovary and make their way down the canal, one of them finally conjugating with the ovum and converting it into an oosperm.

The early stages in the development of the embryo remind us, in their general features, of what we found to occur in mosses (p. 337). The oosperm first divides by a plane parallel to the neck of the ovary, forming two cells, an anterior nearest the growing or distal end of the prothallus, and a posterior towards its proximal end. Each of these divides again by a plane at right angles to the first, there being now an upper and a lower anterior, and an upper and a lower posterior cell : the lower in each case being that towards the downwardly directed neck of the ovary. Each of the four cells undergoes fission, the embryo then consisting of eight cells, two upper anterior (right and left), two lower anterior, two upper posterior, and two lower posterior. We thus get a multicellular but undifferentiated stage, the polyplast.

It will be remembered that in mosses the polyplast forms an apical cell, and develops directly into the sporogonium (p. 337). In the fern the later stages are more complex. One of the upper anterior cells remains undeveloped, the other (Fig. 83, I and K, *st*) takes on the form of a wedge-shaped apical cell, and, dividing in the usual way, forms a structure like the apex of the fern-stem, of which it is in fact the rudiment. The two upper posterior cells divide and subdivide, and form a multicellular mass called the *foot* (*ft*), which becomes embedded in the prothallus, and serves the growing embryo for the absorption of nutriment. One of the lower posterior cells remains undeveloped, the other (*rt*) takes on the form of the apical cell of a root, *i.e.*, of a wedge-shaped cell, which not only produces three sets of segmental cells from its sides but also cap-cells from its base (p. 354) : division of this cell goes on very rapidly, and a root is produced which at once grows downwards into the soil. Finally the two lower anterior cells undergo rapid fission, and develop into the first leaf of the embryo or *cotyledon* (*ct*), which soon begins to grow upwards towards the light.

Thus at a comparatively early stage of its development the fern-embryo has attained a degree of differentiation far beyond anything which occurs in the moss-embryo. The scarcely differentiated polyplast has passed into a stage which may be called the *phyllula*, distinguished by the possession of those two characteristic organs of the higher plants, the leaf and root.

Notice how early in development the essential features of animal or plant manifest themselves. In Polygordius the polyplast is succeeded by a gastrula distinguished by the possession of a digestive cavity: in the fern no such cavity is formed, but the polyplast is succeeded by a stage distinguished by the possession of a leaf and root. In the one case the characteristic organ for holozoic, in the other the characteristic organs for holophytic nutrition make their appearance, and so mark the embryo at once as an animal or plant. We may say then that while the oosperm and the polyplast stages of the embryo are common to the higher plants and the higher animals, the correspondence goes no further, the next step being the formation in the animal of an enteron, in the plant of a leaf and root. In other words the phyllula is the correlative of the gastrula.

The cotyledon increases rapidly in size, and emerges between the lobes of the kidney-shaped prothallus (L): the root at the same time grows to a considerable length, the result being that the phyllula becomes a very obvious structure in close connection with the prothallus, and indeed appearing to be part of it. The two are actually, however, quite distinct, their union depending merely upon the fact that the foot of the phyllula is embedded in the tissue of the prothallus like a root in the soil. Hence the phyllula is related to the prothallus in precisely the same way as the sporogonium to the moss plant (compare Fig. 83, K, with Fig. 81, c^2, and Fig. 83, L, with Fig. 81, c^4).

The rudiment of the stem (L, *st*) continues to grow by the

production of fresh segments from its apical cell : leaves (*l*) are developed from the segments, and grow upwards parallel with the cotyledon. The leaves first formed are small and simple in structure, but those arising later become successively larger and more complicated, until they finally attain the size and complexity of the ordinary leaves of the fern. In the meantime new roots are formed ; the cotyledon, the foot, and the prothallus wither, and thus the phyllula, by the successive formation of new parts from its constantly growing stem, becomes a fern-plant.

We see that the life-history of the fern resembles in essentials that of the moss. In both, alternation of generation occurs, a gamobium or sexual generation giving rise, by the conjugation of ovum and sperm, to an agamobium or asexual generation, which, by an asexual process of spore-formation, produces the gamobium. But in the relative proportions of the two generations the difference is very great. What we know as the moss plant is the gamobium, and the agamobium is a mere spore-producing structure, never getting beyond the stage of a highly differentiated polyplast, and dependent throughout its existence upon the gamobium, to which it is permanently attached. What we know as the fern plant is the agamobium, a large and complex structure dependent only for a brief period of its early life upon the small and insignificant gamobium. Thus while the gamobium is the dominant phase in the life-history of mosses, the agamobium appearing like a mere organ, in ferns the positions are more than reversed—the agamobium may assume the proportions of a tree, while the gamobium is so small that its very existence is unknown to a large proportion of fern-collectors.

It follows from what has just been said that the various organs of a fern do not severally correspond with those of a moss. The leaves of a moss are not homologous with those of a fern, but are rather comparable to lobes of the prothallus : in the same way the rhizoids of a moss correspond not with the complicated roots of the fern but with the rhizoids of the prothallus.

LESSON XXX[1]

THE GENERAL CHARACTERS OF THE HIGHER PLANTS

In the 27th Lesson (p. 307) it was pointed out that a thorough comprehension of the structure and development of Polygordius would enable the student to understand the main features of the organization of all the higher animals.

In the same way the study of the fern paves the way to that of the higher groups of plants, all of which indeed, differ far less from the fern than do the various animal types considered in Lesson XXVII from Polygordius. We saw that the differences between these included matters of such importance as the presence or absence of segmentation and of lateral appendages, the characters of the skeleton, and the structure and position of the nervous system. In the higher plants, on the other hand, the essential organs—root, stem, and leaves—are, save in details of form, size, &c., practically the same in all: the tissues always consist of epidermis, ground-parenchyma, and vascular bundles, the latter being divisible into phloëm and xylem: the growing point both of stem and of root is formed of meristem, from which the permanent tissues arise; and the growing point of the root is

---

[1] Readers who have not studied botany, or at least examined types of the chief groups of plants, will derive little benefit from this lesson.

always protected by a root-cap, that of the stem being simply over-arched by leaves. Moreover an alternation of generations can be traced in all cases.

Plants may be conveniently divided into the following chief groups or phyla:

>*Algæ.*
>*Fungi.*
>*Muscineæ.*
>*Vascular Cryptogams.*
>>Filicinæ.
>>Equisetaceæ.
>>Lycopodineæ.
>*Phanerogams.*
>>Gymnosperms.
>>Angiosperms.

The *Algæ* are the lower green plants. They may be unicellular, or may take the form of linear, superficial, or solid aggregates: they never exhibit more than a limited amount of cell-differentiation. This group has been represented in the foregoing pages by Zooxanthella, diatoms, Vaucheria, Caulerpa, Monostroma, Ulva, Laminaria, and Nitella.[1]

The *Fungi* are the lower plants devoid of chlorophyll: some are unicellular, others are linear aggregates: in none is there any cell-differentiation worth mentioning. Saccharomyces, Mucor, Penicillium, and the mushroom belong to this group.

The position of some of the lower forms which have come under our notice is still doubtful. Bacteria, for instance, are considered by some authors to be Fungi, by others Algæ,

---

[1] By some authors Nitella is placed near the Muscineæ.

while others place them in a group apart. Diatoms also are sometimes placed in a distinct group. It must, moreover, be remembered that most botanists include Hæmatococcus and Volvox among Algæ, and place the Mycetozoa either among Fungi or in a separate group of chlorophyll-less plants (p. 181).

The *Muscineæ* are the mosses and liverworts, the former of which were fully described in Lesson XXVIII.

The *Vascular Cryptogams* are flowerless plants in which vascular bundles are present. Together with the Phanerogams they constitute what are known as vascular plants, in contradistinction to the non-vascular Algæ, Fungi, and Muscineæ, in which no formation of vessels takes place. The group contains three subdivisions.

The first division of Vascular Cryptogams, the *Filicinæ*, includes the ferns, an account of which has been given in the previous lesson. It will be necessary, however, to devote some attention to an aquatic form, called *Salvinia*, which differs in certain important particulars from the more familiar members of the group.

The *Equisetaceæ* include the common horsetails (genus *Equisetum*), a brief account of which will be given, as they form an interesting link in their reproductive processes between the ordinary ferns and Salvinia.

The *Lycopodineæ*, or club-mosses, are the highest of the Cryptogams or flowerless plants. A short description of one of them, the genus *Selaginella*, will illustrate the most striking peculiarities of the group.

The *Phanerogams*, or flowering plants, are so called from the fact that their reproductive organs take the form of specially modified shoots, called cones or flowers. They are sometimes called by the more appropriate name of *Spermaphytes*, or seed-plants, from the fact that they alone among plants, reproduce by means of seeds, structures which differ from spores in the fact that each contains an embryo plant in the phyllula stage.

The *Gymnosperms*, or naked-seeded Phanerogams, include the cone-bearing trees, such as pines, larches, cypresses, &c., as well as cycads and some other less familiar forms. A general account of this group will be given.

The *Angiosperms*, or covered-seeded Phanerogams, include all the ordinary flowering plants, as well as such trees as oaks, elms, poplars, chestnuts, &c. A brief description of the general features of this group will conclude the lesson.

Equisetum

A horsetail consists of an underground creeping stem from which vertical shoots are given off. Some of these bear only leaves and branches, others are peculiarly modified and produce sporangia.

A fertile or sporangium-bearing shoot terminates distally in a conical body (Fig. 84, A), formed of closely-fitting hexagonal scales (*sp. ph*). Each scale (B, *sp. ph*) is attached by a stalk to the axis of the shoot, and bears on its inner surface a number of sporangia (*spg*). The scales are modified leaves, and since they alone produce sporangia they are distinguished from the ordinary foliage-leaves as *sporophylls*.

The spores, which have the same general structure as those of ferns, are liberated by the bursting of the sporangia, and germinate, giving rise to prothalli. But instead of the prothalli being all alike in form and size and all monœcious, some (C) remain small and simple, and produce only spermaries (*spy*), others (D) attain a complicated form and a length of over a centimetre, and produced only ovaries (*ovy*). Thus although there is no difference in the spores, the prothalli produced from them are of two distinct kinds, the smaller being exclusively male, the larger female.

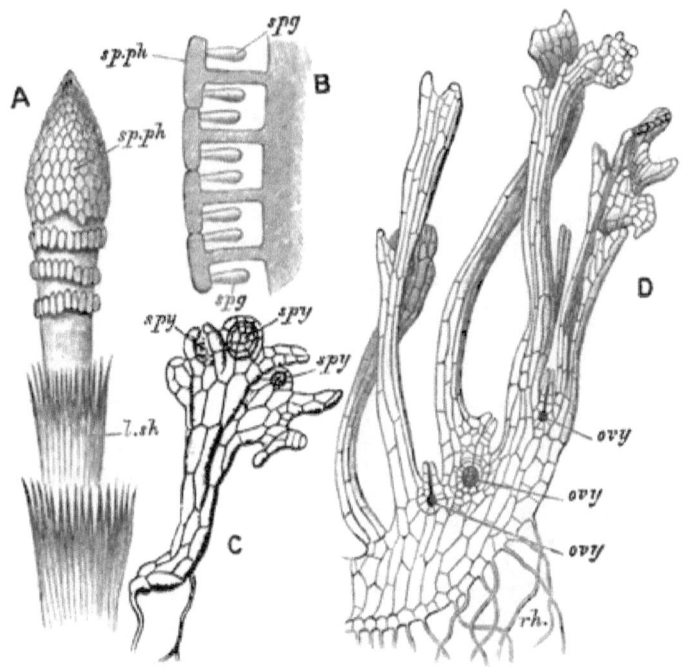

FIG. 84.—Reproduction and Development of *Equisetum*.
A, distal end of a fertile shoot, showing two leaf-sheaths (*l. sh*), and the cone formed of hexagonal sporophylls (*sp. ph*). (Nat. size.)
B, diagrammatic vertical section of a portion of the cone, showing the sporophylls (*sp. ph*) attached by short stalks to the axis of the cone, and bearing sporangia (*spg*) on their inner surfaces.
C, a male prothallus bearing three spermaries (*spy*). (× 100.)
D, portion of a female prothallus bearing three ovaries (*ovy*), those to the right and left containing ova, that in the middle a polyplast; *rh*, rhizoids. (× 30.)
(A, after Le Maout and Decaisne; C and D, after Hofmeister.)

The oosperm develops in much the same way as in ferns: it divides and forms a polyplast, which, by formation of a stem, root, foot, and two cotyledons, becomes a phyllula and grows into the adult plant.

As in the fern, the Equisetum plant, reproducing as it does by asexual spores, is the agamobium, the gamobium being represented by the prothallus. The peculiarity in the present case is that the gamobium is sexually dimorphic, some prothalli producing only male, others only female gonads.

SALVINIA

Salvinia is a fresh-water plant, consisting of a long floating stem bearing at intervals whorls of leaves. Of these some have the ordinary character while others hang downwards into the water and have the form and function of roots. True roots are absent.

The sori or groups of sporangia (Fig. 85, A) are borne on the proximal ends of the submerged leaves, each being enclosed in a globular case corresponding to the indusium of ordinary ferns. They differ from the sori of the typical ferns in being dimorphic, some containing a comparatively small number of large sporangia (*mg. spg*) others a much larger number of small ones (*mi. spg*). The larger kind, distinguished as *megasporangia*, contain each a single large spore, or *megaspore:* the smaller kind, or *microsporangia*, contain a large number of minute spores, like those of an ordinary fern, and called *microspores*. It is this striking dimorphism of the sori, sporangia, and spores which forms the chief distinction between Salvinia and its allies and the true ferns.

The microspore germinates (B), while still enclosed in its sporangium, by sending out a filament, the end of which (*spy*) becomes separated off by a septum and then divided into two cells. The protoplasm of each of these divides into four sperm-mother-cells, and from these, spirally-twisted sperms are produced in the usual manner. It is obvious that the two cells in which the sperms are developed represent a greatly simplified spermary: the single proximal cell

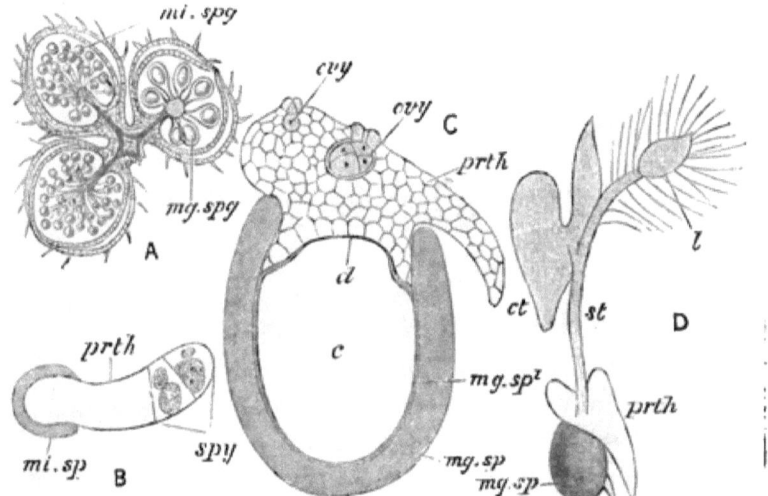

FIG. 85.—Reproduction and Development of *Salvinia*.

A, portion of a submerged leaf, showing three sori in vertical section, two containing microsporangia (*mi. spg*) and one megasporangia (*mg. spg*). (× 10.)

B, a germinating microspore (*mi. spg*), showing the vestigial prothallus (*prth*) and spermary (*spy*). (× 150.)

C, diagrammatic vertical section of a germinating megaspore, showing the outer (*mg. sp*) and inner (*mg. sp′*) coats of the spore, and its cavity (*c*) containing plastic products, separated by a septum (*d*) from the prothallus (*prth*), in which two ovaries (*ovy*) are shown, that to the left containing an ovum, that to the right a polyplast. (× 50.)

D, megaspore (*mg. sp*) with prothallus (*prth*) and phyllula just beginning to develop into the leafy plant: *st*, stem; *ct*, cotyledon; and *l*, outermost leaf of the terminal bud. (× 20.)

(A and B, after Sachs; D, after Pringsheim.)

(*prth*) of the filament arising from the microspore, a still more simplified prothallus. Both prothallus and spermary are vestigial structures; the prothallus is microscopic and unicellular instead of being a solid aggregate of considerable size, as in the two preceding types; the spermary is bicellular instead of being formed of a distinct wall and an internal mass of cells; and the number of sperms is reduced to eight.

The contents of the megaspore are divisible into a comparatively small mass of protoplasm at one end and of starch grains, oil-globules, and proteid bodies, which fill up the rest (C, c) of the spore. The megaspore has in fact attained its large size by the accumulation of great quantities of plastic products, which serve as nutriment to the future prothallus and embryo.

The protoplasm of the megaspore (c) divides and forms a prothallus (*prth*) in the form of a three-sided multicellular mass projecting from the spore, which it slightly exceeds in size. Several ovaries (*ov.y*) are formed on it, having much the same structure as in ordinary ferns. Thus the reduction of the prothallus produced from the megaspore, although obvious, is far less than in the case of that arising from the microspore.

We see that sexual dimorphism has gone a step further in Salvinia than in Equisetum: not only are the prothalli differentiated into male and female, but also the spores from which they arise.

Impregnation takes place in the usual way, and the oosperm divides to form a polyplast, which, by differentiation of a stem-rudiment, a cotyledon, and a foot, passes into the phyllula stage: no root is developed in Salvinia. By the gradual elongation of the stem (D, *st*) and the successive formation of whorls of leaves (*l*), the adult form is assumed.

Thus the life-history of Salvinia resembles that of the fern, but with two important differences: the spores are dimorphic, and the gamobium, represented by the male and female prothalli, is greatly reduced.

Selaginella

Selaginella, one of the club-mosses, consists of a long branching stem bearing numerous close-set leaves. It thus resembles in external appearance a moss, but the essential difference between the two is seen from a study of their histology, Selaginella having a distinct epidermis and vascular bundles like the other Vascular Cryptogams.

The branches terminate in cones (Fig. 86, A) formed of small leaves (*sp. ph*) which overlap in somewhat the same way as the scales of a pine-cone. Each of these leaves is a sporophyll, and bears on its upper or distal side, near the base, a globular sporangium. The sporangia are fairly uniform in size, but some are megasporangia (*mg. spg*) and contain usually four megaspores, others are microsporangia (*mi. spg*) containing numerous microspores.

The microspore (B) cannot be said to germinate at all. Its protoplasm divides, forming a small cell (*prth*), which represents a vestigial prothallus, and a large cell, the representative of a spermary. The latter (*spy*) undergoes further division, forming six to eight cells in which numerous sperm-mother-cells are developed.

A similar but less complete reduction of the prothallus is seen in the case of the megaspore (c). Its contents are divided, as in Salvinia, into a small mass of protoplasm at one end, and a large quantity of plastic products filling up the rest of its cavity. The protoplasm divides and forms a small prothallus (*prth*), and a process of division also takes place in the remaining contents (*prth'*) of the spore, producing a large-celled tissue, the *secondary prothallus*.

By the rupture of the double cell-wall of the megaspore the prothallus is exposed to the air, but it never protrudes through the opening thus made, and is therefore, like the corresponding male structure, purely endogenous. A few ovaries (*ovy*) are formed on it, each consisting of a short neck, an ovum, and two canal-cells afterwards converted into

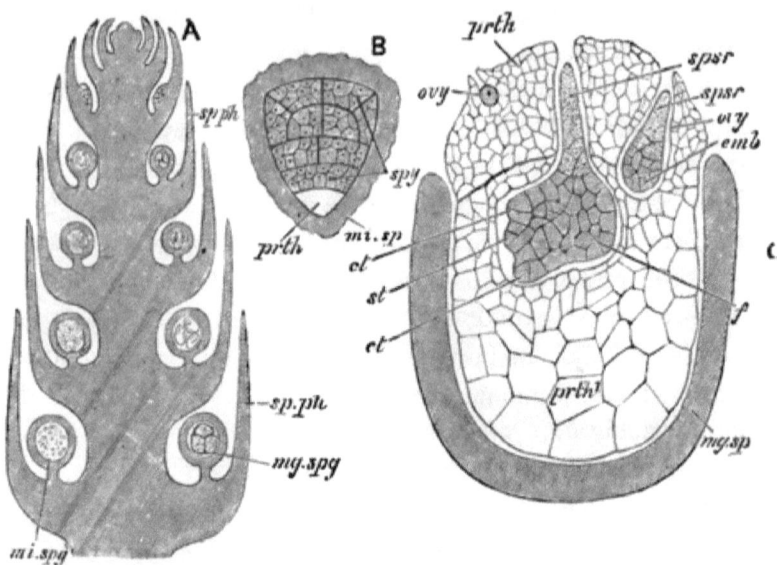

FIG. 86.—Reproduction and Development of *Selaginella*.

A, diagrammatic vertical section of a cone, consisting of an axis bearing close-set sporophylls (*sp. ph*), on the bases of which microsporangia (*mi. spg*) and megasporangia (*mg. spg*) are borne.

B, section of a microspore, showing the outer coat (*mi. sp*), prothallial cell (*prth*), and multicellular spermary (*spy*).

C, vertical section of a megaspore, the wall of which (*mg. sp*) has been burst by the growth of the prothallus (*prth*): its cavity (*prth'*) contains a large-celled tissue, the secondary prothallus: in the prothallus are three ovaries (*ovy*), that to the left containing an ovum, that to the right an embryo (*emb*) in the polyplast stage, and that in the centre an embryo in the phyllula stage, showing stem-rudiment (*st*), foot (*f*), and two cotyledons (*ct*): both embryos are provided with suspensors (dotted) (*spsr*), and have sunk into the secondary prothallus.

(Altered from Sachs.)

mucilage: there is no venter, and the neck consists of only two tiers of cells.

The oosperm divides by a plane at right angles to the neck of the ovary, forming the earliest or two-celled stage of the polyplast. The upper cell undergoes further division, forming an elongated structure, the *suspensor* (dotted in c) : the lower or embryo proper (*emb*) is forced downwards into the secondary prothallus by the elongation of the suspensor, and soon passes into the phyllula stage by the differentiation of a stem-rudiment (*st*), two cotyledons (*ct*), a foot, (*f*) and subsequently of a root.

A further reduction of the gamobium is seen in Selaginella : both male and female prothalli are quite vestigial, never emerging from the spores : and the spermary and ovary are greatly simplified in structure.

Gymnosperms

Such common Gymnosperms as the pines and larches have the character of forest trees, the stem being a strong, woody trunk. The numerous, close-set branches bear small, needle-like leaves, and the root is large and extensively branched.

On the branches are borne structures of two kinds, the male and female cones or flowers (Fig. 87, A and C). Both are to be considered as abbreviated shoots consisting of an axis bearing numerous sporophylls (*sp. ph*). Frequently, as in the pines, several male cones are aggregated together, forming an *inflorescence*, or group of flowers.

In the male cone (A) the sporophylls (stamens, *sp. ph.* ♂) are more or less leaf-like structures, each bearing on its under or proximal side two or more microsporangia (pollen-sacs, *mi. spg*). The mother-cells of these divide each into four microspores (pollen-grains), which are liberated by the rupture of the microsporangia in immense quantities. The microspore (B) is at first an ordinary cell consisting of protoplasm with a nucleus and a double cell-wall, but upon being

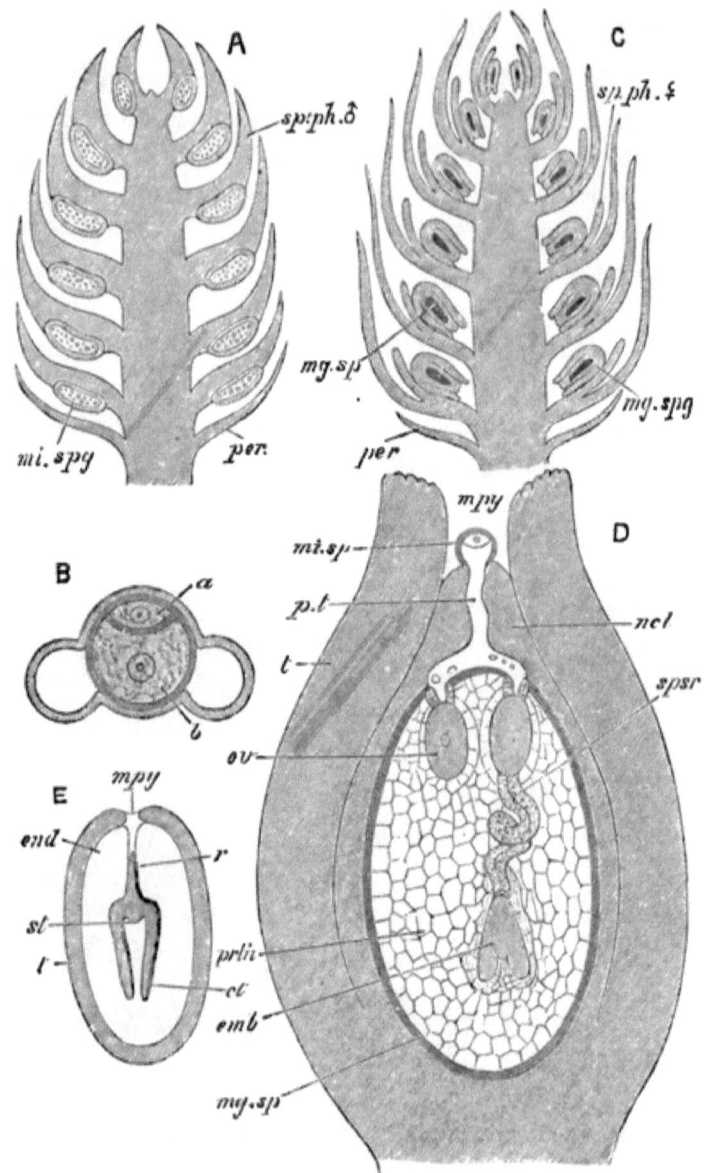

FIG. 87.—Reproduction and Development of *Gymnosperms*.
A, diagrammatic vertical section of male cone, showing axis with male sporophylls (*sp. ph.* ♂) bearing microsporangia (*mi. spg*) : *per*, scale-like leaves forming a rudimentary perianth.

B, a single microspore, showing bladder-like processes of outer coat, and contents divided into small prothallial cell (*a*) and large cell (*b*), from which the pollen-tube arises.

C, diagrammatic vertical section of female cone, showing axis with female sporophylls (*sp. ph.* ♀) bearing megasporangia (*mg. spg*), each of which contains a single megaspore (*mg. sp*) : *per*, the scale-like perianth leaves.

D, diagrammatic vertical section of a megasporangium, showing cellular coat (*t*), and nucellus (*ncl*), micropyle (*mpy*), and megaspore (*mg. sp*) : the latter contains the prothallus (*prth*) in which are two ovaries, that to the left showing a large ovum (*ov*) and neck-cells, while that to the right has given rise to an embryo (*emb*) which is in the phyllula stage, and has sunk into the tissue of the prothallus by the elongation of the long suspensor (*spsr*).

A microspore (*mi. sp*) is seen in the micropyle sending off a pollen-tube (*p. t*), the end of which is applied to the necks of the two ovaries.

E, diagrammatic vertical section of a seed, showing coat (*t*), micropyle (*mpy*), and endosperm (*end*), in which is embedded an embryo in the phyllula stage, consisting of stem-rudiment (*st*), cotyledons (*ct*), and root (*r*).

(A and B, altered from Strasburger ; D and E, altered from Sachs.)

liberated the protoplasm divides, as in Selaginella, into two cells, a small one (*a*) the vestige of the male prothallus, and a large one (*b*) which does not develop sperms, but under favourable circumstances undergoes changes which will be described presently.

In the female cone (c) each sporophyll (carpel, *sp. ph*, ♀) bears on its upper or distal side two megasporangia (so-called ovules, *mg. spg*) the structure of which is peculiar. Each consists of a solid mass of small cells called the *nucellus* (D, *ncl*), attached by its proximal end to the sporophyll, and surrounded by a wall or *integument* (*t*) also formed of a small-celled tissue. The integument is in close contact with the nucellus, but is perforated distally by an aperture, the *micropyle* (*mpy*), through which a small area of the nucellus is exposed.

Each megasporangium contains only a single megaspore (embryo sac, C and D, *mg. sp*) in the form of a large ovoidal body embedded in the tissue of the nucellus. It has at first the characters of a single cell, but afterwards, by division of its protoplasm, becomes filled with small cells representing a prothallus (*prth*). As in Vascular Cryptogams, single superficial cells of the prothallus are converted into ovaries which are extremely simple in structure, each consisting of a large ovum (*ov*), and of a variable number of neck-cells.

The pollen, liberated by the rupture of the microsporangia, is carried to considerable distances by the wind, some of it falling on the female cones of the same or another tree. In this way single microspores (pollen-grains) find their way into the micropyle of a megasporangium (D, *mi. sp*). This is the process known as *pollination*, and is the necessary antecedent of fertilization.

The microspore now germinates: the outer coat bursts, and the larger of the two cells (B, *b*) protrudes in the form of a filament resembling a hypha of Mucor, and called a *pollen-tube* (D, *p.t*). This forces its way into the tissue of the nucellus, like a root making its way through the soil, and finally reaches the megaspore in the immediate neighbourhood of an ovary. A process then grows out from the end of the tube, passes between the neck-cells, and comes in contact with the ovum.

In the meantime the nucleus of the large cell (*b*) of the microspore—that from which the pollen-tube grows—has travelled to the end of the pollen-tube and divided into two. Protoplasm collects round each of the daughter nuclei, converting them into cells, one of which remains undivided, while the other divides, and its substance passes from the pollen-tube into the ovum, where it forms a cell-like body, to which the name of male pronucleus (see p. 263) has been applied. This conjugates with the nucleus of the ovum, or female pronucleus, and thus effects the process of fertilization, or the conversion of the ovum into the oosperm.

The mode of formation of cells described in the preceding paragraph should be specially noted. Instead of the ordinary process of fission hitherto met with, the products of division of a nucleus become surrounded by protoplasm, cells being produced which lie freely in the interior of the mother-cell. This is called *free cell-formation*.

The development of the oosperm is a very complicated process, and results in the formation not of a single polyplast but of four, each at the end of a long suspensor (D, *spsr*), in the form of a linear aggregate of cells, which by its elongation carries the embryo (*emb*) down into the tissue of the prothallus. As a rule only one of these embryos comes to maturity: it develops a rudimentary stem, root, and four or more cotyledons, and so becomes a phyllula.

While these processes are going on the female cone increases greatly in size and becomes woody. The megasporangia also become much larger, their integuments (E, *t*), becoming brown and hard, and the megaspore in each enlarges so much as to displace the nucellus: at the same time the cells of the prothallus filling the megaspore develop large quantities of plastic products, such as fat and albuminous substances, to be used in the nutrition of the embryo: the tissue thus formed is the *endosperm* (*end*). The megasporangium is now called a *seed* (see p. 365).

Under favourable circumstances the seed germinates. By absorption of moisture its contents swell and burst the seed-coat, and the root of the phyllula (*r*) emerges, followed before long by the stem (*st*) and cotyledons (*ct*). The phyllula thus becomes the seedling plant, and by further growth and the successive formation of new parts is converted into the adult.

In Gymnosperms we see an even more striking reduction of the gamobium than in Selaginella. The female prothallus is permanently inclosed in the megaspore, and the megaspore in the megasporangium: the ovaries also are greatly simplified. The male prothallus is represented by the smaller cell of the microspore, and no formation of sperms takes place, fertilization being effected by cells developed in the extremity of a tubular prolongation of the larger cell of the microspore, and resulting from a modification of its nucleus.

It is worthy of notice that Phanerogams alone among the higher organisms, have abandoned the ordinary method of fertilization by the conjugation of ovum and sperm. In this respect they are the most specialized of living things.

ANGIOSPERMS

In this group the general relations of the main parts of the plant—stem, leaves, roots, &c.—are the same as in Gymnosperms.

The flowers, in which, as in Gymnosperms, the organs of reproduction are contained, have a very characteristic structure, which, although presenting almost infinite variety in detail, is the same in its essential features throughout the group.

A typical angiospermous flower (Fig. 88, A) is a greatly abbreviated shoot, consisting of a short axis ($fl. r$) of limited growth bearing four whorls of leaves, of which those of the two distal whorls are sporophylls.

The axis of the floral shoot (A. $fl. r$) is usually broad and more or less conical in form and is called the *floral receptacle*. The leaves of the lower or proximal whorl (per^1), usually from three to five in number, are small green bodies which cover the other parts in the unopened flower: they are called *sepals* and together constitute the *calyx*.

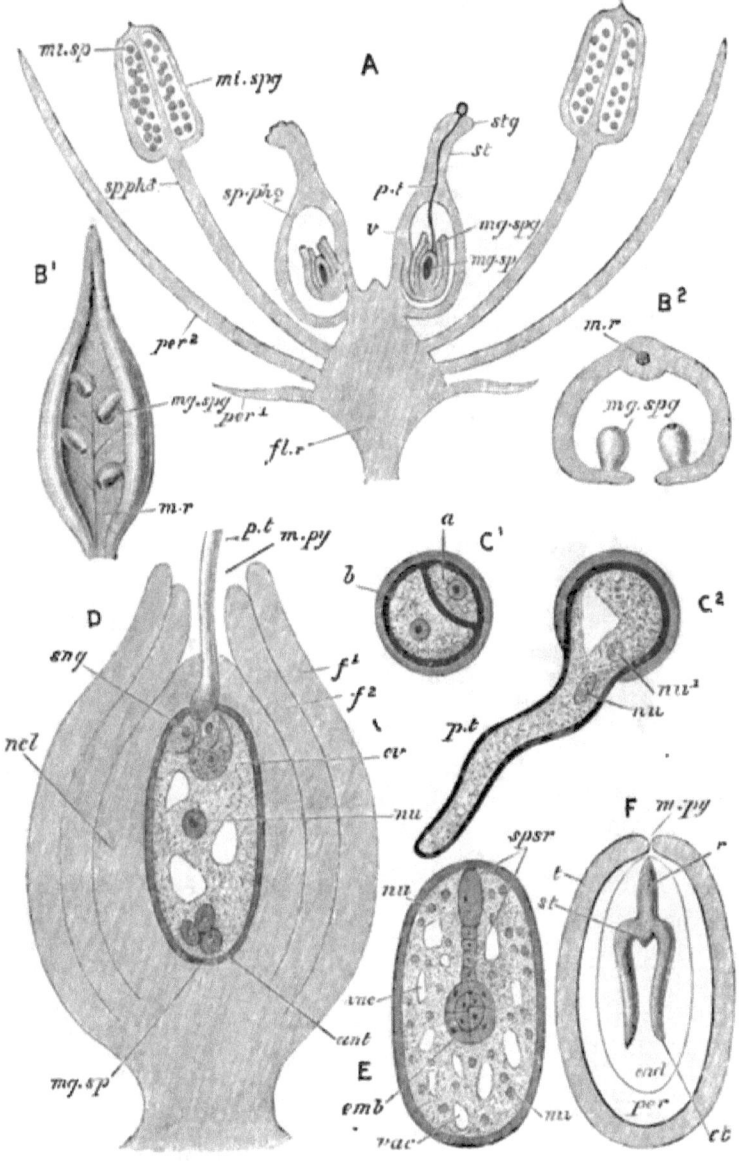

FIG. 88.—Reproduction and Development of Angiosperms.
A, diagrammatic vertical section of a flower consisting of an abbreviated axis or floral receptacle (*fl. r*) bearing a proximal (*per*¹) and a distal (*per*²) whorl of perianth leaves (sepals and petals), a whorl of male

sporophylls or stamens (*sp. ph.* ♂), and one of female sporophylls or carpels (*sp. ph.* ♀).

The male sporophyll bears microsporangia (*mi. spg*) containing microspores (*mi. sp*).

The female sporophyll consists of a solid style (*st*) terminated by a stigma (*stg*), and of a hollow venter (*v*) containing a megasporangium (*mg. spg*) in which is a single megaspore (*mg. sp*).

On the right side a microspore is shown on the stigma, and has sent off a pollen-tube (*p. t*) through the tissue of the style to the micropyle of the megasporangium.

B^1, diagram of a female sporophyll from the distal aspect, and B^2, the same in transverse section, showing the folding in of its edges to form the cavity or venter in which the megasporangia (*mg. spg*) are enclosed: *m.r*, the midrib.

C^1, a microspore, showing the two cells (*a* and *b*) into which its contents divide.

C^2, the same, sending out a pollen-tube (*p. t*): *nu*, *nu*^{1}, the two nuclei.

D, diagrammatic vertical section of a megasporangium, showing the double integument (*f*^{1}, *f*^{2}), nucellus (*ncl*), micropyle (*m.py*), and megaspore (*mg. sp*): the latter contains the secondary nucleus (*nu*) in the centre, three antipodal cells (*ant*) at the proximal end, and two synergidæ (*sng*) and an ovum (*ov*) at the distal end.

A pollen-tube (*p. t*) is shown with its end in contact with the synergidæ.

E, semi-diagrammatic section of the megaspore of a young seed, showing an embryo (*emb*) in the polyplast stage with its suspensor (*spsr*): also numerous vacuoles (*vac*) and nuclei (*nu*).

F, diagrammatic vertical section of a ripe seed, showing the seed-coat (*t*), micropyle (*mpy*), perisperm (*per*) derived from the tissue of the nucellus, and endosperm (*end*) formed in the megaspore and containing an embryo in the phyllula stage with stem-rudiment (*st*), cotyledons (*ct*), and root (*r*).

(B^1 after Behrens; C^1, C^2, and E altered from Howes.)

Above the sepals comes a whorl of leaves (*per*^{2}), usually of large size and bright colour, forming in fact the most obvious part of the flower. These are the *petals* and together constitute the *corolla*. The calyx and corolla together are conveniently called the *perianth*, because they inclose the sporophylls or essential part of the flower. The presence of a well-marked perianth is characteristic of the majority of Angiosperms, and distinguishes them from Gymnosperms, in which this part of the flower is quite rudimentary (see Fig. 87, A. and B, *per*).

The third whorl is called collectively the *andrœcium*, and consists of a variable number of stamens or male sporophylls (*sp. ph.* ♂). Each stamen is a long narrow leaf bearing at its distal end four microsporangia (pollen sacs, *mi. spg*) united into a lobed knob-like body, the *anther*. The microspores (c^1) are at first simple cells with double cell-walls, but subsequently the protoplasm becomes divided into two cells, as in Gymnosperms, a smaller (*a*) and a larger (*b*). The two are not, however, separated by a firm septum of cellulose, and the smaller cell frequently comes to lie freely in the protoplasm of the larger. Moreover it appears that the nucleus of the smaller is the active agent in fertilization, and that the larger must therefore be considered as representing the vestigial prothallus.

The fourth or distal whorl of the flower is called collectively the *gynæcium* or *pistil*, and consists of one or more carpels or female sporophylls (*sp. ph.* ♀), which are modified in a characteristic manner. In some cases each carpel (B^1, B^2) becomes folded longitudinally along its midrib (*m.r*), and its two edges, thus brought into contact, unite so as to inclose a cavity. Concrescence only affects the proximal part of the carpel, which thus becomes a hollow capsule, the *venter* (so-called ovary, A, *v*): its distal portion usually takes the form of a slender rod-like body, the style (*st*), terminated by an enlarged extremity, the stigma (*stg*) which is covered with hairs and is frequently sticky. In some flowers, on the other hand, all the carpels of the gynæcium unite with one another by their adjacent edges, so as to inclose a cavity common to all : in this case also the hollow portion or venter is formed by the proximal part only of the carpels, their distal portions forming a simple or multiple style and stigma.

The megasporangia (ovules, A and B, *mg. spg*) are usually

borne on the edges of the carpels, and, owing to the union of the latter, become inclosed in the cavity of the venter, and are thus completely shut off from all direct communication with the external world. It is this inclosure of the megasporangia in a cavity formed by the sporophylls on which they are borne which constitutes the chief character distinguishing Angiosperms from Gymnosperms.

The megasporangia (D) differ from those of Gymnosperms chiefly in having a double integument: both coats (t^1, t^2) as well as the nucellus (*ncl*), or central mass of tissue, are composed of small cells: and the megaspore (embryo-sac, *mg. sp*) is a single cell of great size embedded in the nucellus.

No prothallus is formed in the megaspore, but its nucleus divides, the products of division pass to opposite ends of the spore, and each divides again and then again, so that four nuclei are produced at each extremity. Three of the nuclei at the proximal end—that furthest from the micropyle—become surrounded by protoplasm and take on the character of cells (D, *ant*): the fourth remains unchanged. Similarly of the four nuclei at the distal or micropylar end, one remains unchanged and three assume the form of cells by becoming invested with protoplasm (see p. 376). Of these three, two lie near the wall of the megaspore and are called *synergidæ* (*sng*): the third, more deeply placed, is the ovum (*ov*). The two unaltered nuclei now travel to the centre of the megaspore and unite with one another, forming the *secondary nucleus* (*nu*) of the spore.

There is thus a single ovum produced in each megaspore, but no ovary and no prothallus: the female portion of the gamobium is reduced to its simplest expression.

Pollination may take place, as in Gymnosperms, by the agency of the wind, but usually the microspores are carried by insects, which visit the flowers for the sake of obtaining nectar, a saccharine fluid' secreted by certain parts. The microspores are deposited on the stigma (A), where they germinate, each sending off a pollen-tube (A and c^2, *p. t*), which grows downwards through the tissue of the stigma and style to the cavity of the venter, where it reaches a megasporangium, and entering at the micropyle (D, *p. t*), continues its course through the nucellus, finally applying itself to the distal end of the megaspore in the immediate neighbourhood of the synergidæ.

In the meantime the nuclei of the microspore (c^2, *nu*, *nu*[1]) have passed into the end of the pollen-tube. The nucleus of the larger cell undergoes degeneration, becoming shrivelled and unaffected by dyes; that of the smaller cell divides by karyokinesis. One of the two daughter-nuclei thus formed also degenerates, the other, accompanied by its directive spheres, passes through the softened cell-wall of the swollen end of the pollen-tube and enters the ovum, uniting with its nucleus in the usual way.

The ovum is thus converted into an oosperm or unicellular embryo: it acquires a cell-wall and almost immediately divides into two cells, of which that nearest the micropyle becomes the suspensor (E, *spsr*), the other, or embryo proper (*emb*), forming a solid aggregate of cells, the polyplast. By further differentiation rudiments of a stem (F, *st*), a root (*r*) and either one or two cotyledons (*ct*) are formed, and the embryo passes into the phyllula stage.

While the early development of the embryo is going on, the secondary nucleus of the megaspore divides repeatedly, and the products of division (E, *nu*) becoming surrounded by protoplasm, a number of cells are produced, which, by further multiplication, fill up all that part of the megaspore which is not occupied by the embryo. The tissue thus formed is called the endosperm (F. *end*), and occupies precisely the position of the vestigial prothallus of Gymnosperms (Fig, 87, p. 374, D, *prth*, and E, *end:* and p. 376), differing from it in the fact that it is only formed after fertilization. We have here a case of retarded development: the degeneration of the prothallus has gone so far that it arises, by free cell-formation, long after the formation of the ovum which, in both Gymnosperms and Vascular Cryptogams, is a specially modified prothallial cell.

The phyllula continues to grow and remains inclosed in the megasporangium, which undergoes a corresponding increase in size and becomes the seed. One or more seeds also remain inclosed in the venter of the pistil, which grows considerably and constitutes the *fruit*. Finally the seeds are liberated, the phyllula protrudes first its root, and then its stem and cotyledons through the ruptured seed-coat, and becomes the seedling plant.

We learn from the present lesson that there is a far greater uniformity of organization among the higher plants than among the higher animals, not only in anatomical and histological structure, but also in the fact that alternation of generations is universal from Nitella and the mosses up to the highest flowering plants. But as we ascend the series, the gamobium sinks from the position of a conspicuous leafy plant to that of a small and insignificant prothallus, becoming finally so reduced as to be only recognizable as such by comparison with the lower forms.

SYNOPSIS

A.—AN ACCOUNT OF THE STRUCTURE, PHYSIOLOGY, AND LIFE-HISTORY OF A SERIES OF TYPICAL ORGANISMS IN THE ORDER OF INCREASING COMPLEXITY.

I.—THE SIMPLER UNICELLULAR ORGANISMS.

1. *Amœba.*

    PAGE

    Cell-body amœboid or encysted: cell-wall nitrogenous (?): nutrition holozoic: reproduction by simple or binary fission 1

2. *Hæmatococcus.*

    Cell-body ciliated or encysted: cell-wall of cellulose: nutrition holophytic: reproduction by binary fission . . 23

3. *Heteromita.*

    Cell-body ciliated: nutrition saprophytic: asexual reproduction by binary fission: sexual reproduction by conjugation of equal and similar gametes followed by multiple fission of the protoplasm of the zygote, forming spores . 36

4. *Euglena.*

    Cell-body ciliated or encysted: cell-wall of cellulose mouth and gullet present: nutrition holophytic and holozoic: reproduction by binary and multiple fission . . 44

5. *Protomyxa.*

    Cell-body amœboid, ciliated or encysted: plasmodia formed by concrescence of amœbulæ: cell-wall nitrogenous (?): nutrition holozoic: reproduction by multiple fission of encysted plasmodium 49

6. *Mycetozoa.*
   Like Protomyxa, but owing to the presence of nuclei the relation of the individual cell-bodies to the plasmodium is more clearly seen : cell wall of cellulose 52

7. *Saccharomyces.*
   Cell-body encysted : cell-wall of cellulose : nutrition saprophytic : reproduction by gemmation or by internal fission : acts as an organized ferment 70

8. *Bacteria.*
   Cell-body ciliated or encysted : cell-wall of cellulose : nutrition saprophytic : reproduction by binary fission or by spore-formation : act as organized ferments : the simplest and most abundant of organisms 82

II.—UNICELLULAR ORGANISMS IN WHICH THERE IS CONSIDERABLE COMPLEXITY OF STRUCTURE ACCOMPANIED BY PHYSIOLOGICAL DIFFERENTIATION.

a. Complexity attained by differentiation of cell-body.

9. *Paramæcium.*
   Medulla, cortex, and cuticle : trichocysts : complex contractile vacuoles : nucleus and paranucleus : mouth, gullet, and anal spot : conjugation temporary, no zygote being formed, but interchange of nuclear material during temporary union 106

10. *Stylonychia.*
    Extreme differentiation or heteromorphism of cilia . . . 116

11. *Oxytricha.*
    Fragmentation of nucleus 120

12. *Opalina.*
    Multiplication of nuclei ; parasitism and its results ; necessity for special means of dispersal of an internal parasite 121

13. *Vorticella.*
    A stationary organism : limitation of cilia to defined regions : muscle-fibre in stalk : necessity for means of dispersal in a fixed organism : conjugation between free-swimming micro- and fixed mega-gamete : zygote indistinguishable from a zooid of the ordinary kind 126

14. *Zoothamnium.*
    A compound organism or colony with dimorphic (nutritive and reproductive) zooids : begins life as a single zooid 135

SYNOPSIS

　　　　　　　　　　　　　　　　　　　　　　　　　　　PAGE
b. Complexity attained by differentiation of cell-wall or by formation of skeletal structures in the protoplasm.

   15. *Foraminifera.*
       Calcareous shells (cell-walls) of various and complicated form 148

   16. *Radiolaria.*
       Membranous perforated shell (cell-wall) and external silicious skeleton often of great complexity: symbiotic relations with Zooxanthella 152

   17. *Diatoms.*
       Silicious, two-valved, highly-ornamented shells 155

c. Complexity attained by simple elongation and branching of the cell.

   18. *Mucor.*
       A branching filamentous fungus: necessity for special reproductive organs in such an organism: they may be sporangia producing asexual spores, or equal and similar gametes producing a resting zygote 158

   19. *Vaucheria.*
       A branched filamentous alga: clear distinction between the gametes or conjugating bodies and the sexual reproductive organs or gonads in which they are produced: gonads differentiated into male (spermary) and female (ovary): gametes differentiated into male (sperm) and female (ovum): zygote an oosperm 169

   20. *Caulerpa.*
       Illustrates maximum differentiation of a unicellular plant: stem-like, leaf-like, and root-like parts 175

III.—ORGANISMS IN WHICH COMPLEXITY IS ATTAINED BY CELL-MULTIPLICATION, ACCOMPANIED BY NO OR BUT LITTLE CELL-DIFFERENTIATION.

a. Linear aggregates.

   21. *Penicillium.*
       A multicellular, filamentous, branched fungus: mycelial, submerged, and aërial hyphæ: apical growth: abundant production of spores by constriction of aërial hyphæ . . 184

   22. *Agaricus.*
       Complexity attained by interweaving of hyphæ in a definite form: illustrates maximum complexity of a linear aggregate . . . .

23. *Spirogyra.*
   A multicellular filamentous unbranched alga: interstitial growth: gonads equal and similar, but gametes show first indication of sexual differentiation 194

b. Superficial aggregate.

24. *Monostroma.*
   Cell-division takes place in two dimensions 202

c. Solid aggregates.

25. *Ulva.*
   Like Monostroma, but cell-division takes place in three dimensions 203

26. *Laminaria.*
   Illustrates maximum size and complexity of a solid aggregate of comparatively slightly differentiated cells . 203

IV.—SOLID AGGREGATES IN WHICH COMPLEXITY IS INCREASED BY A LIMITED AMOUNT OF CELL-DIFFERENTIATION.

27. *Nitella.*
   Segmented axis: nodes and internodes: appendages—leaves and rhizoids: apical growth by binary fission of apical cell accompanied by immediate division and differentiation of newly-formed segmental cells: complex gonads (ovaries and spermaries): alternation of generations, a gamobium or sexual generation (the leafy plant) alternating with an agamobium or asexual generation (the pro-embryo) 206

28. *Hydra.*
   Example of a simple diploblastic animal: cells arranged in two layers (ecto- and endoderm) inclosing an enteron which opens externally by the mouth: combination of intra-cellular with extra-cellular or enteric digestion . . 221

29. *Bougainvillea.*
   Example of a colony with diploblastic zooids which are nutritive (hydranths) and reproductive (medusæ): differentiation of a rudimentary mesoderm producing imperfect triploblastic condition: central and peripheral nervous system: alternation of generations, a gamobium (the medusa) alternating with an agamobium (the hydroid colony); significance of developmental stages—oosperm (unicellular), polyplast (multicellular but undifferentiated), and planula (diploblastic) 237

SYNOPSIS

30. *Diphyes.*
    A free-swimming colony with polymorphic (nutritive, reproductive, protective, and natatory) zooids 250
31. *Porpita.*
    Extreme polymorphism of zooids giving the colony the character of a single physiological individual 253

V.—SOLID AGGREGATES IN WHICH CELL-DIFFERENTIATION, ACCOMPANIED BY CELL FUSION, TAKES AN IMPORTANT PART IN PRODUCING GREAT COMPLEXITY IN THE ADULT ORGANISM.

32. *Polygordius.*
    A triploblastic, cœlomate animal with metameric segmentation: prostomium, peristomium, metameres, and anal segment: besides ecto- and endoderm there is a well developed mesoderm divided into somatic and splanchnic layers separated by the cœlome: differentiation of cells into fibres, &c.: muscle-plates formed as cell-fusions: necessity for distributing system for supply of food to parts of the body other than the enteric canal, and for the removal of waste matters:—circulatory, respiratory, and excretory systems: high development of nervous system—brain and ventral cord, afferent and efferent nerves: characteristic developmental stages—oosperm, polyplast, gastrula (diploblastic), trochosphere (diploblastic with stomodæum and proctodæum), late trochosphere (triploblastic but acœlomate) 271

33. *Mosses.*
    Cell-differentiation very slight, but the type necessary to lead up to ferns: sclerenchyma and axial bundle: distributing system rendered necessary by carbon dioxide being taken in by the leaves, water and mineral salts by the rhizoids: alternation of generations—the leafy plant is the gamobium, the agamobium being represented by the spore-producing sporogonium: developmental stages—oosperm and polyplast, the latter becoming highly differentiated to form the sporogonium 332

34. *Ferns.*
    Extensive cell-differentiation: formation of fibres (elongated cells) and vessels (cell-fusions); general differentiation of tissues into epidermis, ground-parenchyma, and vascular bundles: presence of true roots: the leafy plant is the amagobium and produces spores from which the gamobium, in the form of a small prothallus, arises: developmental stages—oosperm, polyplast, and phyllula (leaf- and root-bearing stage) 344

VI.—Brief Descriptions of Types of the Higher Groups of Animals and Plants in Terms of Polygordius and of the Fern respectively.

PAGE

a. Animals.
All are triploblastic and cœlomate.

35. *Starfish.*
Radially symmetrical: discontinuous dermal exoskeleton: characteristic organs of locomotion (tube feet) in connection with ambulacral system of vessels 309

36. *Crayfish.*
Metamerically segmented: segmented lateral appendages: differentiation of metameres and appendages: continuous cuticular exoskeleton discontinuously calcified: gills as paired lateral offshoots of the body-wall: heart as muscular dilatation of dorsal vessel: cœlome greatly reduced and its place taken by an extensive series of blood-spaces: nervous system sunk in the mesoderm and consisting of brain and ventral nerve-cord 314

37. *Mussel.*
Non-segmented: mantle formed as paired lateral outgrowths of dorsal region: foot as unpaired median outgrowth of ventral region: cuticular exoskeleton in the form of a calcified bivalved shell: gills as paired lateral outgrowth of body-wall: heart as muscular dilatation of dorsal vessel: cœlome reduced to pericardium: nervous system consists of three pairs of ganglia sunk in the mesoderm 320

38. *Dogfish.*
Metamerically segmented: differentiated into head, trunk, and tail: trunk alone cœlomate in adult: appendages as median (dorsal, ventral, and caudal) and paired (pectoral and pelvic) fins: discontinuous dermal exoskeleton and extensive endoskeleton of partially calcified cartilage, including a chain of vertebral centra below the nervous system replacing an embryonic notochord: gills as pouches of pharynx opening on exterior: heart as muscular dilatation of ventral vessel: hollow dorsal nervous system not perforated by enteric canal 324

b. Plants.
All exhibit alternation of generations and the series shows the gradual subordination of the gamobium to the agamobium.

SYNOPSIS

39. *Equisetum.*
    Sporangia borne on sporophylls arranged in cones: spores homomorphic: prothalli dimorphic (male and female) 366

40. *Salvinia.*
    Spores dimorphic; microspore produces vestigial male prothallus: megaspore produces greatly reduced female prothallus 368

41. *Selaginella.*
    Microspore produces unicellular prothallus and multicellular spermary, both endogenously: female prothallus formed in megaspore and is almost endogenous: embryo provided with suspensor 371

42. *Gymnosperms.*
    Cones dimorphic (male and female), with rudimentary perianth: no sperms formed but microspore gives rise to pollen tube, nuclei in which are the active agents in fertilization: single megaspore permanently inclosed in each megasporangium: female prothallus purely endogenous: embryo (phyllula) remains inclosed in megasporangium which becomes a seed 373

43. *Angiosperms.*
    Cone modified into flower by differentiation of sporophylls and perianth: female sporophyll forms closed cavity in which megasporangia are contained: megaspore produces a single ovary represented simply by an ovum and two synergidæ: formation of prothallus retarded until after fertilization 378

B.—SUBJECTS OF GENERAL IMPORTANCE DISCUSSED IN SPECIAL LESSONS.

I.—CELLS AND NUCLEI.

a. The higher plants and animals contain cells similar in structure to entire unicellular organisms, and like them existing in either the amœboid, ciliated, encysted, or plasmodial condition 56

b. Minute structure of cells:—cell-protoplasm, cell-membrane, nuclear membrane, achromatin, chromatin 62

c. Direct and indirect nuclear division 65

d. The higher plants and animals begin life as a single cell, the ovum 68

II.—BIOGENESIS.

a. Definition of biogenesis and abiogenesis: brief history of the controversy 95

b. Crucial experiment with putrescible infusions: sterilization: germ-filters: occurrence of abiogenesis disproved under known existing conditions 98

III.—HOMOGENESIS.

Definition of homogenesis and heterogenesis; truth of the former firmly established 102

IV.—ORIGIN OF SPECIES.

a. Meaning of the term Species: the question illustrated by a consideration of certain species of Zoothamnium 137

b. Definition of Creation and Evolution: hypothetical histories of Zoothamnium in accordance with the two theories . . 141

c. The principles of Classification: natural and artificial classifications 140

d. The connection between ontogeny and phylogeny 146

V.—PLANTS AND ANIMALS.

a. Attempt to define the words plant and animal, and to place the previously considered types in one or other kingdom 176

b. Significance of the "third kingdom," Protista 182

VI.—SPERMATOGENESIS AND OOGENESIS.

Origin of sperms and ova from primitive sex-cells; differences in structure and development of the sexual elements . . 255

VII.—MATURATION AND IMPREGNATION.

a. Formation of first and second polar cells and of female pronucleus 259

b. Entrance of sperm and formation of male pronucleus . . 263

c. Conjugation of pronuclei 263

SYNOPSIS

VIII.—UNICELLULAR AND DIPLOBLASTIC ANIMALS.

PAGE

a. In plants there is a clear transition from unicellular forms to solid aggregates, but in animals the connection of the gastrula with unicellular forms is uncertain 264

b. Hypothesis of the origin of multicellular forms from a colony of unicellular zooids 265

C.—Other matters of general importance, such as the composition and properties of protoplasm, cellulose, chlorophyll, starch, &c. : metabolism : holozoic, holophytic, and saprophytic nutrition : intra- and extra-cellular digestion : amœboid, ciliary, and muscular movements : the elementary physiology of muscle and nerve : parasitism and symbiosis : asexual and sexual generation : and the elements of embryology —are discussed under the various types, and will be most conveniently referred to by consulting the Index.

INDEX AND GLOSSARY

A

Abiogen'esis (a, not: βίος, life: γένεσις, origin), the origin of organisms from not-living matter: former belief in, 96

Absorption by root-hairs, 341

Accre'tion (*ad*, to: *cresco*, to grow), increase by addition of successive layers, 14

Achrom'atin (a, not: χρῶμα, colour), the constituent of the nucleus which is unaffected or but slightly affected by dyes, 7, 63

Acœlom'ate (a, not: κοίλωμα, a hollow), having no cœlome (*q.v.*): 301

Adduct'or muscles, Mussel, 322

Aërob'ic (ἀήρ, air: βίος, life), applied to those microbes to which free oxygen is unnecessary, 93

Agamob'ium (a, not: γάμος, marriage: βίος, life), the asexual generation in organisms exhibiting alternation of generations (*q.v.*)

AGAR'ICUS (mushroom):—Figure, 192: general characters, 191: microscopic structure, 193: spore-formation, 193

Algæ (*alga*, sea-weed), 169

Alternation of Generations, meaning of the phrase explained under Nitella, 220: Bougainvillea, 250: Moss, 340: Fern, 361: Equisetum, 367, 368: Salvinia, 370: Selaginella, 373: Gymnosperms, 377: Angiosperms, 383, 384

Ambula'cral (*ambulacrum*, a walking place) **system**, starfish, 313

AMŒB'A (ἀμοιβός, changing):—Figure, 2: occurrence and general characters, 1: movements, 4, 10: species of, 8: resting condition, 10, 11: nutrition, 11: growth, 13: respiration, 17: metabolism, 17, reproduction, 19: immortality, 20: conjugation, 20: death, 20, 21: conditions of life, 21: animal or plant? 180

Amœb'oid movements, 4

Amœb'ula (diminutive of Amœba), the amœboid germ of one of the lower organisms. 51-54

Anab'olism (ἀναβολή, that which is thrown up). See Metabolism, constructive.

Anaerob'ic (a, not ἀήρ, air: βίος, life), applied to those microbes to which free oxygen is unnecessary, 93

An'al (*ānus*, the vent) **segment**, Polygordius, 273

An'al spot, Paramœcium, 113

An'astates (ἀνάστατος, from ἀναστῆναι, to rise up, 18. See Mesostates, anabolic.

Anatomy (ἀνατέμνω, to cut up), the study of the structure of organisms as made out by dissection.

Andrœ'cium (ἀνήρ, a male: οἶκος, a dwelling), the collective name for the male sporophylls in the flower of Angiosperms, 381

AN'GIOSPERMS (ἀγγεῖον, a vessel: σπέρμα, seed):—Figure, 379: general characters, 378-381: structure of flower, 378: reduction of gamobium, 381, 382: pollination and fertilization, 382, 383: formation of fruit and seed, and development of the leafy plant, 383, 384

Animal, definition of, 176

Animals, classification of, 307

Animals and Plants, comparison of type forms, 176, 177: discussion of doubtful forms, 180

Animals, Protists, and Plants, boundaries between artificial, 181-183

Anther, 381

Antherid'ium. See Spermary.

Antherozo'id. See Sperm.

Antip'odal cells, 382

An'us (*anus*, the vent), the posterior aperture of the enteric canal, 273

Ap'ical cell:—Penicillium, 190: Nitella, 211: Moss, 335: stem of Fern 350: root of Fern, 353: prothallus of Fern, 357

Ap'ical cone, Fern, 350

A'pical growth, 190, 351

A'pical mer'istem, a mass of meristem (*q.v.*) at the apex of a stem or root, 350
Appen'dages, lateral :—crayfish, 314 : dogfish, 324
Archegon'ium (ἀρχή, beginning : γόνος, production), the name usually given to the ovary of the higher plants
Aristotle, abiogenesis taught by, 96
Arteries, in the crayfish, 318
Arthropoda, the, 308
Arthospore (ἄρθρον, a joint : σπορά, a seed), in Bacteria, 89
Artificial reproduction of Hydra, 234
Asexual generation. See Agamobium.
Asexual reproduction. See Fission, Budding, Spore.
Asparagin, 338
Assimila'tion (*assimilo*, to make like). the conversion of food materials into living protoplasm, 13
At'rophy (a, without : τροφή, nourishment), a wasting away, 118
Au'ricle. See Heart.
Autom'atism (αὐτόματος, acting of one's own will), 10, 246
Axial bundle, Moss, 335
Axial fibre, Vorticella, 429
Axil (*axilla*, the arm-pit), 208
Axis, primary and secondary, 209

BOUGAINVILLEA (after L. A. de Bougainville, the French navigator):—Figures, 238, 241 : occurrence and general characters, 237 : microscopic structure, 239 : structure of medusa, 240 : structure and functions of nervous system, 245 : organs of sight, 246 : reproduction and development, 247, 248 : alternation of generations, 250
Bract (*bractea*, a thin plate), 251
Brain :—Polygordius. 286 : trochosphere, 299 : Crayfish, 319 : Dogfish, 330
Branch, Nitella, 209
Branch'ial (βράγχια, *branchiæ*, gills) **apertures**, Dogfish, 324, 329
Browne, Sir Thomas, on abiogenetic origin of mice, 96
Buc'cal (*bucca*, the cheek) **groove**, Paramœcium, 109
Bud, budding, Saccharomyces, 73 : comparison of with fission, 73 : Hydra, 233
Bundle-sheath, 349

C

Calyp'tra (καλύπτρα, a veil), 339
Cal'yx (κάλυξ, the cup of a flower), the outer or proximal whorl of the perianth in the flower of Angiosperms, 378
Canals, radial and circular, medusa, 241
Canal-cells of ovary, 337, 358
Cap-cells of roots, 354
Carbon dioxide, decomposition of by chlorophyll bodies, 29
Car'pel (καρπός, fruit), a female sporophyll, 381
Car'tilage, 324
Cauler'pa (καυλός a stem : ἕρπω, to creep), 174 (Figure)
Cell (*cella*, a closet or hut, from the first conception of a cell having been derived from the walled plant-cell) :—meaning of term, 60 : minute structure of (Figure), 62 : varieties of (Figure), 57
Cell-aggregate, meaning of term, 188
Cell-colony :—temporary, Saccharomyces, 73: permanent, Zoothamnium, 135, 136
Cell-division, 64-67
Cell-fusion 305, 351
Cell-layer, 277
Cell-membrane or **wall**, 11, 27, 28, 63
Cell-multiplication and **differentiation**, 218 : Polygordius, 305 : Fern, 351
Cell-plate, 67
Cell-protoplasm, 60
Cell'ulose, composition and properties of, 28
Central capsule, Radiolaria, 152
Central particle or **Centrosome** (κέντρον, centre : σῶμα, the body), 65, 256 (Figure). See also Directive sphere.
Ceph'alothor'ax, Crayfish, 314
Cerebral ganglion. See Brain.
Cerebro-pleural ganglion, Mussel, 323

B

BACIL'LUS (*bacillum*, a little staff), 85 : Figure, 87
BACTE'RIA (βακτήριον, a little staff) or **MICROBES** (μικρός, small : βίος. life) :—occurrence. 82 : structure of chief genera, 84-87 : reproduction, 87-89 : nutrition, 90 : ferment-action, 91 : parasitism, 92 : conditions of life, 92-94 : presence in atmosphere, 101-102 : animals or plants? 182
BACTER'IUM termo (Figures) 83, 84
Baer, von, Law of Development, 43
Barnacle-geese, supposed heterogenetic production of, 103
Bast. See Phloëm.
Binom'ial nomenclature, 8, 139
Biogen'esis (βίος, life : γένεσις, origin), the origin of organisms from pre-existing organisms, 96 : early experiments on, 96, 97 : crucial experiment on, 97-100
Biol'ogy (βίος, life : λόγος, a discussion), the science which treats of living things
Blast'ocœle (βλαστός, a bud ; κοῖλον, a hollow), the larval body-cavity, 298
Blood, Polygordius, 283
Blood-corpuscles : colourless, see Leuo ocytes : red, 56 : Figures, 57
Blood-vessels, Polygordius, 282: development of, 302
Body-cavity. See Blastocœle and Cœlome.
Body-segments. See Metameres.

Cerebro-spinal cavity, Dogfish, 325
CHARA (χαρά, delight), development and alternation of generations, 219, 220
Chlor'ophyll (χλωρός, green : φύλλον, a leaf), the green colouring matter of plants, properties of, 26 ; occurrence in Bacteria, 87 ; in Hydra, 231
Chrom'atin (χρῶμα, a colour), the constituent of the nucleus which is deeply stained by dyes, 7, 63 ; male and female in nucleus of oosperm, 263
Chrom'atophore (χρῶμα, colour : φέρω, to bear), a mass of proteid material impregnated with chlorophyll or some other colouring matter, 26, **46,** 197, 215, 231
Chromosome (χρῶμα, colour : σῶμα, body), 65, 66, **257, 262,** 263
Cil'ium (*cilium*, an eye-lash), defined, **25n** : comparisons of with pseudopod, 34, 52 : absence of cilia in Arthropoda, 319
Cil'iary movement, 25 ; a form of contractility, 34
Cil'iate Infusor'ia, 107
Classification, natural and **artificial, 141** : natural, a genealogical tree, **142**
Cnid'oblast (κνίδη, a nettle : βλαστός, a bud), the cell in which a nematocyst (*q.v.*) is developed, 230
Cnid'ocil (κνίδη and *cilium*), the "trigger-hair" of a cnidoblast, **230**
Cœlenterata, the, 308
Cœlome (κοίλωμα, a hollow), the body-cavity :—Polygordius, 273 ; Starfish, 311 : Crayfish, 319 : Mussel, 322 : Dogfish, 325 ; development of, Polygordius, 302
Cœlom'ata, provided with a cœlome, 276
Cœlomic epithelium. See Epithelium.
Cœlomic fluid, Polygordius, 281
Col'loids (κόλλα, glue : εἶδος, form), **properties of, 6**
Colony, Colonial organism, meaning of term, 135, 247 : formation of temporary colonies, Hydra, 234
Columel'la (a little column), 162
Com'missure (*commissūra*, a band), 282
Compound organism. See Colony.
Concres'cence (*cum,* together : *cresco,* to grow), the union of parts during growth
Cone, an axis bearing sporophylls :—Equisetum, 366 : Selaginella, 371 : Gymnosperms. 373
Conjuga'tion (*conjugātio*, a coupling), the union of two cells, in sexual reproduction :—Amœba, 20 : Heteromita, 41, 42 : Paramœcium, 114—116 : Vorticella, 132 : Mucor, 165 : Spirogyra, 198 : of ovum and **sperm,** 263 : monœcious and diœcious, 199 : comparison with plasmodium-formation, 54
Connective, œsophageal, 286
Connective tissue, 312
Contractile vac'uole (*vacūus*, empty):—Amœba, 8, 9 : Euglena, 47 : Paramœcium, 111

Contractil'ity (*contractĭo*, a drawing together), nature of, 10, 34 : muscular, 130
Contraction, physical and biological, 10
Corol'la (*corolla*, a little wreath), the inner or distal whorl of the perianth in the flower of Angiosperms, 380
Corpuscles. See Blood-corpuscles, and Leucocytes
Cortex, cor'tical layer (*cortex*, bark), constitution of, 59 : Infusoria, 110
Cotton-wool as a germ-fitter, 99
Cotyle'don (κοτυληδών, **a** cup or socket), the first leaf or leaves of the phyllula (*q.v.*) in vascular plants, 359
Cran'ium (*krānion*, the skull), 328
CRAYFISH :—Figure, 316: general characters, 314, 315 : limited number and concrescence of metameres, 314 : appendages, 314 : exoskeleton, 311 : enteric canal, 315 : gills, 318 : blood-system, 318 : kidney, 318 : nervous system, 319
Creation (*creo*, to produce), definition of, 141 : illustrated in connection with species of Zoothamnium (Diagram), 142
Cross-fertilization : applied to the sexual process when the gametes spring from different individuals, 199
Cryst'alloids (κρύσταλλος, crystal : εἶδος, form), properties of, 6
Cut'icle (*cuticula*, the outer skin), nature of in unicellular animals, 45, 109 : in multicellular animals, **239**
Cyst (κύστις, a bag), **used for** cell-wall in many cases, 11, 54

D

Dallinger, Dr. W. H., observations on an apparent case of heterogenesis, 103
Daughter-cells, cells formed by the fission or gemmation of a mother-cell, 35, 67
Death, phenomena attending, **20,** 21, 166, 167
Decomposition, nature of, 6
Dermis (δέρμα, skin), the deep or connective tissue layer of the skin, 312
Descent, doctrine of. See Evolution.
Development, meaning of the term, 43. For development of the various types see under their names
Dextrin, 113
Diastase, 81
Diast'ole (διαστέλλω, to separate), the phase of dilatation of a heart, contractile vacuole, &c., 111
DIATOMA'CEÆ (διατέμνω, to cut across, because of the division of the shell into two valves), 155 : Figure, 156
Diat'omin, the characteristic yellow colouring matter of diatoms, 155
Dichot'omous (διχοτομέω, to cut in two), applied to branching in which the stem divides into two axes of equal value, 138

Differentia′tion (*differo*, to carry different ways), explained and illustrated, 34, 119

Diges′tion (*digero*, to arrange or digest), the process by which food is rendered fit for absorption, 12, 13: intra- and extra-cellular, 232: contrasted with assimilation, 233

Digest′ive gland, 317

Dimorph′ism, dimorph′ic (δίς, twice: μορφή, form), existing under two forms, 35, 136, 243, 368, 370

Diœ′cious (δίς, twice: οἶκος, a dwelling), applied to organisms in which the male and female organs occur in different individuals, 199

DIPH′YES (διφυής, double): Figure, 252: occurrence and general characters, 250: polymorphism, 251

Diploblast′ic (διπλόος, double: βλαστός, a bud), two-layered: applied to animals in which the body consists of ectoderm and endoderm, 244: derivation of diploblastic from unicellular animals (Figures), 266, 268, 269

Directive sphere, 65, 261, 263. See also Centrosome.

Disc, Vorticella, 128

Dispersal, means of: in internal parasite, 124: in fixed organisms, 133-136

Distal, the end furthest from the point of attachment or organic base, 126

Distribution of food-materials:—in a complex animal, 281: in a complex plant, 341

Divergence of character, 145

Division of physiological labour, 34

DOGFISH:—Figure, 326: general characters, 324: fins, 324: exoskeleton, 325: endoskeleton, 325: enteric canal, 328: gills, 329: blood-system, 329: kidney, 330: gonads, 330: nervous system, 330

Dry-rigor, stiffening of protoplasm due to abstraction of water, 21

E

Echinodermata, the, 308

Ect′oderm (ἐκτός, outside: δέρμα, skin), the outer cell-layer of diploblastic and triploblastic animals, 225-230, 278

Ect′osarc (ἐκτός, outside: σάρξ, flesh), the outer layer of protoplasm in the lower unicellular organisms, distinguished by freedom from granules, 4

Egest′ion (*egero*, to expel), the expulsion of waste matters, 12

Egg-cell. See Ovum.

Em′bryo (ἔμβρυον, an embryo or fœtus), the young of an organism before the commencement of free existence.

Em′bryo-sac. See Megaspore.

Encysta′tion, being enclosed in a cyst (*q.v.*)

End′oderm (ἔνδον, within: δέρμα, skin) the inner cell-layer of diploblastic and triploblastic animals, 225, 231, 278

End′oderm-lamella, Medusa, 241

Endog′enous (ἔνδον, within: γίγνομαι, to come into being), arising from within, *e.g.* the roots of vascular plants, 354

End′osarc (ἔνδον, within: σάρξ, flesh), the inner, granular protoplasm of the lower unicellular organisms, 4

Endoskel′eton (ἔνδον, within, and *skeleton*, from σκέλλω, to dry), the internal skeleton of animals, 325

End′osperm (ἔνδον, within: σπέρμα, seed), nutrient tissue formed in the megaspore of Phanerogams, 377, 383

Endospore (ἔνδον, within: σπορά, a seed), a spore formed within a vegetative cell, 89

Energy, conversion of potential into kinetic, 15: source of, in chlorophyll-containing organisms, 31

Enter′ic (ἔντερον, intestine), **canal**, the entire food-tube from mouth to anus:—Polygordius, 273, 279: Starfish, 312: Crayfish, 315: Mussel, 322: Dogfish, 328

Ent′eron or **Enteric cavity**, the simple digestive chamber of diploblastic animals, 225

Epiderm′is (ἐπί upon: δέρμα, the skin): in animals synonymous with deric epithelium (*q.v.*, under Epithelium): in vascular plants a single external layer of cells, 348, 352, 353

Epithel′ial cells: columnar, 58: ciliated, 59

Epithel′ium (ἐπί, upon: θηλή, the nipple), a cellular membrane bounding a free surface, of cells: **cœlomic**, 277, 304: **deric**, 272: **enteric**, 276, 278

Equator′ial plate, 67

EQUISE′TUM (*equus*, a horse: *seta*, a bristle):—Figure, 367: general characters, 366: cone and sporophylls, 366: male and female prothalli, 366, 367: alternation of generations, 367, 368

Equiv′ocal generation. See Abiogenesis.

EUGLEN′A (εὔγληνος, bright-eyed):—Figure, 45: occurrence and general characters, 44: movements, 44: structure, 45: nutrition, 46: resting stage, 47: reproduction, 48: animal or plant? 178

Euglen′oid movements, 45

Ev′olution (*evolvo*, to roll out), **organic**: definition, 143: illustration of in connection with species of Zoothamnium (Diagram), 144

Excre′tion (*excerno*, to separate), the separation of waste matters derived from the destructive metabolism of the organism, 16, 284

Exog′enous (ἐξ, out of: γίγνομαι, to come into being), arising from the exterior, *e.g.* leaves, 354

INDEX AND GLOSSARY

Exoskel´eton (ἔξω, outside, and *skeleton*, from σκέλλω to dry), the external or skin-skeleton: **cuticular**, 239, 277-279; **cuticular and calcified**, 315, 320: **epidermal** (hair and nails): **dermal**, 312, 325
Eye-spots or **Ocel´li**:—Medusa, 241: Polygordius, 290, 299

F

Fæces (*faex*, dregs), solid excrement, consisting of the undigested portions of the food, 16
Ferm´ent (*fermentum*, yeast, from *ferveo*, to boil or ferment), a substance which induces **fermenta´tion**, *i.e.* a definite chemical change, in certain substances with which it is brought into contact, without itself undergoing change: **unorganized** and **organized ferments** 80: alcoholic, 76-81: **acetous**, 91: **diastatic** or **amylolytic**, 81: **lactic**, 91: **peptonizing** or **proteolytic**, 81; **putrefactive**, 91: ferment-cells of Mucor, 168
FERNS:—Figures, 346, 356: general characters 314, 345: histology of stem, leaf, and root, 344-354: nutrition, 354: spore-formation, 355: prothallus and gonads, 357-358: development, 359: alternation of generations, 361
Fertiliza´tion (*fertilis*, bearing fruit); the process of conjugation of a sperm or sperm-nucleus with an ovum, whereby the latter is rendered capable of development: a special case of conjugation (*q.v.*), 199: details of process, 263: in Vaucheria, 173: in Gymnosperms, 376, in Angiosperms, 383
Filtering air, method of, 99
Fins, Dogfish 324
Fiss´ion (*fissio*, a cleaving), **simple** or **binary**, the division of a mother-cell into two daughter-cells: in Amœba, 19; Heteromita, 40: animal- and plant-cells generally, 65-67: Paramœcium, 114: Vorticella 131
Fission, multiple, the division of a mother-cell into numerous daughter-cells:—in Heteromita, 42: Protomyxa, 51: Saccharomyces, 74
Fission, process intermediate between simple and multiple, Opalina, 124
Flagella. See Cilium.
Flag´ellate Infusoria, 107
Flagell´ula (diminutive of *flagellum*), the flagellate germ of one of the lower organisms (often called zoospores, 51, 54
Flagell´um (*flagellum*, a whip): defined, 25: transition to pseudopod, 52, 231
Floral receptacle, the abbreviated axis of an angiospermous flower 378
Flower, a specially modified cone (*q.v.*), having a shortened axis, which bears perianth-leaves as well as sporophylls, 378: often applied to the cone of Gymnosperms, 373
Food-current, Mussel 320, 322
Food-vacuole, a temporary space in the protoplasm of a cell containing water and food-particles, 11, 112
Foot: of Mussel, 320: of phyllula of fern 359
FORAMINIF´ERA (*forāmen*, a hole: *fero* to bear), 148: Figures, 149, 150, 151
Fragmenta´tion of the nucleus, 120
Free cell formation, 377
Fruit of Angiosperms, 384
Func´tion (*functio*, a performing), meaning of the term, 9

G

Gam´ete (γαμέω, to marry), a conjugating cell, whether of indeterminate or determinate sex:—Heteromita, 41: Mucor, 165: Spirogyra, 198: Vaucheria 173
Gamob´ium (γάμος, marriage: βίος, life), the sexual generation in organisms exhibiting alternation of generations (*q.v.*): progressive subordination of, to agamobium in vascular plants, 361, 384
Ganglion (γάγγλιον, a tumour), a swelling on a nerve-cord in which nerve-cells are accumulated, 319, 323
Gastric juice (γαστήρ, the stomach), properties of, 12
Gast´rula (diminutive of γαστήρ, the stomach), the diploblastic stage of the animal embryo in which there is a digestive cavity with an external opening: characters and Figure of, 265: contrasted with phyllula,, 360
Gemma´tion (*gemma*, a bud). See Budding.
Genera´tion, asexual, See Agamobium.
Sexual. See Gamobium.
Genera´tions, Alternation of. See Alternation of generations.
Gen´eralized, meaning of term, 140
Ge´nus (*genus*, a race), generic name, generic characters, 8, 139
Germ-filter, 99
Ger´minal spot, the nucleolus of the ovum, 259
Germina´tion (*germinatio*, a budding), the sprouting of a spore, zygote, or oosperm to form the adult plant: for germination of the various types see under their names.
Gill, an aquatic respiratory organ, 318, 323, 324, 329
Gland (*glans*, an acorn), an organ of secretion (*q.v.*): gland-cells, 231, 232, 285, 286
Glochid´ium, 323
Gon´ad (γόνος, offspring, seed), the essen-

tial organ of sexual reproduction, whether of indeterminate or determinate sex, *i.e.* an organ producing either undifferentiated gametes, ova, or sperms; see under the various types, and especially 172, 198, 209, 214
Gon'aduct (*gonad*, and *dūco*, to lead), a tube carrying the ova or sperms from the gonad to the exterior, 295, 313, 319, 323, 380
Grapping-lines, Diphyes, 251
Growing point: Nitella, 211, Moss, 535: Fern, 350
Growth, 13
Guard-cells of stomates, 353
Gullet, the simple food-tube of Infusoria, 47, 110: or part of the enteric canal of the higher animals, 280
GYMNOSPERMS (γυμνός, naked. σπέρμα seed): Figure, 374: general characters, 373: structure of cones and sporophylls, 373, 375, : reduction of gamobium (prothalli and gonads), 376,: pollination and fertilization, 376, 377: formation of seed and development of leafy plant, 377
Gynœc'ium (γυνή, a female : οἶκος, a dwelling), the collective name for the female sporophylls in the flower of Angiosperms, 381

H

Hæm'atochrome (αἷμα, blood : χρῶμα, colour), a red colouring matter allied to chlorophyll, 26
HÆMATOCOCCUS (αἷμα, blood : κόκκος a berry):—Figure, 24: general characters, 23: rate of progression, 23: ciliary movements, 25, 33: colouring matters, 26: motile and stationary phrases, 28: nutrition, 28: source of energy, 30: reproduction, 35: dimorphism, 35: animal or plant? 180
Hæmoglob'in (αἷμα, blood : *globus*, a round body, from the circular red corpuscles of human blood), 58: properties and functions of, 283
Head-kidney: trochosphere, 299
Heart:—Crayfish, 318 : Mussel, 323 : Dogfish, 329
Heat, evolution of, by oxidation of protoplasm, 17
Heat-rigor (*rigor*, stiffness), heat-stiffening, 21
Heliotropism, 168
Hered'ity (*hereditas*, heirship), 147
Hermaph'rodite (ἑρμαφρόδιτος, from Hermes and Aphrodite). See Monœcious.
Heterogen'esis (ἕτερος, different : γένεσις, origin), meaning of term, 102 : supposed cases of, 103 : not to be confounded with metamorphosis or with evolution, 104
HETEROMITA (ἕτερος, different μίτος,

a thread) :—Figure 38 : occurrence and general characters, 36 : movements, 37 : nutrition, 37 : asexual reproduction, 40 : conjugation, 41 : development and lifehistory, 42, 43 : animal or plant? 181
High and **low organisms**, 106
Higher (triploblastic) **animals**, uniformity in general structure of, 307
Higher (vascular) **plants**, uniformity in general structure of, 363
Histol'ogy (ἱστίον, a thing woven : λόγος, a discussion), minute or microscopic anatomy.
Holophyt'ic (ὅλος, whole : φυτόν, a plant), **nutrition**, defined, 31
Holozo'ic (ὅλος, whole : ζῶον, an animal), **nutrition**, defined, 31
Homegen'esis (ὁμός, the same : γένεσις origin), meaning of the term, 102
Homol'ogous (ὁμόλογος, agreeing), applied to parts which have had a common origin, 243
Homomorph'ism homomorph'ic (ὁμός the same : μορφή, form), existing under a single form, 139
Host, term applied to the organism upon which a parasite preys, 123
HYDRA (ὕδρα, a water-serpent) : Figures, 222, 226, 228, 235 : occurrence and general characters, 221 : species, 223 : movements, 223, 224 : mode of feeding, 224 : microscopic structure, 225: digestion, 232 : asexual, artificial, and sexual reproduction 233, 235 : development, 236
Hydr'anth (ὕδρα, a water-serpent : ἄνθος a flower), the nutritive zooid of a hydroid polype, 239, 243
Hydroid (ὕδρα, a water-serpent : εἶδος form) **Polypes** (πολύπους, many-footed), compound organisms, the zooids of which have a general resemblance to Hydra, 237
Hyper'trophy (ὑπέρ, over : τροφή, nourishment), an increase in size beyond the usual limits, 118
Hyph'a (ὑφαίνω, to weave) applied to the separate filaments of a fungus : they may be **mycelial** (see mycelium), **submerged**, or **aërial** : Mucor, 160, 163, Penicillium, 185, 188
Hyp'odermis (ὑπό, under : δέρμα, skin), Fern, 345, 348
Hyp'ostome (ὑπό, under : στόμα, mouth), 223, 239

I

Immortality, virtual, of lower organisms, 21
Income and **expenditure** of protoplasm, 18
Individual. See Zooid.
Individuation, meaning of the term. 233, 254

INDEX AND GLOSSARY

Indu'sium (*indusium*, an under-garment), 354
Inflores'cence (*floresco*, to begin to flower), an aggregation of cones or flowers, 373
Infusor'ia (so called because of their frequent occurrence in infusions), 107
Ingesta (*ingero*, to put into) and Egesta (*egero*, to expel), balance of, 32
Ingestion (*ingero*, to put into), the taking in of solid food, 11, 58
Insola'tion (*insolo*, to place in the sun), exposure to direct sunlight, 94
Integ'ument (*integumentum*, a covering) of megaspore: Gymnosperms, 375: Angiosperms, 382
Inter-cellular spaces, 347
Inter-muscular plexus (πλέκω, to twine), 288
Internode (*inter*, between: *nodus*, a knot), the portion of stem intervening between two nodes, 208
Intersti'tial (*interstitium*, a space between) cells, Hydra, 227: growth, Spirogyra, 198
Intest'ine (*intestinus*, internal), part of the enteric canal of the higher animals, 280
Intus-suscep'tion (*intus*, into: *suscipio*, to take up), addition of new matter to the interior, 13
Iodine, test for starch, 27
Irritabil'ity (*irritabilis*, irritable), the property of responding to an external stimulus, 10

J

Jaws: Crayfish, 315: Dogfish, 324, 328

K

Karyokines'is (κάρυον, a kernel or nucleus: κίνησις, a movement), indirect nuclear division, 67
Katab'olism (καταβολή, a laying down), 18. See Metabolism, destructive.
Kat'astates (καταστῆναι, to sink down), 18. See Mesostates, katabolic.
Kidney :—Crayfish, 318: Dogfish, 330

L

LAMINAR'IA (*lamina*, a plate), 203 (Figure), 204
Labial palps, Mussel, 322
Larva, the free-living young of an animal in which development is accompanied by a metamorphosis, 299
Larval stages, significance of, Polygordius, 301

Leaf, structure of :—Nitella, 208, 209, 213: Moss, 335: Fern, 344, 352: limited growth of, 214
Leaflet, Nitella, 209
Lept'othrix (λεπτός, slender: θρίξ, a hair), filamentous condition of Bacillus, 89: Figure, 87
LESSONIA (after Lesson, the French naturalist), 204 (Figure)
Leuc'ocyte (λευκός, white: κύτος, a hollow vessel, cell), a colourless blood corpuscle : —structure of, in various animals (Figures), 57 : ingestion of solid particles by, 58 : fission of, 58 : formation of plasmodia by, 58
Leuwenhoek, Anthony van, discoverer of Bacteria, 97
Life, origin of. See Biogenesis.
Life-history, meaning of the term, 43
Lignin (*lignum*, wood), composition and properties of, 348
Linear aggregate, an aggregate of cells arranged in a single longitudinal series, 188
Linnæus, C., introducer of binomial nomenclature, 8, 139
Liver, Dogfish, 328

M

MACROCYSTIS (μακρός, long: κύστις, bladder), 204
Mad'reporite (from its similarity to a *madrepore* or stone-coral), 311
Mantle, Mussel, 320
Manub'rium (*manubrium*, a handle) of Medusa, 241
Matura'tion of ovum, 259
Maximum temperature of amœboid movements, 21
Medul'la or medul'lary substance (*medulla*, marrow): in Infusoria, 110
Medus'a (Μέδουσα, name of one of the Gorgons), the free-swimming reproductive zooid of a hydroid polype, 239-243: derivation of a, from hydranth (Figure), 241
Medus'oid, a reproductive zooid having the form of an imperfect Medusa, Diphyes, 251
Meg'agamete (μέγας, large: γαμέω, to marry), a female gamete (*q.v.*) distinguished by its greater size from the male or microgamete, 132
Meg'anucleus (μέγας, large: *nucleus*, a kernel), 111, 128
Meg'asporan'gium (μέγας large σπορά, seed: ἀγγεῖον, a vessel), the female sporangium in plants with sexually dimorphic sporangia, usually distinguished by its greater size from the male or micro-sporangium :—Salvinia, 368 : Sela-

D D

ginella, 371 : Gymnosperms, 375 : Angiosperms, 381.

Meg'aspore (μέγας, large : σπορά, a seed), the female spore in plants with sexually dimorphic spores, always distinguished by its large size from the male or microspore :—Salvinia, 368 : Selaginella, 371 : Gymnosperms, 375 : Angiosperms, 382

Megazo'oid (μέγας large : ςῷον, animal : εἶδος, form), the larger zooid in unicellular organisms with dimorphic zooids, 35, 132

Mer'istem (μερίστημα, formed from μερίζω, to divide), indifferent tissue of plants from which permanent tissues are differentiated, 350

Mes'entery (μέσος, middle : ἔντερον, intestine), a membrane connecting the enteric canal with the body-wall, 279 : development of, 302

Mes'oderm (μέσος, middle : δέρμα, skin), the middle cell-layer of triploblastic animals : Polygordius, 278 : development of, 299 : splitting of to form somatic and splanchnic layers, 302

Mesoglœ'a (μέσος, middle : γλοία, glue), a transparent layer between the ecto- and endo-derm of Cœlenterates :—in Hydra, 225 : in Bougainvillea, 244

Mes'ophyll (μέρος, middle : φύλλον, a leaf), the parenchyma of leaves, 352

Mes'ostates (μέσος, middle : στῆναι, to stand), intermediate products formed during metabolism (*q.v.*) and divisible into (*a*) **anabolic mesostates** or **anastates**, products formed during the conversion of food-materials into protoplasm ; and (*b*) **katabolic mesostates** or **katastates**, products formed during the breaking down of protoplasm, 18

Metab'olism (μεταβολή, a change), the entire series of processes connected with the manufacture of protoplasm, and divisible into (*a*) **constructive metabolism** or **anabolism**, the processes by which the substances taken as food are converted into protoplasm, and (*b*) **destructive metabolism** or **katabolism**, the processes by which the protoplasm breaks down into simpler products, excretory or plastic, 17

Met'amere (μέτα, after : μέρος, a part), a body-segment in a transversely segmented animal such as Polygordius, 271, 273 : development of, 301 : limited number and concrescence of in Crayfish, 314

Metamorphos'is (μεταμόρφωσις), a transformation applied to the striking change of form undergone by certain organisms in the course of development after the commencement of free existence :—Vorticella, 133 : Polygordius, 306

Mic'robe (μικρός, small : βίος, life). See eria.

MICROCOC'CUS (μικρός, small : κόκκος, a berry) (Figure), 86

Microgam'ete (μικρός, small : γαμέω, to marry), a male gamete (*q.v.*), distinguished by its smaller size from the female or megagamete, 132

Micro-millimetre, the one-thousandth of a millimetre, or 1-25,000th of an inch, 84

Micro-organism. See Bacteria.

Micronucleus (μικρός, small : *nucleus*, a kernel), 111, 128

Micropyle (μικρός, small : πύλη, an entrance), 375

Micro-sporan'gium (μικρός, small : σπορά, a seed : ἀγγεῖον, a vessel), the male sporangium in plants with sexually dimorphic sporangia, usually distinguished by its smaller size from the female or mega-sporangium :—Salvinia, 368 : Selaginella, 371 : Gymnosperms, 376 : Angiosperms, 381

Mic'rospore (μικρός, small : σπορά, a seed), the male spore in plants with sexually dimorphic spores, always distinguished by its small size from the female or mega-spore :—Salvinia, 368 : Selaginella, 371 : Gymnosperms, 376 Angiosperms. 381

Microzo'oid (μικρός, small : ζῷον, an animal : εἶδος, form), the smaller zooid in unicellular organisms with dimorphic zooids, 35, 132

Midrib of leaf, Moss, 335

Minimum temperature for amœboid movements, 21

Mollusca, the, 309

Monœc'ious (μόνος, single : οἶκος, a house), applied to organisms in which the male and female organs occur in the same individual, 199, 234

Monopod'ial (μόνος, single : πούς, a foot), applied to branching in which the main axis continues to grow in a straight line and sends off secondary axes to the sides, 138

MONOSTROMA (μόνος, single : στρῶμα, anything spread out), 202 (Figure)

Morphol'ogy (μορφή, form : λόγος a discussion), the department of biology which treats of form and structure, 9

Mor'ula (diminutive of *mōrum*, a mulberry) See Polyplast.

MOSSES :—Figures, 333, 338 : general characters, 332 : structure of stem, 334 : leaf, 335 : rhizoids, 335 : terminal bud, 335 : reproduction, 336 : development of sporogonium, 337 : of leafy plant, 339, 340 : alternation of generations, 340 : nutrition, 340, 341

Mouth :—Euglena, 47 : Paramœcium, 110 : Hydra, 223 : Medusa, 241 : Polygordius, 271 : backward shifting of in Crayfish, 315

MUCOR (*mucor*, mould) :—Figure, 159 :

occurrence and general characters, 158 : mycelium and aërial hyphæ, 160-163 : sporangia and spores,160-162-165 ; transition from uni- to multi-cellular condition, 162 : development of spores, 163 ; conjugation, 165 ; death, 166 : nutrition, 167 ; parasitism. 167 : ferment-cells, 168

Mucous membrane, 58

Multicellular, formed of many cells, 61, 162

Muscle (*músculus*, a little mouse, a muscle), nature of, 130, 131

Muscle-fibres, Bougainvillea, 244

Muscle-plate, Polygordius, development of, 305

Muscle-process, Hydra, 227, 232

Mushroom. See Agaricus.

MUSSEL (same root as *muscle*), Freshwater :—Figure, 321 : general characters, 320 : mantle, shell, and foot, 320 : food-current, 320 : enteric canal, 322 : gills and blood-system, 323 : nephridia, gonads, and nervous system, 323

Mycelial hyphæ, the hyphæ interwoven to form a mycelium.

Mycel'ium (μύκης, a fungus), a more or less felt-like mass formed of interwoven hyphæ :—Mucor, 160 : Penicillium, 185

MYCETOZOA (μύκης, a fungus : ζῶον, an animal) :—Figure, 53 : occurrence and general characters, 52 : nutrition, 54 : reproduction and life-history, 54, 55 : animals or plants ? 181

My'ophan (μῦς, mouse, muscle : φαίνω, to appear), 110

Myxomyce'tes (μύξα, mucus : μύκης, a fungus). See Mycetozoa.

N

Nem'atocyst (νῆμα, a thread : κύστις, a bag), 229

Nephrid'iopore (νεφρός, a kidney : πόρος, a passage), the external opening of a nephridium, 285

Nephrid'ium (νεφρός, a kidney), structure of, Polygordius, 285 (Figure) : development of, 304 : Mussel, 323 : Dogfish, 330

Neph'rostome (νεφρός, a kidney : στόμα, a mouth), the internal or cœlomic aperture of a nephridium, 285

Nerve, afferent and efferent, functions of, 288

Nerve-cell, 230, 245

Nervous system, central and peripheral : —Medusa, 245 : Polygordius, 286 : functions of, 288 : Starfish, 313 : Crayfish, 319 : Mussel, 323 : Dogfish, 330

Neur'ocœle (νεῦρον, a nerve : κοίλη, a hollow), the central cavity of the vertebrate nervous system, 330

NITELL'A (*niteo*, to shine) :—Figures, 207, 212, 215, 217. 219 : occurrence and general characters, 206 : microscopic structure, 209 : terminal bud, 211 : structure and development of gonads, 209, 214 : development, 219 : alternation of generations, 220

Node (*nodus*, a knot), the portion of a stem which gives rise to leaves, 208

Not'ochord (νῶτος, the back : χορδή, a string), 328

Nucel'lus (diminutive of *nucleus*, the name formerly applied), 375, 382

Nuclear division, indirect : 64 (Figure) : 65, 67 : direct, 67

Nuclear membrane, 62,

Nuclear protoplasm. See Achromatin.

Nuclear spindle, 65, 66

Nucle'olus (diminutive of nucleus), 8

Nu'cleus (*nucleus*, a kernel), minute structure of, 63 ; Amœba, 7, 8 : Paramœcium, 111, 114 : Opalina, 121 : Vorticella, 128 : Nitella, 210, 213 : fragmentation of, 120

Nucleus, secondary, of megaspore, Angiosperms, 382

Nutrient solution, artificial, principles of construction of, 78,

Nutrition :—Amœba (holozoic), 11 : Hæmatococcus (holophytic), 28 : Heteromita (saprophytic), 37 : Opalina (type of internal parasite), 123 : Mucor 167 : Penicillium, 190 : Polygordius (type of higher animals), 273, 281 : Moss (type of higher plants), 340

O

Ocel'lus (*ocellus*, a little eye), structure and functions of, Medusa, 241, 246

Œsoph'agus (οἰσοφάγος, the gullet). See Gullet.

Ontog'eny (ὄντος, being : γένεσις, origin), the development of the individual : a recapitulation of phylogeny (*q.v.*), 146

Oogen'esis (ᾠόν, an egg : γένεσις, origin), the development of an ovum from a primitive sex-cell, 255, 258

Oogon'ium (ᾠόν, egg : γόνος, production), the name usually given to the ovary of many of the lower plants.

Oosperm (ᾠόν, egg : σπέρμα, seed), a zygote (*q.v*), formed by the ovum and sperm : a unicellular embryo, 173 : origin of nucleus of, 263

Oosphere (ᾠόν, an egg : σφαῖρα, a sphere), a name frequently given to the ovum of plants.

Oospore (ᾠόν, an egg : σπορά, a seed), a name frequently applied to the oosperm of plants.

OPALIN'A (from its *opalescent* appearance) :—Figure, 122 : occurrence and general characters, 121-123 : structure and division of nuclei, 121: parasitic nutrition, 123: reproduction, 124: means of dispersal, 124: development, 125

Opt'imum (*optimus*, best) temperature for amœboid movements, 21: for saprophytic monads, 40

Organ (ὄργανον, an instrument), a portion of the body set apart for the performance of a particular function, 291

Or'ganism, any living thing, whether animal or plant, 5

Oss'icle (diminutive of ὄς, a bone), 311

Ov'ary (*ovum*, an egg), the female gonad or ovum-producing organ; see under the various types and especially Vaucheria, 172: atrophy of, in Angiosperms, 382. The name is also incorrectly applied to the venter of the pistil of Angiosperms, 381

Ovi'duct (*ovum*, an egg: *duco*, to lead), a tube conveying the ova from the ovary to the exterior, 295

Ov'um (*ovum*, an egg), the female or megagamete in its highest stage of differentiation: general structure of, 68, 69: minute structure and maturation of, 258, 259: see also under the various types and especially Vaucheria, 172: formation of, in Angiosperms, 382

Ov'ule (diminutive of *ovum*), the name usually applied to the megasporangium of Phanerogams.

Oxidation of protoplasm, 15

OXYTRICH'A (ὀξύς, sharp: θρίξ, a hair), 120 (Figure)

P

Pancreas (παγκρέας, sweetbread), 328
Pandorina, 266 (Figure), 267
Param'ylum (παρά, beside: ἄμυλον, fine meal, starch), 46
PARAMŒ'CIUM :—Figures, 108, 115: structure, 107: mode of feeding, 112: asexual reproduction, 114: conjugation, 114

Par'asite, parasitism (παράσιτος, one who lives at another's table) :—Opalina, 123: Bacteria, 92: Mucor, 167

Paren'chyma (παρέγχυμα, anything poured in beside, a word originally used to describe the substance of the lungs, liver, and other soft internal organs), applied to the cells of plants the length of which does not greatly exceed their breadth and which have soft non-lignified walls, 60: ground-parenchyma, 345, 347

Pari'etal (*paries*, a wall), applied to the layer of cœlomic epithelium lining the body-wall, 277, 278

Parthenogen'esis (παρθένος, a virgin: γένεσις, origin), development from an unfertilized ovum or other female gamete, 200

Parthenogenet'ic ova, characteristics of, 262

Pasteur, Louis, researches on yeast, 78-80

Pasteur's solution, composition of, 76
Pedal (*pes*, the foot) ganglion, Mussel, 323

PENICILL'IUM (*penicillum*, a painter's brush, from the form of the fully-developed aërial hyphæ) :—Figure, 186: occurrence and general characters, 184: mode of growth, 185: microscopic structure, 185: formation and germination of spores, 189: sexual reproduction, 190: nutrition, 190: vitality of spores, 191

Peps'in (πέπτω, to digest), the proteolytic or pepsonizing ferment of the gastric juice, 12, 80

Peptones, 12

Perianth (περί, around: ἄνθος, a flower), the proximal infertile leaves of a flower, 380

Perisperm (περί, around: σπέρμα, seed), nutrient tissue developed in the nucleus of the seed, 380 (description of figure)

Peristom'e (περί, around: στόμα, the mouth), Vorticella, 128

Peristom'ium (περί, around: στόμιον, a little mouth), the mouth-bearing segment of worms, 273, 297

Peritone'um (περιτόναιον), the membrane covering the viscera, 325

Pet'als (πέταλον, a leaf), the inner or distal perianth leaves in the flower of Angiosperms, 380

Phar'ynx (φάρυγξ, the throat) :—Polygordius, 280: Dogfish, 328

Phloem (φλοιός, bark or bast), the outer portion of a vascular bundle, 349

Phyla (φῦλον, a tribe) of the animal kingdom, 307: of the vegetable kingdom, 364

Phyll'ula (diminutive of φύλλον, a leaf), the stage in the embryo of vascular plants at which the first leaf and root have appeared, 360: contrasted with gastrula, 360

Phylog'eny (φῦλον, a race: γένεσις, origin), the development of the race, 147

Physiol'ogy (φύσις, the nature or property of a thing: λόγος, a discussion), the department of biology which treats of function, 9 *et seq.*

Pigment-spot, Euglena, 47
Pileus (*pileus*, a cap), Agaricus, 191
Pinna (*pinna*, a feather), of leaf, 352
Pistil (*pistillum*, a pestle, from *pinso*, to pound.) See Gynœcium.

Plan'ula (diminutive of πλάνος, a wandering about), the mouthless diploblastic larva of a hydroid, 248

Plant, definition of, 176

INDEX AND GLOSSARY 405

Plants, classification of, 364
Plas'ma (πλάσμα, anything moulded), of blood, 56
Plasmo'dium (πλάσμα, anything moulded), 52-55: comparison of with zygote, 54
Plastic (πλαστικός, formed by moulding) products, products of katabolism which remain an integral part of the organism, 33
Pod'omere (πούς, a foot : μέρος, a part), a limb-segment, 314
Polar cells, formation of, 262
Pollen grain (*pollen*, fine flour), a name given to the microscope of Phanerogams.
Pollen-sac, a name given to the microsporangium of Phanerogams.
Pollen-tube, 376, 383
Pollina'tion, 376, 383
POLYGORD'IUS (πολύς, many : Γόρδιος, King of Phrygia, inventor of the Gordian knot):—Figures, 272, 274, 285, 287, 294, 296, 298, 300, 303: occurrence and general characters, 271, 274 : metameric segmentation, 271-273 : mode of feeding, 273: enteric canal, 273, 277 ; cell-layers, 276-278 ; coelome, 273, 277 : distribution of food, 281 ; blood-system, 282 : nephridia, 284 : nervous system, 286 : differentiation of definite organs and tissues, 291 : reproduction, 293 : development and metamorphosis, 299-306
Polymorph'ism (πολύς, many : μορφή, form), existing under many forms, 251
Pol'yplast (πολύς, many: πλαστός, formed, modelled), the multicellular stage of the embyro before the differentiation of cell-layers or organs :—Hydroids, 248 : Moss, 337 : Fern, 359
PORPITA (πόρπη, a brooch), 253 : (Figure), 253
Primor'dial utricle, 196, 210
Proctodæ'um (πρωκτός, the anus : ὁδαῖος, belonging to a way), an ectodermal pouch which unites with the enteron and forms the posterior end of the enteric canal, its external aperture being the permanent anus, 298
Pro-embryo, chara, 219 (Figure)
Pro-nucleus, *female*, 262 : *male*, 263 : conjugation of male and female, 263
Prostom'ium (πρό, before : στόμιον, a little mouth), the first or pre-oral segment in worms, &c., 271, 296
PROT'AMŒBA (πρῶτος, first : ἀμοιβός, changing), 9 (Figure).
Prothal'lus (πρό, before : θαλλός, a twig), the gamobium of vascular plants :—Fern, 355 : dimorphism of in Equisetum, 367 ; reduction of in Salvinia, 369 ; Selaginella, 371, and Gymnosperms, 376, 378 ; retarded development of in Angiosperms, 384
Prothallus, secondary, Selaginella, 371

Prot'eids (πρῶτος, first), composition of, 5
Protist'a (πρώτιστος, the first of all), the lowest organisms intermediate between the lowest undoubted animals and plants, 182
Protococ'cus (πρῶτος, first : κόκκος, berry). See Hæmatococcus.
PROTOMYX'A (μρῶτος, first: μύξα, mucus): Figure, 50 : occurrence and general characters, 49 : life-history, 51 ; animal or plant? 181
Protonem'a (πρῶτος, first : νῆμα, a thread), Moss, 336, 339
Prot'oplasm (πρῶτος, first : πλάσμα, anything moulded), composition of, 5 : properties of, 5-7 : micro-chemical tests for, 7, 8 : minute structure of, 62, 63: continuity of in Fern, 350 : in Polygordius, 292 : intra- and extra-capsular, Radiolaria, 152
Protozoa, the, 308
Prox'imal (*proximus*, nearest), the end nearest the point of attachment or organic base, *e.g.* in the stalk of Vorticella, 126
Pseud'opod' (ψευδής, false : πούς, foot), described, 4 : comparison of with cilium, 34, 52 : in columnar epithelium, 59 : in endoderm cells of Hydra, 231
Pteris. See Ferns.
Punctum vegetationis. See Growing point.
Putrefac'tion (*putrefacio*, to make rotten) nature of, 82 : a process of fermentation, 91 : conditions of temperature, moisture, &c., 93, 94
Putres'cent (*putresco*, to grow rotten) solution, characters of, 37, 82
Putres'cible infusion, sterilization of, 99-102
Pyren'oid (πυρήν, the stone of stone-fruit : εἶδος, form), a small mass of proteid material invested by starch, 27

R

Radial symmetry, starfish, 309
RADIOLAR'IA (*radius*, a spoke or ray):—Figures, 152, 153 : occurrence and general characters, 152 : central capsule, 152 : intra- and extra-capsular protoplasm, 152 : silicious skeleton, 152 : symbiotic relations with Zooxanthella, 154
Rect'um (intestinum rectum, the straight gut), the posterior or anal division of the enteric canal, 281
Redi, Francisco (Italian *savant*), experiments on biogenesis, 97
Reducing division, 257, 262
Reflex action, 289
Reproduction, necessity for, 19
Reproductive organ. See Gonad.

Reservoir of contractile vacuole, Euglena, 47
Respiration :—Amœba, 17 : Polygordius, 284
Respiratory cæca, Starfish, 312
Rhiz'oid (ῥίζα, root : εἶδος, form):—Nitella, 206, 214 ; Moss, 335 : prothallus of Fern, 355
Root, Fern, 344, 353
Root-cap, 354
Root-hairs, 353, 357
Ross, Alexander, on abiogenetic origin of mice, insects, &c., 96
Rotation of protoplasm, 210
Rudiment, rudimentary (*rudimentum*, a beginning), the early stage of a part or organ : often used for a structure which has undergone partial atrophy, but in such cases the word vestige (*q. v.*) is more suitable.

S

SACCHAROMY'CES (σάκχαρον, sugar : μύκης, fungus):—Figure, 72 : occurrence, 71 : structure, 71 : budding, 73 : internal fission, 74 : nutrition, 75 : alcoholic fermentation caused by, 75, 79, 80 : experiments on nutrition of, 78-80 : animal or plant? 182
SALVIN'IA :—Figure, 369 : general characters, 368 : mega- and micro-sporangia and spores, 368 : male and female prothalli and gonads, 369, 370 : development and alternation of generations, 370
Saprophyt'ic (σαπρός, putrid : φυτόν, a plant) **nutrition**, defined, 39
Schulze's solution, test for cellulose, 28 : for lignin, 348, 349
Scleren'chyma (σκληρός, hard : ἔγχυμα, infusion):—Moss, 334 : Fern, 345, 348, 351
Secre'tion (*secretus*, separate), nature of, 231 : formation of cell-wall a process of, 14
Seed, formation of, 377, 384 : germination of, 377
Seg'ment (*segmentum*, a piece cut off), in plants a node together with the next proximal internode, 208 : in animals the name is variously applied. See Metamere, Podomere.
Segment'al cell : Nitella, 211 : Moss, 335 : Fern, 350
Segmentation, metameric. See Metamere.
SELAGINELL'A (σελαγέω, to shine):—Figure, 372 ; general characters, 371 : cone, sporangia, and spores, 371 : prothalli and gonads, 371 : development and alternation of generations, 372, 373
Self-fertilization, applied to the sexual process when the gametes spring from the same individual, 199
Sep'als (*separ*, separate), the outer or proximal perianth-leaves in the flower of Angiosperms, 378
Sep'tum (*septum*, a barrier):—In plants 187 : in Polygordius, 280 : development of, 302
Set'a (*seta*, a bristle), 290
Sex-cells, primitive, 255 : origin of in Hydroids, 247 : in Polygordius, 293
Sexual differentiation, illustrated by Vaucheria, 172 : by Spirogyra, 199
Sexual generation. See Gamobium.
Sexual reproduction, nature of, 42
Shell, Mussel, 320
Shoot, in plants, an axis of the second or any higher order with its leaves, 209
Sieve-tubes and **plates**, 350
Sinus (*sinus*, a hollow), a spacious cavity 318
Skeleton. See Endo- and Exo-skeleton.
Slime-fungi. See, Mycetozoa.
Solid aggregate, 203
Somat'ic (σῶμα, the body), applied to the layer of mesoderm which is in contact with the ectoderm and with it forms the body-wall, 278
Sor'us (σωρός, a heap), an aggregation of sporangia, 354, 368
Species (*species*, a kind), meaning of term illustrated, 8, 137 ; definition of, 139 : origin of, 141, 144
Specific characters, specific name, 8, 139
Specialized, meaning of, 140
Sperm (σπέρμα, seed), the male or microgamete in its highest stage of differentiation : structure and development of, 255 : see also under the various types, and especially Vaucheria, 172, 173
Spermatozo'id, spermatozo'on (σπέρμα, seed : ζῶον, animal, from the actively moving sperms of animals having been supposed to be parasites), synonyms of sperm.
Spermary (σπέρμα, seed), the male gonad or sperm-producing organ : see under the various types, and especially Vaucheria, 172
Sperm'iduct (σπέρμα, seed : *duco*, to lead), a tube conveying the sperm from the spermary to the exterior, 295
Spermatogen'esis (σπέρμα, seed ; γένεσις, origin), the development of a sperm from a primitive sex-cell, 255, 256 (Figure).
Spinal cord, Dogfish, 330
Spiral vessel. See Vessel.
SPIRILL'UM (*spira*, a coil) 86, 88 (Figure)
SPIROGYRA (*spira*, a coil : *gyrus*, a revotion):—Figure, 195 : occurrence and general characters, 194 : microscopic structure, 194 : growth, 197 : conjugation, 198 : development, 200 : nutrition, 200

Splanch'nic (σπλάγχνον, intestine or viscus), applied to the layer of mesoderm which is in contact with the endoderm and with it forms the enteric canal, 278
Spontaneous generation. See Abiogenesis.
Sporan'gium (σπορά, seed: ἀγγεῖον, a vessel), a spore-case:—Mucor, 160 : Vaucheria, 171 : Fern, 354. See also Mega- and Micro-sporangium.
Spore (σπορά, a seed), an asexual reproductive cell : see under the various types and especially Heteromita, 42 : Saccharomyces, 74 ; Bacteria, 89 : vitality of in Bacteria, 99, 101 : Penicillium, 189 : Moss, 339 : Fern, 355. See also Mega- and Micro-spore.
Sporogon'ium σπορά seed : γόνος, production), the agamobium of a moss, 337
Spor'ophyll (σπορά, seed : φύλλον, leaf), a sporangium-bearing leaf :—Equisetum, 366 : Selaginella, 371 : Gymnosperms, 373, 375 : Angiosperms, 381
Stamen (*stamen*, a thread), a male sporophyll, 373, 381
Starch, composition and properties of, 27
STARFISH :—Figure, 310 : general characters, 309-311 : **radial symmetry**, 309 : tube-feet and ambulacral system, 311, 313 : exoskeleton, 312
Stem, structure of :—Moss, 334 ; Fern, 345
Sterig'ma (στήριγμα, a support): Penicillium, 188 ; Agaricus, 193
Sterilization of putrescible infusions, 99-102
Stigma (στίγμα, a spot), the receptive extremity of the style, 381
Stimulus, various kinds of, 289
Stock. See Colony.
Stom'ate (στόμα, mouth), 353
Stomodæ'um (στόμα, mouth : ὁδαῖος, belonging to a way), an ectodermal pouch which unites with the enteron and forms the anterior end of the enteric canal, its aperture being the permanent mouth, 298
Stone-canal, Starfish, 313
Style (*stylus*, a column), the distal solid portion of the female sporophyll or of the entire gynœcium in Angiosperms, 381
STYLONYCH'IA (στῦλος, a column : ὄνυξ, a claw), Figure, 117 : occurrence and general characters, 116 : polymorphism of cilia, 118-119
Sub-apical cell. See Segmental cell.
Superficial aggregate, 202
Supporting lamella. See Mesoglœa.
Suspensor : Selaginella, 373 ; Gymnosperms, 377 ; Angiosperms, 383
Sweet Wort, composition of, 75
Swimming-bell, Diphyes, 251
Symbio'sis (συμβίωσις, a living with), an intimate and mutually advantageous association between two organisms, 154

Syner'gidæ (συνεργός, a fellow worker), 382
Sys'tole (συστολή, a drawing together, contraction), the phase of contraction of a heart, contractile vacuole, &c., 111

T

Teeth, Dogfish, 328
Temperature, effects of on protoplasmic movements, 20, 21
Tentacles :—Hydra, 223 ; Bougainvillea, 239 ; Polygordius, 271 : development of, 302
Term'inal bud :—Nitella, 208, 210 ; Moss 335
Testis (the Latin word), generally used for the spermary in animals.
Thermal death-point. See Ultra-maximum temperature.
Tissues, differentiation of :—Polygordius, 291 ; Fern, 353
Tracheides (τραχύς, rough : εἶδος, form). See Vessels of Plants, 349
Transpiration, the giving off of water from the leaves of plants, 341
Trich'ocyst (θρίξ, a hair : κύστις, a bag), 113
Triploblast'ic (τριπλόος, triple : βλαστός, a bud), three-layered : applied to animals in which the body consists of ectoderm, mesoderm, and endoderm, 244, 278
Troch'osphere (τροχός, a wheel, in reference to the circlet of cilia : σφαῖρα, a sphere), the free-swimming larva of Polygordius, &c. :—characters of, 296 (Figure) ; origin of from gastrula, 297, 298 ; metamorphosis of, 299
Tube-feet, Starfish, 311, 313

U

Ultra-maximum temperature, for amœboid movements, 21 ; for monads, 40 ; for Bacteria, 93
ULVA (*ulva*, an aquatic plant), 203
Umbell'ate (*umbella*, a sun-shade, umbrella) applied to branching in which the primary axis is of limited growth and sends off a number of secondary axes from its distal end, 138
Unicell'ular, formed of a single cell, 61 ; connection of uni- with multi-cellular organisms, 264-270
Ureter (οὐρητήρ, the Greek name), the duct of the kidney, 330

V

Vac'uole (*vacuus*, empty), contractile, 11 111 : non-contractile, 71
Variability, 147
Variation, individual, 140, 147

INDEX AND GLOSSARY

Variety, an incipient **species**, 147
Vasc'ular (*vasculum*, a small vessel) **bundles**, 345, 348
Vascular plants, 365
VAUCHERIA (after J. P. E. Vaucher, a Swiss botanist):—Figure, 170: occurrence and general characters, 169 : **minute structure**, 169 ; **asexual** reproduction, 171: sexual reproduction, 172 : nutrition, 175
Veins of Dogfish, 329 : of leaves, 352
Vel'um (*velum*, a veil) of medusa, 241
Vent, the aperture of the cloaca, 324
Venter (*venter*, the belly), of ovary of Moss, 336, and Fern, 358 : of the female sporophyll or of the entire gynœcium of Angiosperms (so-called ovary) 381
Ventral nerve-cord :—Polygordius, 286 : development of, 301 : Crayfish, 319
Ventricle. See Heart.
Vermes, the, 308
Ver'tebral (*vertebra*, a joint) **centra** and **column**, Dogfish, 328
Vertebrata, the, 309
Vessels :—of plants, spiral and scalariform, 348, 349 : of animals, see Blood-vessels.
Vestige, vestigial (*vestigium*, a trace), applied to any structure which has become atrophied or undergone reduction beyond the limits of usefulness, 118
Vib'rio (*vibro*, to vibrate), 86, 88, (Figure)
Vis'ceral (*viscus*, an internal organ), applied to the layer of cœlomic epithelium, or of peritoneum, covering the intestine and other internal organs, 277
Visceral ganglion, Mussel, 323
Vitelline (*vitellus*, yolk) **membrane**, the cell-membrane of the ovum, 259
Volvox (*volvo*, to roll), 267, 268, 269. (Figures)
VORTICELLA (diminutive of *vortex*, a eddy):—Figure, 127 : occurrence and general characters, 126 : structure, 126 : asexual reproduction, 131 : conjugation, 132 : means of dispersal, 132-136 : encystation, spore-formation, development, and metamorphosis, 133

W

Waste-products, 33
Water of organization, 5, 29
Whorl of leaves, 208
Wood. See Xylem.
Work and Waste, 14

X

Xylem ($\xi\dot{\upsilon}\lambda o\nu$, wood), the inner portion o. vascular bundle, 349

Y

Yeast, 71
Yeast-plant. See Saccharomyces.
Yellow-cells of Radiolaria, 154
Yolk-granules or **spheres**, 68, 235, 258

Z

Zooglœ'a ($\zeta\tilde{\omega}o\nu$, an animal : $\gamma\lambda o\iota\alpha$, glue), 85
Zooid ($\zeta\tilde{\omega}o\nu$, an animal; $\epsilon\tilde{\iota}\delta o s$, form), a single individual of a compound organism, 137, 237
Zootham'nium ($\zeta\tilde{\omega}o\nu$, an animal : $\theta\acute{\alpha}\mu\nu o s$, a bush):—Figures, 134, 138 : occurrence and general characters, 135 : dimorphism of zooids, 135 : means of dispersal, 136 : characters and mutual relations of species, 135-139
Zooxanthell'a ($\zeta\tilde{\omega}o\nu$ an animal : $\xi\alpha\nu\theta\acute{o}s$, yellow), 154
Zyg'ospore ($\zeta\upsilon\gamma\acute{o}\nu$, a yoke : $\sigma\pi o\rho\acute{\alpha}$, a seed), applied to a resting zygote formed by the conjugation of similar gametes, 166
Zygote ($\zeta\upsilon\gamma\omega\tau\acute{o}s$, yoked), the products of conjugation of two gametes :—Heteromita, 41 : Vorticella, 133 : Mucor, 165 : Vaucheria, 174 : Spirogyra, 198-200.

THE END

www.ingramcontent.com/pod-product-compliance
Lightning Source LLC
Chambersburg PA
CBHW051733300426
44115CB00007B/541